The Physics of Bacteria

Bacteria are the most ubiquitous life-forms on Earth, and are studied extensively to gain insight into their function and understand how they interact with their environment. In recent years, bacterial biophysics has added a new dimension to this research by using the tools of physics to investigate the quantitative principles that underpin these cellular systems. This book provides a modern and cohesive introduction to bacterial biophysics, with a focus on biofilms, slimes and capsules. In the first of three parts, key techniques and models from the physical sciences that can be applied to bacterial problems are presented. Part II then provides a bacterial microbiology primer for physical scientists and an examination of single-cell phenomena. The final section explores interacting bacteria and biofilms from a physical perspective. Ideal for physics graduates interested in this important field, this book is also relevant for researchers in physical chemistry, bioengineering, mathematics and microbiology.

Thomas Andrew Waigh is Reader in Biological Physics in the Department of Physics and Astronomy at the University of Manchester, UK. He has written three books: *Applied Biophysics* (2002), *The Physics of Living Processes* (2014) and *Some Critical Questions in Biological Physics* (2017) and has published over 120 articles in peer reviewed journals. He has worked in the field of bacterial biophysics for over 10 years and organized many meetings on 'The physics of microorganisms'. He currently works with industrialists on antimicrobial detergents.

The Physics of Bacteria

From Cells to Biofilms

THOMAS ANDREW WAIGH
University of Manchester

Shaftesbury Road, Cambridge CB2 8EA, United Kingdom

One Liberty Plaza, 20th Floor, New York, NY 10006, USA

477 Williamstown Road, Port Melbourne, VIC 3207, Australia

314–321, 3rd Floor, Plot 3, Splendor Forum, Jasola District Centre, New Delhi – 110025, India

103 Penang Road, #05–06/07, Visioncrest Commercial, Singapore 238467

Cambridge University Press is part of Cambridge University Press & Assessment,
a department of the University of Cambridge.

We share the University's mission to contribute to society through the pursuit of
education, learning and research at the highest international levels of excellence.

www.cambridge.org
Information on this title: www.cambridge.org/9781009313520

DOI: 10.1017/9781009313506

When citing this work, please include a reference to the DOI 10.1017/9781009313506

First published 2025

A catalogue record for this publication is available from the British Library

A Cataloging-in-Publication data record for this book is available from the Library of Congress

ISBN 978-1-009-31352-0 Hardback

For Emily and Sally. One who is a big fan of bacteria and the other who isn't.
Thanks also go to the indifferent members of my family
(Roger, Cathy, Paul, Bronwyn and Oliver).

Contents

Part II Single Bacteria

Part III Interacting Bacteria and Biofilms

Preface

There is a gap in the scientific literature for biophysics books on bacteria. The classic book is by Howard C. Berg, *Random Walks in Biology* (1993), and was ground breaking at the time it was published.[1] However, the book only strongly focused on motility in a single species of bacteria, *Escherichia coli* (*E. coli*), and helped stress the area of stochastic processes in biology. This was followed by another very useful, but fairly specialised, book on K-12 *E. coli* motility (2004) by the same author.[2] Other than this, the bacterial biophysics textbook literature is almost empty with the exception of a few general biophysics textbooks. These have the occasional bacterial example, but they tend to concentrate on eukaryotic phenomena.[3–5] Why have not more books been published on this topic, given the huge impact of bacteria on our daily lives? It is believed that prospective authors have all been scared off by the immense size of the task in terms of the number of bacterial species i.e. even taxonomy is still tricky with modern microbiology techniques due to the immense numbers of organisms involved. The current book thus demonstrates either a great deal of foolishness or bravery on the part of the author, depending on the reader's perspective and charitable disposition.

Biophysics as a topic keeps moving in and out of fashion, and it has only been in recent years that quantitative biology went from being a background task required during instrument development to the forefront of life sciences research (with the occasional high profile exception e.g. Max Delbrück's work on genetics[6] or Crick and Watson's model for DNA[7]). Particularly, the glut of data available from modern high-throughput molecular biology tools, such as fluorescence microscopy, mass spectrometry, electron microscopy, crystallography, nuclear magnetic resonance, electrophoresis, fluorescence in situ hybridisation and DNA microarrays, is driving the need for better tools in quantitative biology and biophysics.

A number of multi-authored books have been published on biofilms,[8–10] but what they have in authority, they tend to lack in presenting the field as a cohesive whole. Some other relevant books are focused on measurement techniques, such as *Single-Molecule Techniques* by Mark Leake,[11] but the bacteria take a back seat in this presentation. The engineering community has also made many useful contributions to the field, but they lack some of the more modern physical tools that could be important for a fundamental understanding. From an engineering perspective, a recommended book on biofilms is *Fundamentals of Biofilm Research* by Lewandowski and Beyenal.[12] The hope is that physics has something more to offer to the field.

Applied mathematicians have made some important progress in the fluid mechanics of microorganisms. The book by Lauga is readable by non-specialists[13] and covers some fruitful quantitative theories for some standard dilute microorganisms in simple geometries. Effective theories for interacting bacteria are more hotly debated.

A successful time-tested approach in physics is reductionist, and thus a first step in understanding bacterial biophysics is to look at the behaviour of single bacterial cells. Thus, after introducing a range of modern physics tools to tackle bacterial problems in Part I, Part II concentrates on the physical behaviour of single cells. Part II includes a crash course on bacterial microbiology for physical scientists and considers the growth of bacteria, motility, chemotaxis and molecular machines from a physical perspective. Chapter 20 considers capsules and slimes, since they are important virulence factors in disease.

A modern shift of emphasis in bacterial biophysics has been to consider assemblies of bacterial cells in *biofilms*, and this is the main subject of Part III. Bacterial biofilms are an everyday phenomenon, and mankind will have been indirectly aware of them for the entire existence of our species e.g. plaque on people's teeth, the distinctive colours of lakes and thermal springs, the smell of the countryside after a rain shower (volatile compounds in aerosols emitted by *Actinomycetes*[14]) and the spoilage of beer.[15] However, in the 1970s, the first direct evidence of distinct phenotypic and genetic differences in bacteria was observed when they occurred in biofilms[16] (the term *biofilm* was coined by Bill Costerton in 1978). The bacteria run a distinct genetic programme once they commit themselves to creating a biofilm, and thus it becomes possible to rigorously quantify whether a single bacterium is in the biofilm state. During the 1980s and 1990s, research on bacterial biofilms thus began in earnest, driven by the definitive evidence that a biofilm is a well-defined state of matter.

Palaeontological evidence of biofilms is the oldest on record for a living organism, at 3.48 billion years old in Western Australia.[17] Some of the earliest organisms thus appear to have existed in the biofilm state, and the fossils resemble biomats formed by present-day marine bacteria. A huge number of present-day organisms still use biofilms.

Biofilms touch on some of the most important issues currently facing mankind. Specifically, biofilm research relates to global warming (e.g. carbon dioxide or methane sequestration, changing environments and ecology[18]), recycling (e.g. degradation of synthetic plastics, such as polyethylene terephthalate[19] and polyurethane[20]), food security (e.g. animal husbandry, nitrogen fixation and fermentation products), antibiotic resistance (e.g. immunocompromised individuals who have undergone heart transplant and HIV patients), renewable energy (e.g. fuel cells and photosynthesis) and biodiversity (most of the DNA diversity on planet Earth is directly contained in bacterial chromosomes or in viruses that infect bacteria). Antibiotic resistance continues to be a huge global problem facing humanity. Deaths due to sepsis (an immunological response to bacterial infections) in 2017 were estimated to be 11 million, which is even higher than the global deaths due to cancer (eight million).[21] Thus bacterial biophysics continues to be of high importance to mankind.

Approximately 107 billion modern humans are thought to have lived on planet Earth over the last 50000 years, and a large proportion of them (probably the majority) will have died from microbial diseases. For example, when Robert Koch did his groundbreaking research during the 1870s, establishing the field of microbiology, a seventh of the people alive were dying from tuberculosis caused by a single variety of bacteria (*Mycobacterium tuberculosis*). It was therefore a huge step forward when Koch discovered that the causative agent was a type of bacterium, and it was instrumental in reducing the mortality rate to the ~1/8000 of the people alive today dying from the disease. The most fatal pandemic in history was also due to a bacterium, *Yersinia pestis*. The Black Death killed around half of Europe's population in the mid-1300s.

Archaea (which together with bacteria make up the prokaryotes) also commonly form biofilms, as opposed to bacteria, which tend to be the prokaryotes that people most commonly consider (archaea are a relatively recent discovery from the 1970s[22]). Biofilms also occur with some eukaryotic microorganisms,[23] such as fungi and microalgae, which also present significant health problems e.g. the fungus *Candida albicans* causes morbidity in people with either cystic fibrosis or chronic obstructive pulmonary disorder (a condition often seen with smokers). Therefore, two out of three branches of life on planet Earth (archaea and bacteria) tend to use biofilms as a default mode of growth, and the third branch (eukaryotes) has some very common examples that do so e.g. yeast. Multicellularity mitigates much of the need for biofilms in more advanced eukaryotic life forms, since cells are protected by their neighbours, although some analogous functions are often performed by the extracellular matrix e.g. collagen in human skin or mucus in the intestines. Unfortunately, in what follows, there is no room for biofilms formed by non-bacterial microorganisms, but it is expected that many of the research ideas are transferable due to the similar timescales, length scales and molecular architectures involved, which all relate to similar physical phenomena e.g. the mesoscopic forces and hydrodynamics that guide biofilm morphogenesis in prokaryotes and eukaryotes are all similar. Eukaryotic biofilms are also underrepresented in the biophysics literature, which makes an accurate overview of them more challenging.

Slightly more open to debate is the subject of viral biofilms. There are some data that indicate human T-cell leukaemia viruses[24] hijack the metabolism of host eukaryotic cells to create favourable aggregates of extracellular matrix molecules that resemble biofilms. This study is now 10 years old and remains the most persuasive in the literature. Speculation would indicate that this could be a strong virulence factor for infections in humans, allowing viruses to hide from immune responses embedded in the extracellular matrix, but electron microscopy measurements to rigorously check this hypothesis are very challenging due to the small length scales and the extreme levels of molecular heterogeneity that are present. Synergy of viral infections with bacterial or fungal biofilms is also a distinct possibility, and it seems very likely that bacterial biofilms act as a reservoir for new viral infections in a number of diseases of vertebrate organisms. Furthermore, there is a large literature on mucosal interactions with viruses[25] (a broad range of external barriers to infections in humans are protected with a covering of mucus, such as the lungs, the eyes and the gastrointestinal tract), although there is no direct evidence to date that viruses directly modify the expression of mucin gels.[26]

Viral infections of bacterial biofilms, as opposed to viral biofilms, have been studied in detail.[27] Bacteriophages are exquisitely well adapted to life in bacterial biofilms for example they can communicate with one another to synchronise their cycles of infection within biofilms (a viral quorum sensing mechanism[28]) and can target specific capsular strains of bacteria.[29] Bacteriophages are thought to contain the majority of genetic diversity on planet Earth.[30]

Bacterial biofilms are emergent phenomena created from the simultaneous interactions of many thousands of separate individual microbes. In recent years, such generic phenomena have been categorised under the banner of complexity theory in statistical physics.[31–32] Analogous to termite mounds, the structures formed in biofilms are shaped by subtle evolutionary processes that guide the interactions of the individuals e.g. ventilation shafts in termite mounds can be compared with breathing channels in bacterial biofilms. The field thus provides fascinating challenges for statistical analysis, applied mathematics, non-linear physics and computational physics to explain the emergent phenomena e.g. through the development of agent-based models (Chapter 27).[33,34]

Inevitably, there will be gaps in the molecular biology coverage in what follows due to the vast nature of the venture and the expertise of the author (a physicist). The book follows more of a physics approach, and the hope is that physics has lots to offer in addition to traditional descriptive molecular biology methods. A simplistic generic approach is taken in the face of extreme complexity. It is also hoped that the book will act as an effective primer for physical scientists hoping to move into this important field and at worst will act as a place marker to encourage someone else to write a better book.

The book will by necessity be an incomplete survey of bacteria on planet Earth. An emphasis on biofilms, slimes and capsules helps to focus the work, but since the vast majority of bacteria spend the majority of their time in surface-attached biofilms,[35] in practice, it is more of a change in focus within bacterial biophysics than a true subfield. Indeed, the rarer examples of purely planktonic bacteria found in nature can be predominantly considered as a dispersal mechanism for biofilms, since they go on to nucleate new biofilm-based communities when they attach to new surfaces (with evolutionary delays for the occasional non-biofilm creating strains). It is easy to become drowned in the complexity of the problems presented in bacterial biophysics, and the superficial agreement of models with experiments is often undermined by too many artificial assumptions. Therefore, when given the choice, an attempt is made to strip down the problems to their bare bones, sacrificing some complexities to discriminate between alternative generic patterns and thus increase the generality of the approach.

In addition to biofilms, *capsules* and *slime layers* are important virulence factors for bacterial infections. Virulence factors allow the bacteria to proliferate during infections and are thus important in disease. The word *capsule* invokes images of a rigid covering around a bacterium, and its use in this context is unfortunate. The term *capsule* might be better applied to S-layers or peptidoglycan on the surface of some bacteria, which are layers of composite external structures that reinforce the surfaces and provide them with a shear resistance i.e. proper elastic shell-like coatings. What are currently called *bacterial capsules* are actually polymer brushes and thus provide soft repulsive potentials that repel surrounding macroscopic surfaces and nanoparticles (a steric stabilisation effect due to the entropy of the polymeric brushes). Bacterial capsules have little in-plane elasticity i.e. the shear resistance expected for something genuinely capsular. *Slime layers* are also haphazardly defined in the literature. They are formed from shed polymers (often capsular material) that create a viscoelastic layer in which the bacteria are embedded. Slime layers are differentiated from biofilms in that they are often viscoelastic fluids rather than gels (and thus can be more easily washed off), although the complex nature of polymer viscoelasticity can mean that the two possibilities are hard to differentiate in practice. In the literature, slime layers are presumed to be less structured than biofilms, but this may just correspond to a reduction in the number of cross-links between the polymers that constitute the slimes, which means that they have no dedicated mechanism for gelation, although some slimes used in food (e.g. gellan and xanthan) do form weak physical gels. Structuration in slime layers, biofilms and mixtures of the two predominantly remains an open question. It is expected that slime layers and biofilms correspond to separate genetic programmes (which are sometimes coexpressed), with those for slime layers significantly simpler than biofilms e.g. slimes are due to a simple upregulation of biopolymer production via the activation of single genes or operons, rather than the complex array of genes that control biofilm signalling and production.

From a positive perspective, researchers have no fear of running out of problems to explore or new phenomena to discover in bacteria and their biofilms. A vast number of bacterial species are still left to be discovered (including even basic descriptions of their taxonomy, and the vast range of physiological behaviours and phenotypes they display), let alone the fundamental physical phenomena they demonstrate. Current estimates state that approximately 30 000 bacteria have been rigorously classified in detail to date from an estimated 10^{12} total number of species on planet Earth (these numbers are still hotly debated in the literature).[36] There are still a large number of outstanding physical questions with *E. coli*, which so far is the best described bacterium.

The choice of material in many places is directed by the tastes of the author. A necessity was to keep the physics relatively simple so that models could be posed clearly to help build understanding and show that they have been carefully tested against reality.

An attempt has been made to include lots of cartoons (rough schematics) of bacterial phenomena to try to communicate the ideas effectively. These always tend to be a gross simplification, since biology invariably has a huge range of complex phenomena continuously distributed over a diverse hierarchy of different length scales and timescales.

As befitting of a research monograph, an attempt has been made to properly reference the work in the manuscript, but the reference lists are not exhaustive and an apology is made to anyone whose work is omitted.

There are a number of ethical questions associated with medical research, biotechnology and synthetic biology. Clearly, regulatory guidelines should be followed in research into bacterial biophysics, and sloppy ill-defined experiments are the bane of biological research i.e. the reproducibility of results in the biological literature is a constant problem. For physical scientists hoping to move into the field of microbiology, it is recommended that they find collaborators in dedicated microbiology laboratories.

Suggested Reading

Biological Physics

Boal, D., *Mechanics of the Cell.* Cambridge University Press: 2012. An excellent introduction to cellular biophysics. Some interesting calculations on bacterial cell walls are included.

Lewandowksi, Z.; Beyenal, H., *Fundamentals of Biofilm Research.* CRC Press: 2013. A bioengineering approach to the study of biofilms.

Waigh, T. A., *The Physics of Living Processes.* Wiley: 2014. An undergraduate textbook that describes some useful physical tools that can be applied to biology.

Waigh, T. A., *Some Critical Questions in Biological Physics: A Guided Tour around the Bugbears.* Institute of Physics Publishing: 2017. An informal discussion of some important problems in biological physics, which includes biofilms.

Microbiology

Cossart, P., *The New Microbiology: From Microbiomes to CRISPR.* ASM: 2018. A popular science book that provides a modern perspective on new developments in microbiology with an emphasis on molecular biology.

Kim, B. H.; Gadd, G. M., *Prokaryote Physiology and Metabolism*, 2nd ed. Cambridge University Press: 2019. Slightly simpler than the White book and a little less detailed, but marginally more accessible.

Madigan, M. T.; Bender, K. S.; Buckley, D. H.; Sattley, W. M.; Stahl, D. A., *Brock Biology of Microorganisms.* Pearson: 2018. A classic broad brush stroke approach to the whole field of microbiological research.

White, D., *The Physiology and Biochemistry of Prokaryotes.* Oxford University Press: 2011. The field of microbiology is vast, but this textbook is probably the best on aspects of bacterial molecular biology.

References

1. Berg, H. C., *Random Walks in Biology*. Princeton University Press: 1993.
2. Berg, H. C., *E. coli in Motion*. Springer: 2004.
3. Boal, D., *Mechanics of the Cell*. Cambridge University Press: 2012.
4. Philips, R.; Kondev, J.; Theriot, J.; Garcia, H.; Kondev, J., *Physical Biology of the Cell*. Garland: 2012.
5. Waigh, T. A., *The Physics of Living Processes*. Wiley: 2014.
6. Luria, S. E.; Delbruck, M., Mutations of bacteria from virus sensitivity to virus resistance. *Genetics* **1943**, *28* (6), 491–511.
7. Watson, J. D.; Crick, F. H. C., Molecular structure of nucleic acids: a structure for deoxyribose nucleic acids. *Nature* **1953**, *171*, 737–738.
8. Leake, M. C., *Biophysics of Infection*. Springer: 2018.
9. Romero, T., *Bacterial Biofilms*. Springer: 2008.
10. Ghannoum, M.; Parsek, M.; Whiteley, M.; Mukherjee, P. K., *Microbial Biofilms*. ASM Press: 2015.
11. Leake, M. C., *Single-Molecular Cellular Biophysics*. Cambridge University Press: 2013.
12. Lewandowki, Z.; Beyenal, H., *Fundamentals of Biofilm Research*. CRC Press: 2013.
13. Lauga, E., *The Fluid Dynamics of Cell Motility*. Cambridge University Press: 2020.
14. Joung, Y. S.; Buie, C. R., Aerosol generation by raindrop impact on soil. *Nature Communications* **2015**, *6*, 6083.
15. Riedl, R.; Dunzer, N.; Michel, M.; Jacob, F.; Hutzler, M., Beer enemy number one: genetic diversity, physiology and biofilm formation of *L. brevis*. *Journal of the Institute of Brewing* **2019**, *125* (2), 250–260.
16. Davey, M. E.; O'Toole, G. A., Microbial biofilms: from ecology to molecular genetics. *Microbiology and Molecular Biology Reviews* **2000**, *64* (4), 847–867.
17. Noffke, N.; Christian, D.; Wacey, D.; Hazen, R. M., Microbially induced sedimentary structures recording an ancient ecosystem in the ca. 3.48 billion year old dresser formation, Pilbara, Western Australia. *Astrobiology* **2013**, *13* (12), 1103–1124.
18. Pepper, I.; Gerba, C. P.; Gentry, T. J., *Environmental Microbiology*. 3rd ed.; Academic Press: 2015.
19. Austin, H. P.; et al., Characterization and engineering of a plastic-degrading aromatic polyesterase. *Proceedings of the National Academy of Sciences* **2018**, *115* (19), E4350.
20. Pathak, V. M., Navneet, Review of the current status of polymer degradation: a microbial approach. *Bioresources and Bioprocessing* **2017**, *4*, Article Number: 15.
21. Rudd, K. E.; et al., Global, regional and national sepsis incidence and mortality 1990–2017: analysis for the global burden of disease study. *Lancet* 2020 (10219), *395*, 200–211.
22. Woese, C. R.; Fox, G. E., Phylogenetic structure of the prokaryotic domain. *Proceedings of the National Academy of Sciences* **1977**, *74* (11), 5088–5090.

23. Beckwith, J. K.; Ganesan, M.; VanEpps, J. S.; Kumar, A.; Solomon, M. J., Rheology of *Candida albicans* fungal biofilms. *Journal of Rheology* **2022**, *66* (4), 683–697.
24. Pais-Correia, A. M.; Sachse, M.; Guadagnini, S.; Robbiati, V.; Lasserre, R.; Gessain, A.; Gout, O.; Alcover, A.; Thoulouze, M. I., Biofilm-like extracellular viral assemblies mediate HTLV-1 cell-to-cell transmission at virological synapses. *Nature Medicine* **2010**, *16*, 83–89.
25. Barr, J. J.; et al., Bacteriophage adhering to mucus provide a non-host-derived immunity. *Proceedings of the National Academy of Sciences* **2013**, *110* (26), 10771–10776.
26. Smith, P. D.; Blumberg, R. S.; MacDonald, T. T., *Principles of Mucosal Immunology*. Garland: 2020.
27. Vidakovic, L.; Singh, P. K.; Hartmann, R.; Nadell, C. D.; Drescher, K., Dynamic biofilm architecture confers individual and collective mechanisms of viral protection. *Nature Microbiology* **2018**, *3*, 26–31.
28. Erez, Z.; et al., Communication between viruses guides lysis-lysogeny decisions. *Nature* **2017**, *541*, 488–493.
29. Scholl, D.; Adhya, S.; Merril, C., *E. coli* K1's capsule is a barrier to bacteriophage T7. *Applied and Environmental Microbiology* **2005**, *71* (8), 4872–4874.
30. Zrelovs, N.; Dislers, A.; Kazaks, A., Motley crew: overview of the currently available phage diversity. *Frontiers in Microbiology* **2020**, *11*, 579452.
31. Holland, J. H., *Complexity: A Very Short Introduction*. Oxford University Press: 2014.
32. Jensen, H. J., *Complexity Science: The Study of Emergence*. Cambridge University Press: 2023.
33. Grimm, V., *Individual-based Modeling and Ecology*. Princeton University Press: 2005.
34. Mounfield, C. C., *The Handbook of Agent Based Modelling*. Independent Publishing: 2020.
35. Flemming, H. C.; Wuertz, S., Bacteria and archaea on Earth and their abundance in biofilms. *Nature Reviews Microbiology* **2019**, *17*, 247.
36. Locey, K. C.; Lennon, J. T., Scaling laws predict global microbial diversity. *Proceedings of the National Academy of Sciences* **2016**, *113* (21), 5970–5975.

Acknowledgements

I would like to thank colleagues and students for extending my understanding of bacteria and for being a pleasure to work with: Ian Roberts, Andrew McBain, Jian Lu, Nickolay Korabel, Johanna Blee, Jack Hart, Sorasak Phanphak, Haoning Gong, Jane King, Chris van der Waale, Salman Rogers, Abdhul Harrani, Daniela Ciumac, Dick Strugnell, Frances Separovic, Laura Fox, Emmanuel Akabuogu, Victor Martorelli, Lin Zhang, Dirmeier Reinhard, Rok Krašovec, Josh Milstein, Thomas Hutton, Seamus Holden, Stuart Middlemiss, Bartek Waclaw, Mark Leake, Achillefs Kapanidis, Viki Chen, Adam Cohen, Tom Millard, Viki Allen, Philip Woodman, Andrew Harrison, David Kenwright, Sergei Fedetov, Jacob Asmat, Owen Moores, Conor Lewis, Roger Waigh, Nickolai Lavroff and Mathias Enderle. A part of the manuscript was written at the University of Melbourne as part of an exchange scheme with the University of Manchester in August 2019. Another major component was written during the coronavirus lockdown in April 2020.

PART I

PHYSICAL TOOLS

This section introduces some modern physical tools that are useful in bacterial biophysics. The hope is to bridge the gap to the research level challenges presented in Parts II and III, with a 'what can I do to help' approach, and to emphasise the unique perspectives physics can contribute to bacterial research. To future proof this material, only generic techniques and models to interpret the data are considered. Bacteria are extremely complex and physical measurements invariably neglect a huge number of hidden variables that cannot be easily assessed in current experiments. The challenge is to develop hypotheses that are robust to these unknown parameters and careful control experiments are thus mandatory in bacterial biophysics.

1 How to Track Cells and Molecules

Seeing is believing and the development of high-resolution microscopes originally provided the most conclusive evidence for the existence of bacteria or *animalcules* (tiny animals) as they were first described.[1] Microscopy continues to be a central tool in modern bacterial biophysics and, when combined with quantitative image analysis tools, microscopes can provide unambiguous quantitative data to answer many of the questions related to bacterial behaviour.

A simple inverted optical microscope is shown in Figure 1.1. It follows a simple *4f* geometry, where *f* is the focal length of the condenser and objective lenses, and in practice, additional improvements are standard on laboratory microscopes e.g. Köhler illumination (to illuminate specimens in a uniform manner), phase contrast (to provide additional contrast for thin or transparent specimens), fluorescence optics (to allow imaging of fluorescent samples, Figure 13.8) and confocal pin holes (to improve background rejection, Figure 13.7).[2]

Once objects are identified in a microscopy image (the process of *segmentation*), analysing their dynamics by linking objects in consecutive images provides a rich source of additional information i.e. *tracks* are created that can be statistically analysed. Tracks can describe the motion of whole cells on the microscale, single molecules on the nanoscale or organelles on intermediate length scales e.g. the swimming behaviour of *Escherichia coli* (of micrometre length scales), the motion of proteins attached to a membrane (of nanometre length scales) or the transverse fluctuations of the endoplasmic reticulum in eukaryotic cells (of 10–1 000 nm length scales).[3]

1.1 How to Track Cells

Experimentally, tracking single cells is less demanding than single molecules due to their larger size, so it is a good place to start.[4] For strains of readily culturable bacteria, cells can be imaged with standard microscopy techniques using absorption contrast i.e. no complicated sample preparation is required, such as staining techniques. To track cells, first a sequence of well-resolved microscopy images need to be acquired. Standard types of imaging modality that can be used to create movies of dispersed bacteria or early stages of biofilms include *bright-field microscopy* (very high speeds are possible i.e. ~10^5 frames per second), *fluorescence microscopy* (specific labelling is possible, but the technique is relatively slow due to the low photon yield of fluorescent processes) and *confocal microscopy* (allows three-dimensional [3D]

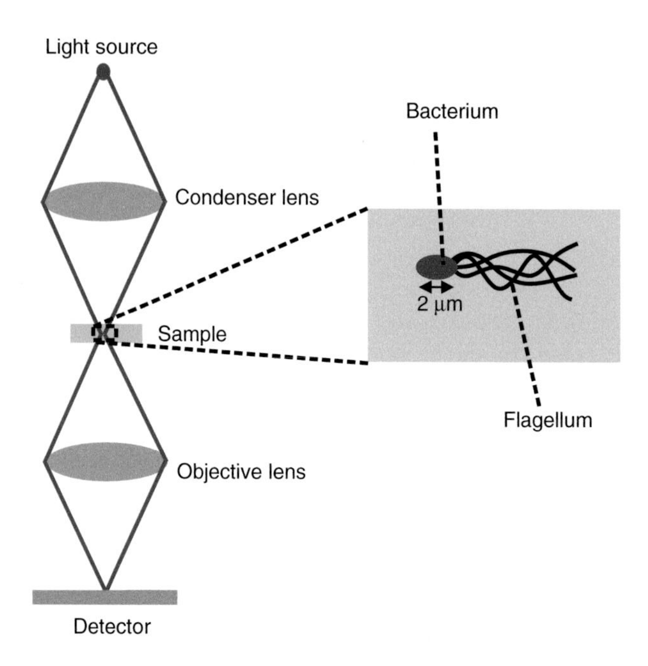

Figure 1.1 A simple *inverted optical microscope* in a *4f* configuration (*f* is the focal length of both the condenser and the objective lenses).[2] An *E. coli* cell is shown, and the flagella would be invisible without a dedicated contrast mechanism (fluorescence or phase contrast is needed).

imaging, but the scanning process of image acquisition often makes it slow). State-of-the-art super-resolution fluorescence microscopy techniques include *lattice sheet microscopy* that can achieve ~50 nm resolution at video rates (~50 frames per second)[5] and MINIFLUX (a variety of *stimulated emission depletion microscopy*, STED, Figure 13.10) that can achieve ~1 nm resolution at 1 000 frames per second.[6] The super-resolution techniques tend to be technically challenging (Section 13.2.4), and bright-field microscopy is much easier for beginner microscopists.

Ideally, movies of cells should be as long as possible, in terms of the number of frames, to maximise the amount of information available in the resultant tracks. Track length can be limited by the depth of focus of the microscope (z sectioning), field of view of the microscope (sampling in x and y), excessive particle speeds, the available memory on the camera (particularly an issue with ultrafast cameras), photobleaching of fluorescent labels and phototoxicity that damages the cells.

Once a movie of the cells has been made, the next challenge is to segment the images of cells using image analysis software. Gaussian trackers can be used to locate the positions of compact symmetrical bacterial cells that may be reasonably approximated by Gaussian functions, but more complicated cellular shapes need more sophisticated forms of segmentation, such as neural networks (NNs) or snakes algorithms[9] (Figures 1.2 and 1.4). Tracks are then made by connecting the centres of the segmented cells together in consecutive frames to form a linked list. Software searches for the closest positions of cells in consecutive images to link the cell centres together.

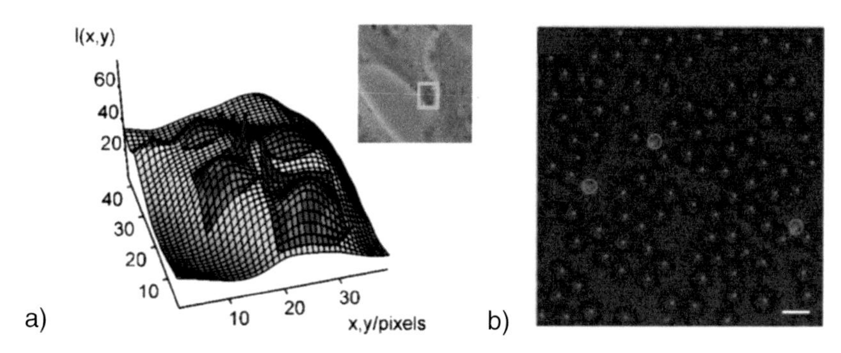

Figure 1.2 (a) A *Gaussian tracker* segments objects[7] in a human epithelial cell.[7] Endosomes (red, ~100 nm in size) are identified within a frame from a bright-field microscope (inset). (b) A *convolutional neural network* (CNN) segments *Bacillus subtilis* cells (red circles) immersed in a suspension of Brownian particles (red dots).[8]

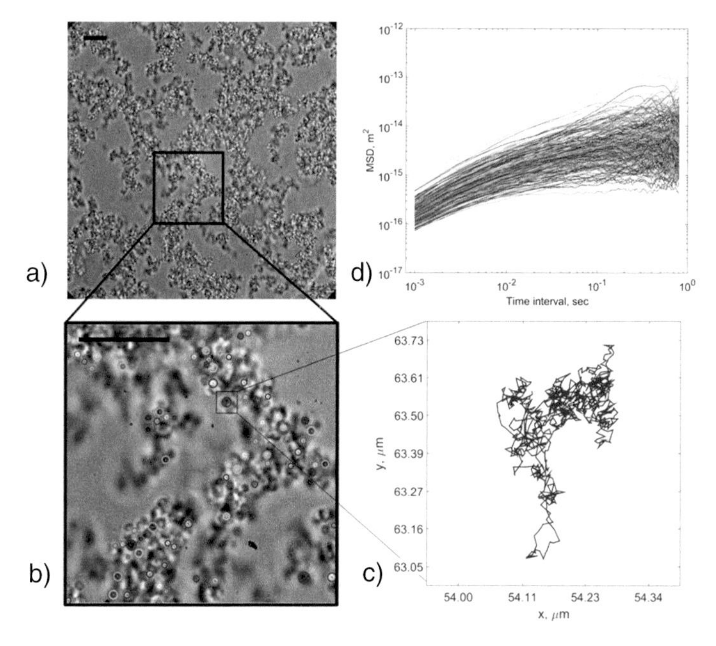

Figure 1.3 (a) An image of an early-stage *Staphylococcus aureus* biofilm from bright-field optical microscopy.[13,14] (b) Zoomed in region where single bacterial cells (1–2 μm in size) can be observed. (c) A track of a single *S. aureus* cell made using Gaussian tracker software (Figure 1.2a). (d) Mean square displacements (MSD) as a function of time interval of hundreds of single *S. aureus* cells in the biofilm calculated from the tracks. *S. aureus* is immotile, so the cells' motions are due to thermal forces modulated by the viscoelasticity of the biofilm.

If particles move substantial distances between consecutive images or the particle concentrations are too high, it can be an impossible task to unambiguously identify which particle contributes to which track. Particle positions can typically be measured with subdiffraction-limit resolution (often an order of magnitude improvement is possible on the diffraction limit) and sub-camera pixel resolution, because the weighted mean of measurements of the optical centre of mass of a particle is used i.e. averages over many pixels are calculated. Thus cell positions can be routinely tracked with ~10 nm resolution at ~10^4 frames per second using standard optical microscopes combined with fast complementary metal oxide semiconductor cameras (Figure 1.3).[10] Higher resolutions have been achieved with quantum metrology using squeezed light.[11,12]

1.2 How to Track Single Molecules

Single molecules can now be routinely tracked both *in vitro* and *in vivo* inside single cells, although it was seen as a dramatic advance when single molecules were first imaged in condensed phases (Nobel Prize 2014).[15,16] Many people were surprised that it was possible to discriminate single molecules in condensed phases against backgrounds containing vast numbers of molecules of the order of Avogadro's number (6×10^{23}) with a sufficient signal-to-noise ratio (SNR). Single-molecule imaging techniques are primarily based on fluorescence microscopy, and this requires specific labelling of the molecules of interest with fluorophores. An emission filter based on the wavelength shift of emitted photons from a fluorophore compared with the wavelength of the excitation light source (the Stokes shift) can be used to discriminate single fluorescent molecules against the background of a huge number of non-fluorescent molecules.

Large catalogues are available for commercial fluorophores that can label biomolecules with varying degrees of specificity, such as proteins, nucleic acids, carbohydrates and lipids. The specificity of the labels needs to be determined in a biological experiment to be certain of what is labelled using careful control experiments due to the large number of factors that affect fluorophore binding. An elegant solution for labelling proteins is to genetically modify them to add an extra fluorescent protein domain to their structure. This can be very effective for *in vivo* studies, but green fluorescent proteins (GFPs) can suffer from fast photobleaching (synthetic fluorophores often are much more photostable), bulky GFPs can perturb protein functionality (control experiments are needed), and there are time lags introduced by the GFP transcription that can limit studies of fast intracellular dynamics.

For molecular imaging, the choice of segmentation algorithm is determined in part by the geometry of the molecule. Extended molecules with extensive labelling (e.g. a large DNA molecule in which all the base pairs are fluorescently labelled) require snakes algorithms (Figure 1.4), whereas molecules with point-like labelling often use Gaussian trackers (Figure 1.2a).[9] AI techniques (e.g. convolutional neural networks [CNNs]) can be more flexible in the types of molecular geometry they can analyse[19] but will suffer from poor SNRs if they are not properly constrained (Figure 1.2b). Often it is best to constrain artificial intelligence (AI) algorithms using simple physical models e.g. the probabilities of particle displacements can be constrained on the basis that particles will not teleport between different locations, which is a Bayesian approach. Current AI techniques often require extensive data sets to perform the training procedure i.e. they involve *supervised learning* (Chapter 14).

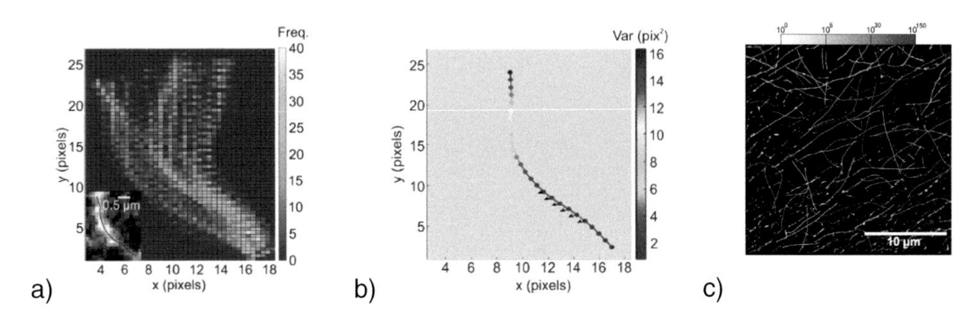

Figure 1.4 *Snakes algorithms* allow tracking of extended objects, such as the endoplasmic reticulum (ER) in human cells[3] or a peptide fibre in a gel.[17] (a) Time dependence of the tracked contours of the ER tubules from fluorescence microscopy. (b) Mean position of the ER tubule, where the variances of the transverse motions are highlighted. The tracked contours of ER tubules indicate active motion due to motor proteins. (c) Peptide fibre positions in a gel from fluorescence microscopy.[17,18]

The choice of algorithm to link the positions of segmented particles together into a track also has a variety of options[20] e.g. *nearest neighbour linking* or *multi-track optimisation* are possible. Particular care is required when particles closely approach one another (they can easily switch labels), and tracks can become fragmented due to low SNRs (they can be stitched together, but often with limited success). Our experience is that the segmentation algorithm plays a more important role than the linking algorithm in the quality of the final tracks, but both are important.

Bayesian tracking techniques (Chapter 14) can be very useful to remove the noise on tracks e.g. Kalman filtering,[21,22] and the methodology has been extensively developed with satellite imaging. However, care must be taken that this noise is random and Markovian (independent noise fluctuations occur with no memory), since it is an assumption used in Kalman filtering. It is particularly an issue when considering non-Markovian processes e.g. the motility of microorganisms or the intracellular motion of molecules are frequently non-Markovian.[23]

1.3 The Statistics of Structures

The static images of molecules and cells from microscopy experiments can provide a range of useful information e.g. calculating their sizes, conformations and relative organisation. Standard freeware software allows the segmentation of bacteria in microscopy images and can provide quantitative descriptors of cell shape.[24] Sophisticated software has also been developed to segment bacterial biofilms and quantify their structures in three dimensions.[25]

Different statistical tools are needed to quantify the relative positions of bacteria or the molecules associated with them. The *Ripley K function* ($K(r)$) quantifies the intuitive notion of whether particles have been placed at random across a surface or they

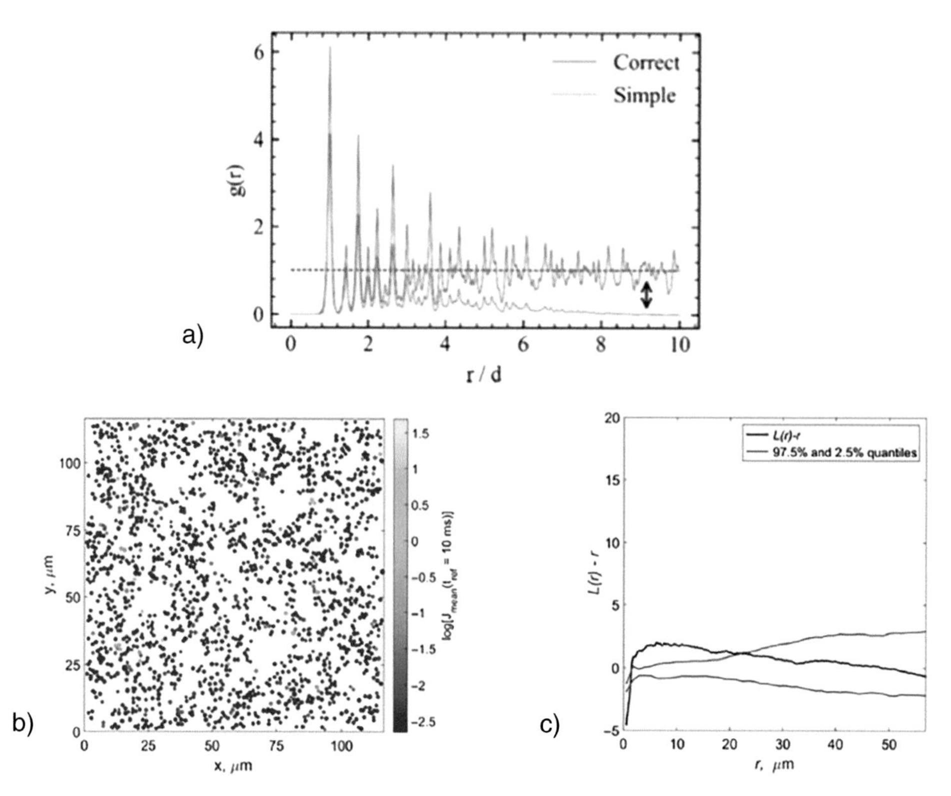

Figure 1.5 (a) The *radial distribution function* $g(r)$ for a random lattice of points[26] (blue). The red curve is from a naïve numerical calculation. (b) *Segmented positions* of *S. aureus* bacteria in a biofilm (coloured with a measure of the linear viscoelasticity via the creep compliance, $J(t)$, Chapter 10) and (c) the *Ripley K function* of the *S. aureus* bacteria in the biofilm[13] shown in (b) (r is the distance from a test point, $K(r)$ is the Ripley K function and $L(r)=\sqrt{\frac{K(r)}{\pi}}$ is the black line). It shows considerable clumping of the *S. aureus*.

are clustered together or dispersed.[27] It is defined so that $K(r)$ is the expected number of additional points within a distance r of a given point. This is a useful tool to understand the distributions of bacteria as they adsorb to surfaces. An alternative measure is given by the pair correlation function ($g(r)$), which is widely used in condensed matter physics, particularly liquid-state theory and models of colloidal matter.[28] $g(r)$ is defined as

$$g(r)=\frac{1}{2\pi r}\frac{\mathrm{d}K(r)}{\mathrm{d}r}, \tag{1.1}$$

where r is again the distance from a test point (Figure 1.5). With a stationary Poisson distribution of points, $g(r)=1$ i.e. a complete random arrangement with no correlations. $g(r)<1$ indicates an anti-correlation between points (dispersion), whereas $g(r)>1$ indicates clustering.[27] $g(r)$ can be related to the interparticle potential if Boltzmann statistics are assumed for systems in thermal equilibrium[28,29] and has been extensively developed in liquid-state theory. In anisotropic systems, $g(r)$ needs to be generalised[26] e.g. correlations along separate lattice directions should be averaged separately to maintain the additional information needed to quantify the degree of anisotropy, such as with liquid crystalline materials.

1.4 How to Analyse Particle Tracking Data

Statistical tools for handling tracking data can be very powerful. They play a central role in modern biological physics, since microscopy methods can provide high-resolution time series of images of living cells, biofilms and single molecules. Robust statistics are needed to test hypotheses on the behaviour of particles e.g. how they move, react, sense and oscillate. Furthermore, the analysis of particle tracking data can be conveniently extended to include tools from machine learning, since they have a common statistical basis, greatly increasing the possibilities for pattern recognition and large-scale automation[22] (Chapter 14).

Tracks of individual bacteria provide a rich source of information on their behaviour e.g. their motility, chemosensing and interactions. Statistical tools need to be applied to the tracks to make sense of them. A wide variety of *ad hoc* bespoke statistical parameters could be defined e.g. a bacterium is motile if its average velocity over 1 s is 1 μm s^{-1}, but they are often unsatisfactory. To choose between alternative possible statistical parameters, standard mathematically elegant methods are preferable, since they provide better prospects for quantitative comparison with both analytical models and simulations. They can also be more robust to varying experimental conditions and thus generalise more easily.

The transport of bacterial cells and molecules in the cells is often *anomalous* e.g. the central limit theorem breaks down and the probability density functions of their displacements are non-Gaussian. Mathematical models have been developed to describe anomalous transport (Chapter 2), although the relative merits of competing models are still debated.[30] A recent innovation is to train NNs on anomalous transport models since NNs can then provide the dynamic segmentation of tracks (Chapter 14).[31] Particle tracks represent a special case of *time series analysis*, which find wide-ranging applications inside and outside biology e.g. forecasting the stock market or diagnosing heart disease based on electrocardiograms.[32] There is thus a huge literature, and a wide range of mathematical tools have been developed.

A central tool for quantifying stochastic motion of particles is the *mean square displacement* (MSD, Chapter 2). For a random walk, the MSD $\left(\left\langle \Delta r^2 \right\rangle\right)$ has a simple scaling dependence on time, $\left\langle \Delta r^2 \right\rangle \sim t^1$ e.g. during Brownian motion (Figure 1.6). Furthermore, scaling of the MSDs is used to define anomalous transport, $\left\langle \Delta r^2 \right\rangle \sim t^{\alpha}$, where $\alpha \neq 0,1,2$, and this is the type of stochastic motion most commonly observed for cellular motility and the motion of larger molecules inside cells.[23,33] Average displacements $\left(\left\langle \Delta r \right\rangle\right)$ of particles are often not a useful measure of stochastic transport, since for symmetric stochastic processes, they average to zero, $\left\langle \Delta r \right\rangle = 0$. Higher moments

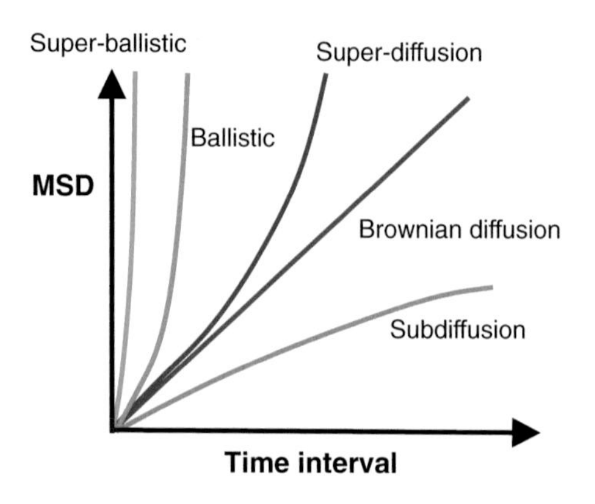

Figure 1.6 MSD as a function of time interval showing *sub-diffusive*, *diffusive* (Brownian), *super-diffusive*, *ballistic* and *super-ballistic* scaling behaviours. Note that super-ballistic scaling is rarely observed in low Reynolds number systems (they are overdamped), although it is possible in turbulent flows.

of the displacement probability distribution based on the third and fourth moments $\left(\langle \Delta r^3 \rangle \text{and} \langle \Delta r^4 \rangle\right)$ are useful for quantifying the skew and the degree of peakedness (the kurtosis), respectively. Moments of probability distributions of the displacements can provide average quantities to describe stochastic motility, which are reasonably robust to noise, but probability distribution functions (pdfs) contain additional information. Mathematically, the moment distribution is insufficient to unambiguously determine a pdf.[34]

For stationary statistical processes,[35] often MSDs are averaged over time, and the MSD is then considered as a function of time interval (TAMSD(τ)) i.e.

$$TAMSD(\tau)=\frac{1}{T-t}\int_0^{T-t}\left(\left(x_i(\tau+t)-x_i(\tau)\right)^2+\left(y_i(\tau+t)-y_i(\tau)\right)^2+\left(z_i(\tau+t)-z_i(\tau)\right)^2\right)dt \tag{1.2}$$

where T is the duration of the track and t is the time. x_i, y_i and z_i are the Cartesian coordinates of the particle i. The calculation of time-averaged MSDs (TAMSDs) can provide a major improvement in the SNR at short time intervals in experiments. If there are n steps in a track, the error bars scale as $(n-1)^{-1/2}$ for the shortest time interval of the TAMSD, $(n-2)^{-1/2}$ for the next shortest time interval and so on. Ensemble averaging of MSDs (EMSDs) over different particles is also possible to improve the SNR i.e. the MSDs are averaged over i in Equation (1.2).

There is a general theorem by Birkhoff from dynamical systems theory[36,37] that states TAMSD = EMSD for an *ergodic process*, and it can be used as a diagnostic for ergodicity breaking e.g. whether glassy behaviour occurs in the tracks. MSDs can be calculated in one, two and three dimensions, and their analysis conveniently generalises to different dimensionalities e.g. to describe the motion of a motor along a DNA chain (one-dimensional [1D]), a particle in the plane of focus of a conventional optical microscope (two-dimensional [2D]) or a particle in a confocal microscope (3D).

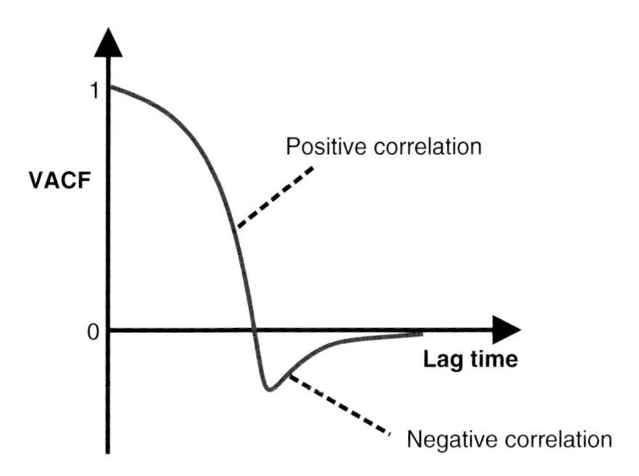

Figure 1.7 *Velocity autocorrelation function* (*VACF*) as a function of lag time for a particle moving inside a cell. The negative values of the *VACF* are due to anti-correlation.

Stochastic aging (SA) is a separate issue to ergodicity and is often observed in biology e.g. stochastic processes are not stationary and evolve with time during the growth of a cell.[37] SA can be diagnosed by delaying tracks by different aging times (i.e. chop off the start of the data corresponding to the aging time) and then comparing the resultant *MSD*s. Aging and glassy phenomena have direct implications in medicine e.g. scarring during wound healing that involves non-ergodic glassy fibrous composites of collagen.

Velocities need to be handled carefully with stochastic processes, since with random walks they depend on the time scale at which they are measured.[38] Often instantaneous velocities are defined in experiments as $\frac{\Delta r}{\Delta t}$ (Δr is the displacement of a particle over a time interval, Δt), but this quantity is sensitive to the choice of Δt e.g. a smaller choice of Δt can correspond to higher values of velocity for sub-ballistic processes. Lots of values of motor protein velocities in the literature are mishandled due to such issues and when faster cameras are manufactured, the quoted motor protein velocities often also increase. More robust methods to quantify velocities are to consider velocity autocorrelation functions (*VACF*) or velocities calculated via first passage probabilities[39] (see later). *VACF*s (Figure 1.7) can be defined as

$$VACF(t) = \int_0^{T-t-\delta} \frac{v(t'+t)v(t')}{T-t-\delta} dt', \tag{1.3}$$

$$v = \frac{r(t+\delta)-r(t)}{\delta}, \tag{1.4}$$

where v is the velocity at time t, δ is the time interval, T is the duration of the experiment and r is the displacement.

The use of probability distributions of *survival times* has its origins in medicine (Figure 1.8a).[40] Histograms of the number of patients in a medical trial can be plotted

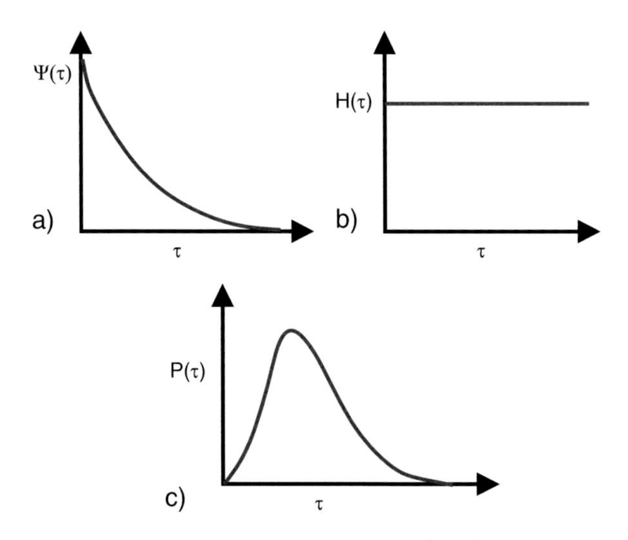

Figure 1.8 Plots of (a) the *survival time distribution* $(\Psi(\tau))$, (b) the *hazard function* $(H(\tau))$ and the *probability density function* $(P(\tau))$ as a function of time (τ) for the run times of a bacterium.

as a function of time. If the death rate occurs at a constant value per unit time, the survival distribution has an exponential decay ($\Psi(\tau)$, a Poisson process). Decays due to more complex processes can be non-exponential, and hazard rates $(H(\tau))$ can be introduced to make it easier to visualise them i.e. a constant hazard rate as a function of time is equivalent to a single exponential decay for the survival time (Figure 1.8b). The hazard rate is the rate of death of a subject of age t.

A practical problem for calculating survival times is when patients leave trials before they die, which biases the data due to a form of censuring. Kaplan–Meier estimators can be used to correct for these biases,

$$\Psi(t_i) = \left(1 - \frac{d_i}{n_i}\right)\Psi(t_{i-1}), \tag{1.5}$$

where $\Psi(t_i)$ is the survival distribution at time t_i, d_i is the number of events that happen at t_i and n_i is the number of events that survive up to t_i. Survival times can be used in the more general context of biological physics using such Kaplan–Meier corrections e.g. the run times of bacteria can be considered as a distribution of survival times. Biases introduced by the finite length of tracks in tracking experiments can be corrected using Equation (1.5). Survival times can also be used to understand the residence times of bacteria on surfaces.[41] Survival times (e.g. for runs, Figure 1.8a) can be simpler than just considering histograms (or pdfs), since they are monotonic decreasing functions (in contrast, the run time pdfs, $P(\tau)$, will be peaked, Figure 1.8c).

The *first passage probability* (FPP) for particle tracks is defined as the probability distribution for the times a particle takes to travel a specific distance for the first time[43] (Figure 1.9a). The *mean* FPP (MFPP, v_{FPP}) is the mean of the FPP distribution.

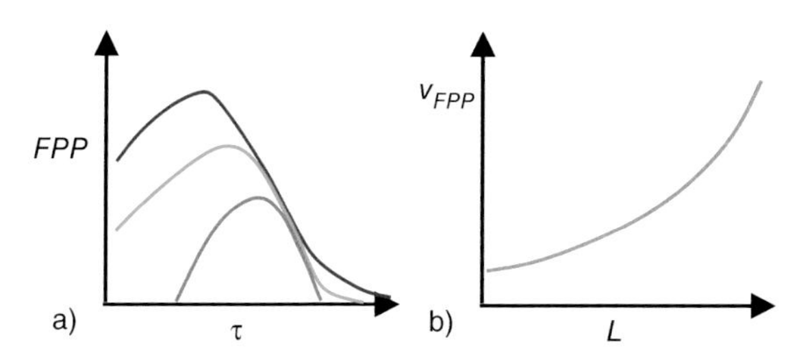

Figure 1.9 (a) *First passage probability* (FPP) of particles moving inside a cell as a function of time (τ), red < blue < navy blue correspond to longer transit lengths. (b) *Mean first passage velocity* (v_{FPP}) of the particles as a function of transit length (L).[39,42]

In numerous biological situations, the FPP is the crucial statistical quantity of interest e.g. times for a chemical reaction to occur or for a particle to leave a maze. MFPPs for particles with multiple scaling regimes as a function of time imply that the reaction kinetics of the particle will also exhibit multiple regimes. Furthermore, FPPs can also provide more robust alternatives to instantaneous velocities to quantify motility and can help separate up the motility of particles using the average FPP velocity (v_{FPP}) as a function of transit length (L, Figure 1.9b) e.g. the question can be asked as to whether long-range transits happen at larger velocities, which is useful with endosomal transport.

MSDs are insensitive to direction (so too are the survival times and the FPPs), and they just provide a measure of the amplitude of motion as a function of time. *Angular correlations* of particle displacements thus provide crucial information to understand particle motility[44] with respect to direction. Analogous to an MSD, the average direction cosine $(\langle \cos\theta(\tau) \rangle)$ for segments along a track can be quantified as a function of time interval (averaged in an analogous manner to a TAMSD, Equation (1.1), although three points are required to define the consecutive displacements Δr_i and Δr_{i+1}), and a scalar product is used,

$$\langle \cos\theta(\tau) \rangle = \frac{\langle \Delta r_i . \Delta r_{i+1} \rangle}{|\Delta r_i||\Delta r_{i+1}|}. \tag{1.6}$$

Cosines are bounded functions; $-1 \le \cos\theta \le 1$. Negative values of the angular correlation function correspond to anti-persistent motion i.e. the particle is constantly changing direction and tends to move back on itself. $\langle \cos\theta \rangle = 0$ corresponds to no average directional bias and is expected for an unbiased random walk. Positive values of $\langle \cos\theta(\tau) \rangle$ correspond to *directional persistence*. Such measures of directionality are useful for the development of models for bacteria, since bacteria act as stochastic swimmers, and for the motility of intracellular cargoes within bacterial cells. Some similar information is encoded in velocity correlation functions (Equation 1.2), but it is useful to have both measures.

More sophisticated statistical measures are needed to describe the correlated motion of particles. Two-point correlation functions are one possibility, $\langle \Delta r_1 \Delta r_2 \rangle$, where Δr_1 and Δr_2 are the displacements for two different particles, which has been studied from the perspective of two-particle microrheology.[45] Velocity cross-correlation functions $(\langle v_1 v_2 \rangle)$ are also useful to study the mutual motion of cells e.g. in chemotactic fields, and provide similar information to $\langle \Delta r_1 \Delta r_2 \rangle$.

Flocking order parameters (Φ) have also been introduced to describe phase transitions during coherent motion in motile particles[46] (such as bacteria, starlings and ants) e.g.

$$\Phi = \frac{1}{Nv}\left|\sum_i v_i\right|, \tag{1.7}$$

where N is the number of particles, v is the average velocity and v_i is the velocity of each particle. Care must be taken in calculating Φ for particles that experience anomalous transport, since the values of v_i will depend on time for non-ballistic particle motion, and other order parameters have been suggested to make them more robust.[47]

A sophisticated modern approach to the motility of both particles inside bacteria and cellular motility follows a framework of *heterogeneous anomalous transport* (HAT).[48] This attempts to quantify the heterogeneity of the anomalous transport of particles in both space and time by considering generalised diffusion coefficients $(D_\alpha(r,t))$ and scaling exponents $(\alpha(r,t))$ that vary in space and time, defined via the MSDs of the distributions using

$$\langle \Delta r^2 \rangle = 2nD_\alpha(r,t)\tau^{\alpha(r,t)}, \tag{1.8}$$

where n is the number of dimensions. Note that D_α has fractional units, which provides some challenges e.g. it is not possible to plot D_α on a single axis. Rescaling D_α by characteristic length and time scales solves many of these problems. Thus, values of both D_α and α are allowed to vary with time and space during the analysis, which corresponds to a multi-fractal model.[23] There is good evidence that HAT occurs for the majority of cellular motility and intracellular motility of large molecules and aggregates.

Experiments with extended linear objects, such as single molecules, aggregates of cells, organelles or individual cells, lend themselves to Fourier analysis of data segmented using snakes algorithms[49] (Figure 1.4). The equipartition theorem can be used to calculate the energy of each Fourier mode assuming the fibres are in thermal equilibrium and simple continuum models are used for the energy of the snakes e.g. all the energy is stored in Hookean bending modes.[49] Similar analysis is also possible with cell membranes in two dimensions.[50] Challenges occur to describe systems in which quenched disorder or active transport affect the conformations of the extended objects.[3,17]

Other less direct methods of calculating stochastic processes occur in the literature. For example, the square of the Fourier transform of the particle displacement as a function of the correlation time is called the power spectral density $(P(\omega))$ and is often measured in optical tweezer experiments. The information content is similar to a MSD as a function of time interval, but MSDs are often simpler to work with.

1.5 Scattering Alternatives

Instead of working directly with images, *scattering experiments* function in reciprocal space. Historically, scattering techniques were used in situations where imaging was not possible e.g. in experiments with hard X-rays or thermal neutrons where it is hard to construct an imaging lens, such as the Braggs' initial work to study the structure of simple crystals, such as sodium chloride. Images tend to be preferred in modern-day biological physics experiments, since they are easier to interpret and tend to be less ambiguous, but scattering data can also be useful. Inelastic scattering experiments (e.g. dynamic light scattering, X-ray photon correlation spectroscopy or quasi-elastic neutron scattering) detect small energy changes in scattered radiation and often provide much faster dynamic information than is currently possible with imaging experiments e.g. point detectors can stream data much faster than pixel arrays. In bacterial biophysics, *fluorescence correlation spectroscopy* (FCS) and *dynamic differential microscopy* (DDM) are commonly used scattering techniques and can be microscope based to improve their spatial sensitivity.

In DDM, a movie of a biological system is made with a microscope and versions are possible using both coherent (e.g. bright-field contrast) and incoherent (e.g. fluorescence) scattering. Software correlators can then be used with stacks of the images to calculate correlation functions that describe the image dynamics (Figure 1.10). The correlation functions allow access to identical information as inelastic scattering experiments, although they are at relatively slow time scales due to the update times of pixel arrays used on standard digital cameras in optical microscopes.

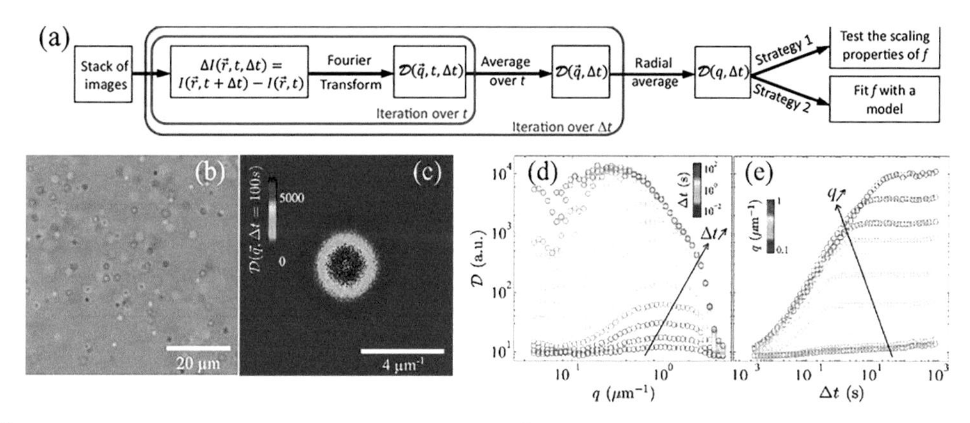

Figure 1.10 (a) A schematic diagram of the algorithm used to analyse DDM[51] experiments. (b) An example of the difference of two images $\left(d(r,t,\tau)\right)$. (c) The square of the Fourier transform of the difference of two images. (d) $D(q,\tau)$ (the square of the Fourier transform of $d(r,t,\tau)$ averaged over t) as a function of momentum transfer, q. (e) $D(q,\tau)$ as a function of time interval ($\Delta t = \tau$). Reprinted from [Germain D., Leocmach M., Gibaud T., Differential dynamics microscopy to characterise Brownian motion and bacteria motility. *American Journal of Physics* 2016, 84, 202], with the permission of AIP Publishing.

Intermediate scattering functions $\left(\text{ISF}, f(q,\tau)\right)$ are a key statistical tool used in inelastic scattering experiments to quantify the dynamics e.g. with light, neutrons and X-ray.[52] DDM can be used to extract the ISF from stacks of images. There needs to be a source of speckle on the images (which is non-necessarily due to coherent scattering) and the images do not need to be particularly well resolved. DDM works well on images from both bright-field (with both laser and light-emitting diode [LED] illumination) and fluorescence microscopy (the calculations are slightly different in each case). Taking differences between images suppresses the noise due to stationary particles and detector heterogeneities[53] (hence the name DDM), which gives

$$d(r,t,\tau) = I(r,t_0+\tau) - I(r,t_0). \tag{1.9}$$

Next, these image differences are Fourier transformed in space (q is the momentum transfer) and squared,

$$D(q,\tau) = \left\langle |d(q,t_0,\tau|^2 \right\rangle_{t_0}. \tag{1.10}$$

The ISF $\left(f(q,\tau)\right)$ is constructed by fitting $A(q)$ and $B(q)$ using

$$D(q,\tau) = A(q)\left[1 - f(q,\tau)\right] + B(q), \tag{1.11}$$

where $A(q)$ and $B(q)$ are assumed arbitrary smooth functions.

In DDM, a major advantage compared with tracks from direct imaging is that the dynamics can be averaged and quantified without segmentation of the images. The neglect of segmentation can be an advantage for complex hierarchical structures e.g. the endoplasmic reticulum in eukaryotic cells, where it can be hard to unambiguously locate the objects' boundaries.[3] Challenges with DDM are that some spatial information is lost in the averaging procedures (e.g. during the calculation of the Fourier transforms), analytic calculations are slightly harder in reciprocal space and it can be more challenging to determine which specific structures are being analysed.

Fluorescence correlation spectroscopy considers the fluctuations in fluorescent emission from a small volume that is illuminated in a sample[54] (Figure 1.11). The fluctuations in fluorescent emission can be related to the motion of the fluorophores through the calculation of correlation functions ($G(\tau)$, where τ is the correlation time) and can be performed relatively quickly (using fast point photodetectors) and thus experiments can be performed with quickly photobleaching fluorophores. If only a single detection volume is used in the sample, it is harder to explore the length dependence of the dynamic processes using FCS (via q, the momentum transfer). This can make the exploration of anomalous transport more challenging (the spatial dependence of the motility is ambiguous) and a partial solution is to use a range of pin hole sizes, so different volumes are illuminated in the sample.[55] In this case, the fluorophores need to be long-lived and the statistical processes must be stationary (they should not evolve with time).

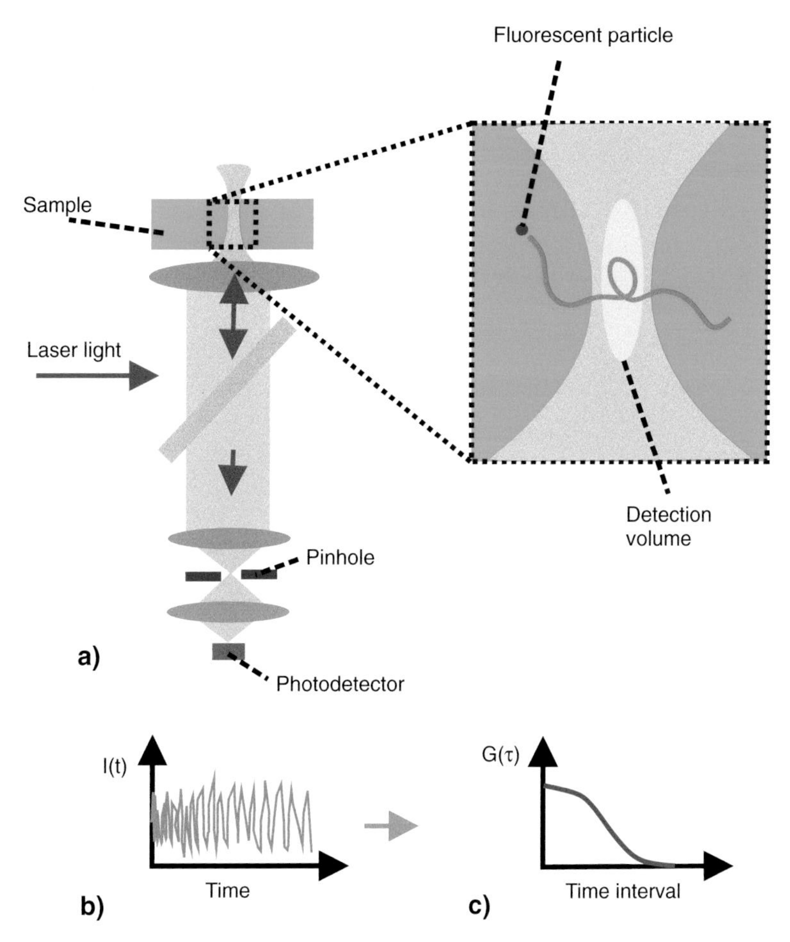

Figure 1.11 (a) Apparatus for FCS is based on a fluorescence microscope, which uses a pinhole to define the detection volume. (b) The intensity of light ($I(t)$) emitted by fluorophores in the detection volume as a function of time. (c) A correlation function ($G(\tau)$) of the intensity fluctuations as a function of time interval (τ) from (b). The correlation function can be used to quantify the motion of the fluorescence particles within the detection volume.

Suggested Reading

Höfling, F.; Franosch, T., Anomalous transport in the crowded world of biological cells. *Reports on Progress in Physics* **2013**, *76*, 046602. Good overview of the experimental evidence for anomalous transport inside cells.

Ibe, O. C. *Elements of Random Walks and Diffusion Processes*. Wiley: 2013. Introduces anomalous transport (e.g. fractional Brownian motion) in an intelligible manner for non-mathematicians. Also acts as a good primer for stochastic processes.

Klafter, J.; Sokolov, I. M., *First Steps in Random Walks: From Tools to Applications*. Oxford University Press: 2011. Short and fairly mathematical introduction to some modern models for anomalous transport.

Waigh, T. A.; Korabel, N., Heterogeneous anomalous transport in molecular and cellular biology. *Reports on Progress in Physics* **2023**, *86*, 126601. Considers some challenges in the modelling of anomalous transport in cellular biology e.g. multi-fractal effects.

References

1. Lane, N., The unseen world: Reflections on Leeuwenhoek (1677) 'Concerning little animals'. *Philosophical Transactions of the Royal Society B* **2015**, *370* (1666), 20140344.
2. Mertz, J., *Introduction to Optical Microscopy*. Cambridge University Press: 2019.
3. Perkins, H. T.; Allan, V. J.; Waigh, T. A., Network organisation and the dynamics of tubules in the endoplasmic reticulum. *Scientific Reports* **2021**, *11* (1), 16230.
4. Dubay, M. M.; Acres, J.; Riebeles, M.; Nadeau, J. L., Recent advances in experimental design and data analysis to characterise prokaryotes motility. *Journal of Microbiological Methods* **2023**, *204*, 106658.
5. Chen, B. C.; et al., Lattice light-sheet microscopy-imaging molecules to embryos at high spatiotemporal resolution. *Science* **2014**, *346* (6208), 1257998.
6. Wolff, J. O.; Scheiderer, L.; Engelhardt, T.; Maththias, J.; Hell, S. W., MINIFLUX dissects the unimpeded walking of kinesin-1. *Science* **2023**, *379* (6636), 1004–1010.
7. Rogers, S. S.; Waigh, T. A.; Zhao, X.; Lu, J. R., Precise particle tracking against a complicated background: Polynomial fitting with Gaussian weight. *Physical Biology* **2007**, *4* (3), 220–227.
8. Helgadottir, S.; Argua, A.; Volpe, G., Digital video microscopy enhanced by deep learning. *Optica* **2019**, *6* (4), 506.
9. Szeliski, R., *Computer Vision: Algorithms and Applications*, 2nd ed. Springer: 2022.
10. Waigh, T. A., Advances in the microrheology of complex fluids. *Reports on Progress in Physics* **2016**, *79* (7), 074601.
11. Xu, C.; Zhang, L.; Huang, S.; Ma, T.; Liu, F.; Yonezawa, H.; Zhang, Y.; Xiao, M., Sensing and tracking enhanced by quantum squeezing. *Photonics Research* **2019**, *7* (6), 14.
12. Taylor, M. A.; Janousek, J.; Daria, V.; Knittel, J.; Hage, B.; Bachor, H. A.; Bowen, W. P., Biological measurement beyond the quantum limit. *Nature Photonics* **2013**, *7* (3), 229–233.
13. Hart, J. W.; Waigh, T. A.; Lu, J. R.; Roberts, I. S., Microrheology and spatial heterogeneity of *Staphylococcus aureus* biofilms modulated by hydrodynamic shear and biofilm-degrading enzymes. *Langmuir* **2019**, *35* (9), 3553–3561.

14. Rogers, S. S.; van der Walle, C.; Waigh, T. A., Microrheology of bacterial biofilms *in vitro*: *Staphylococcus aureus* and *Pseudomonas aeruginosa*. *Langmuir* **2008**, *24* (23), 13549–13555.
15. Moerner, W. E.; Kardor, L., Optical detection and spectroscopy of single molecules in a solid. *PRL* **1989**, *62* (21), 2535–2538.
16. Leake, M. C., *Single-Molecular Cellular Biophysics*. Cambridge University Press: 2013.
17. Cox, H.; Xu, H.; Waigh, T. A.; Lu, J. R., Single-molecule study of peptide gel dynamics reveals states of prestress. *Langmuir* **2018**, *34* (48), 14678–14689.
18. Cox, H.; Cao, M.; Xu, H.; Waigh, T. A.; Lu, J. R., Active modulation of states of prestress in self-assembled short peptide gels. *Biomacromolecules* **2019**, *20* (4), 1719–1730.
19. Newby, J. M.; Schaefer, A. M.; Lee, P. T.; Forest, M. G.; Lai, S. K., Convolutional neural networks automate detection for tracking of submicron-scale particles in 2D and 3D. *Proceedings of the National Academy of Sciences* **2018**, *115* (36), 9026–9031.
20. Chenouard, N.; et al., Objective comparison of particle tracking methods. *Nature Methods* **2014**, *11* (3), 281–289.
21. Wu, P. H.; Agarwal, A.; Hess, H.; Khargonekar, P. P.; Tseng, Y., Analysis of video-based microscopic particle trajectories using Kalman filtering. *Biophysical Journal* **2010**, *98* (12), 2822–2830.
22. Murphy, K. P., *Probabilistic Machine Learning: An Introduction*. MIT: 2022.
23. Waigh, T. A.; Korabel, N., Heterogeneous anomalous transport in cellular and molecular biology. *Reports on Progress in Physics* **2023**, *86* (12), 126601.
24. Dacret, A.; Quardokus, E. M.; Brun, Y. V., MicrobeJ, a tool for high throughput bacterial cell detection and quantitative analysis. *Nature Microbiology* **2016**, *1* (7), 1.
25. Hartmann, R.; et al., Quantitative image analysis of microbial communities with BiofilmQ. *Nature Microbiology* **2021**, *6* (2), 151.
26. Kopera, B. A. F.; Retsch, M., Computing the 3D radial distribution function from particle positions: An advanced analytic approach. *Analytic Chemistry* **2018**, *90* (23), 13909–13914.
27. Holmes, S.; Huber, W., *Modern Statistics for Modern Biology*. Cambridge University Press: 2019.
28. Allen, M. P.; Tildesley, D. J., *Computer Simulation of Liquids*. Oxford University Press: 2017.
29. Hansen, J. P.; McDonald, I. R., *Theory of Simple Liquids: With Applications to Soft Matter*. Academic Press: 2013.
30. Metzler, R.; Klafter, J., The restaurant at the end of the random walk. *Journal of Physics A: General Physics* **2004**, *37* (31), R161–R208.
31. Han, D.; Korabel, N.; Chen, R.; Johnston, M.; Gavrilova, A.; Allan, V. J.; Fedotov, S.; Waigh, T. A., Deciphering anomalous heterogeneous intracellular transport with neural networks. *eLife* **2020**, *9*, e52224.
32. Nielsen, A., *Practical Time Series Analysis: Prediction with Statistics and Machine Learning*. O'Reilly: 2020.

33. Hofling, F.; Franosch, T., Anomalous transport in the crowded world of biological cells. *Reports on Progress in Physics* **2013**, *76*, 046602.
34. Sornette, D., *Critical Phenomena in Natural Sciences*. Springer: 2003.
35. Ibe, O. C., *Elements of Random Walks and Diffusion Processes*. Wiley: 2013.
36. Birkhoff, G. D., Proof of the ergodic theorem. *Proceedings of the National Academy of Sciences* **1931**, *17* (12), 656–660.
37. Korabel, N.; Taloni, A.; Pagnini, G.; Allan, V. J.; Fedotov, S.; Waigh, T. A., Ensemble heterogeneity mimics ageing for endosomal dynamics within eukaryotic cells. *Scientific Reports* **2023**, *13* (1), 8789.
38. Berg, H. C., *Random Walks in Biology*. Princeton University Press: 1993.
39. Rogers, S. S.; Flores-Rodriguez, N.; Allan, V. J.; Woodman, P. G.; Waigh, T. A., The first passage probability of intracellular particle trafficking. *PCCP* **2010**, *12* (15), 3753–3761.
40. Aalen, O.; Borgan, O.; Gjessing, H., *Survival and Event History Analysis: A Process Point of View*. Springer: 2008.
41. Blee, J. A.; Roberts, I. S.; Waigh, T. A., Membrane potentials, oxidative stress and the dispersal response of bacterial biofilms to 405 nm light. *Physical Biology* **2020**, *17* (4), 036001.
42. Flores-Rodriguez, N.; Rogers, S. S.; Kenwright, D. A.; Waigh, T. A.; Woodman, P. G.; Allan, V. J., Roles of dynein and dynactin in early endosome dynamics revealed using automated tracking and global analysis. *PLOS One* **2011**, *6* (9), e24479.
43. Redner, S., *A Guide to First Passage Processes*. Cambridge University Press: 2001.
44. Harrison, A. W.; Kenwright, D. A.; Waigh, T. A.; Woodman, P. G.; Allan, V. J., Modes of correlated angular motion in live cells across three distinct time scales. *Physical Biology* **2013**, *10* (3), 036002.
45. Levine, A. J.; Lubensky, T. C., One- and two-particle microrheology. *Physical Review Letters* **2000**, *85*, 1774.
46. Vicsek, T.; Czirok, A.; Ben-Jacob, E.; Cohen, I.; Shochet, O., Novel type of phase transition in a system of self-driven particles. *Physical Review Letters* **1995**, *75* (6), 1226–1229.
47. Cavagna, A.; Giardina, I.; Grigera, T. S., The physics of flocking: Correlation as a compass from experiments to theory. *Physics Reports* **2018**, *728* (3), 1–62.
48. Itto, Y.; Beck, C., Superstatistical modelling of protein diffusion dynamics in bacteria. *Journal of the Royal Society – Interface* **2021**, *18* (176), 20200927.
49. Gittes, F.; Mickey, B.; Nettleton, J.; Howard, J., Flexural rigidity of microtubules and actin filaments measured from thermal fluctuations in shape. *Journal of Cellular Biology* **1993**, *120* (4), 923–934.
50. Monzel, C.; Sengupta, K., Measuring shape fluctuations in biological membranes. *Journal of Physics D: Applied Physics* **2016**, *49* (24), 243002.
51. Germain, D.; Leocmach, M.; Gibaud, T., Differential dynamic microscopy to characterize Brownian motion and bacteria motility. *American Journal of Physics* **2016**, *84* (3), 202.

52. Berne, B. J.; Percora, R., *Dynamic Light Scattering: With Applications to Chemistry, Biology and Physics*. Dover: 2003.
53. Cerbino, R.; Cicuta, P., Perspective: Differential dynamic microscopy extracts multiscale activity in complex fluids and biological systems. *Journal of Chemical Physics* **2017**, *147* (11), 110901.
54. Rigler, R.; Elson, E. S., *Fluorescence Correlation Spectroscopy: Theory and Applications*. Springer: 2001.
55. Stolle, M. D.; Fradin, C., Anomalous diffusion in inverted variable-lengthscale fluorescence correlation spectroscopy. *Biophysical Journal* **2019**, *116* (5), 791–806.

2 Statistics of Bacterial Motility

2.1 Passive Motility

The simplest model for bacterial motility needs to describe dead or immotile bacteria (e.g. *Staphylococcus aureus*) in dilute suspensions. The motion of the bacteria will follow a random walk in three dimensions driven by thermalised collisions of the bacteria with the surrounding solvent molecules i.e. *Brownian motion*. Einstein's PhD thesis results are very useful for the statistical analysis of the passive motion of bacteria.[1] Since the immotile bacteria only experience Brownian motion, they have no preferred direction of motion and the average displacement $(\langle \Delta r \rangle)$ is zero, $\langle \Delta r \rangle = 0$. The average displacement is therefore not a particularly useful way to quantify the motion and instead the second moment of the displacement probability distribution (Δr^2, related to the variance) is used which is called the mean square displacement (MSD). The MSD $\left(\langle \Delta r^2 \rangle\right)$ has a characteristic linear dependence on the time interval (τ) for a random walk,

$$\langle \Delta r^2 \rangle = 2nD\tau, \tag{2.1}$$

where n is the number of spatial dimensions considered (1, 2 or 3) and D is the diffusion coefficient. The derivation of Equation (2.1) is robust to the exact molecular details of the collisions of the bacteria with their surroundings, as long as they are thermalised and the time scales are sufficiently long (inertia can be neglected). The diffusion coefficient can be calculated from the *fluctuation-dissipation theorem,*[2]

$$D = \frac{kT}{f}, \tag{2.2}$$

where kT is the thermal energy and f is the friction coefficient of the particle that is diffusing. For spherical particles, f is given by the Stokes equation, from the solution of Stokes' hydrodynamic equation (Chapter 9, Equation (9.3)),

$$f = 6\pi\eta r, \tag{2.3}$$

where η is the viscosity and r is the radius of the sphere. MSDs are often preferred over velocity measurements to describe diffusive processes because a single diffusion coefficient can characterise the motion over a wide range of time scales. Defining instantaneous velocities will necessitate that they depend on the time scale considered and their exact functional form then needs to be considered (Section 1.5). Shorter time scales imply higher velocities for diffusion due to the fractal nature of random walks i.e. the physical process has a characteristic non-integer dimensionality.

Einstein's microscopic interpretation of Brownian motion that gives rise to Equation (2.1) can be shown to be equivalent to the macroscopic continuum approach (Fick's 2nd law),

$$\frac{\partial c}{\partial t} = D\nabla^2 c, \tag{2.4}$$

where c is the concentration of diffusing particles (e.g. bacteria) and D is the same coefficient that is observed on the microscopic level, Equation (2.2) i.e. a single diffusion coefficient, describes single particle motion and the collective arrangements of large numbers of dilute bacteria (below a critical bacterial concentration where the distinction between mutual and self-diffusion becomes necessary[2]).

In practice, applying Equations (2.1) to (2.4) for diffusion to experiments on immotile bacteria is not necessarily straightforward. Bacteria are often denser than water and will sediment, limiting the time over which pure diffusion can be observed. Furthermore, bacteria (e.g. *S. aureus*) are often sticky and clump together, so finding the correct friction coefficient (f) for the aggregate needs to be carefully considered.

Since collisions with solvent molecules transfer angular momentum in addition to translational momentum, Brownian motion can also affect the angular motion of molecules and cells through *angular diffusion*. The subject has an illustrious history and is found in the PhD thesis of J. B. Perrin (Nobel Laureate for Brownian motion in 1926), who solved Fick's second law in spherical coordinates (r,θ,φ) and similar work was also performed by Peter Debye.[3] Such an analysis has been extended to micron-sized anisotropic colloidal particles[4] and tested using tracking with three-dimensional (3D) confocal microscopy experiments.

To calculate the angular fluctuations in the direction that a rigid particle is pointing, diffusion can be considered in spherical coordinates on a sphere of unit radius $(r = 1,\theta,\varphi)$. For small angles (and thus small times), an analogous expression to Equation (2.1) for the mean square angle $\left(\langle \Delta\theta^2 \rangle\right)$ is

$$\langle \Delta\theta^2 \rangle \approx 2nD_\theta \tau, \tag{2.5}$$

where $n = 1$ or 2, if one or two angular coordinates (θ,φ) in the spherical geometry are considered respectively, and D_θ is the angular diffusion coefficient. For spheres, the angular diffusion coefficient can be calculated from

$$D_\theta = \frac{kT}{8\pi\eta a^3}. \tag{2.6}$$

Equation (2.5) is a solution of Fick's 2nd law, Equation (2.4), expressed in spherical coordinates with $r = 1$,

$$\frac{\partial c}{\partial t} = D_\theta \frac{1}{\sin^2\theta}\left[\sin\theta \frac{\partial}{\partial\theta}\left(\sin\theta \frac{\partial}{\partial\theta}\right) + \frac{\partial^2}{\partial\varphi^2}\right] c. \tag{2.7}$$

For anisotropic particles it is possible to generalise Fick's 2nd law (Equation 2.4), combining both translational and rotational motion for three dimensions, in a single equation

$$\frac{\partial c}{\partial t} = D_{ij}\frac{\partial^2 c}{\partial q_i \partial q_j}, \tag{2.8}$$

where $D_{ij} = \begin{pmatrix} D_t & D_c^T \\ D_c & D_r \end{pmatrix}$ and q_i are a combination of the Cartesian coordinates $(i = 1,2,3)$ and the Euler angles $(i = 4,5,6)$. D_t is the translational diffusion submatrix, D_r is the rotational diffusion submatrix and D_c is the submatrix that describes coupling between translation and rotation. Numerical calculations can be made for D_{ij} for any rigid particle shapes based on Stokes hydrodynamics.[5]

Care is needed when considering the coordinates to describe the rotational diffusion of particles. If the question is asked, what direction is the rigid particle pointing, then spherical coordinates are convenient, the radius can be neglected and only θ and φ change the results (Figure 2.1a) e.g. Equation (2.7) results with the solution Equation (2.5) for small angles. However, in general, the particle will also be rotating along its axis via the Brownian motion of its third Euler angle[6] (ψ, Figure 2.1b, this is a general result of rigid body dynamics and we know the extra degree of freedom is supplied with thermal energy due to the equipartition theorem) and a more general approach is required as described by Equation (2.8).

Whether the fluctuations in all the Euler angles ((θ,φ,ψ), Figure 2.1b) can be measured in an experiment on angular diffusion depends on the instrumentation e.g. on the particular microscope used and the mechanism that provides image contrast. Recent confocal experiments (Section 13.2.2) on complex nanoparticles do provide information on all three Euler angles and the generalised approach is more informative.[5]

For motile bacteria, a combination of swimming and diffusion often occurs. The prototypical model describes runs of the bacteria (e.g. standard strains of *Escherichia coli*) as they swim in straight lines followed by tumbles in which the direction of motion is actively randomised by motors (Section 2.2). Such active transport is superposed on passive diffusion. Furthermore, in the long time limit, the processes of active transport can have diffusive scaling (they follow Equation (2.1)), but the effective diffusion coefficient is much larger than for a passive thermal mechanism, Equation (2.2), as seen in Equation (2.11). Angular diffusion restricts how bacteria

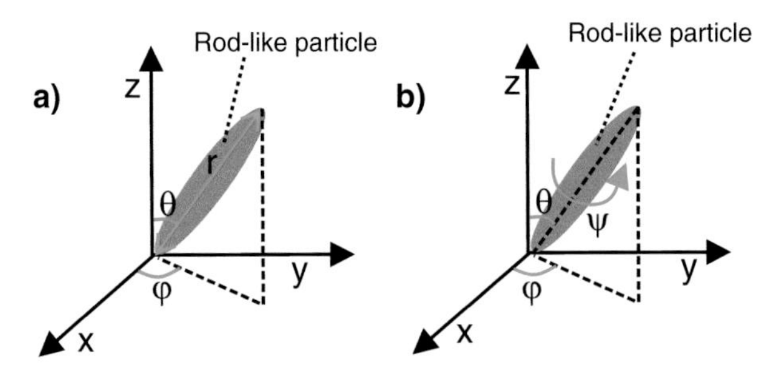

Figure 2.1 Geometries for the calculation of rotational diffusion. (a) Position of a rod-like particle on a sphere of radius r (spherical coordinates, r,θ,φ) and (b) Euler angles (θ,φ,ψ) for the position of the rod-like particle.

can use chemosensing to determine the direction of stimuli, since large angular fluctuations add significant noise to the detection process during their run and tumble motility (Chapter 18).

2.2 Active Motility of Single Bacteria

The standard statistical model for the motility of a single bacterium was developed by Lovely et al. and explored in detail with *E. coli*[8] by H. Berg and coworkers (Figure 2.2). For a bacterial cell swimming at constant speed, v, the cell is assumed to have a sequence of exponentially distributed (Poisson) runs that follow straight lines of average duration, τ. At the ends of the runs, the direction is randomised by a tumbling motion when the direction of the flagellar motor reverses and the flagellar motion becomes incoherent (*E. coli* are propelled by bundles of flagella).

Specifically, the Lovely and Dahlquist model has the following assumptions[7]:

a. The bacterium follows a series of straight-line trajectories joined by instantaneous changes in direction. These *runs* are characterised by their speed, direction and time duration.
b. The *runs* all have the same constant velocity (v).
c. A probability distribution can be defined for the choice of the new direction for a cell's motion and it is azimuthally symmetrical.
d. The probability distribution for the direction chosen for the motion of the cell is Markovian i.e. the direction chosen is independent of the preceding run direction.
e. The duration of a *run* follows an exponential Poisson distribution i.e. it has a constant probability per unit time to stop.

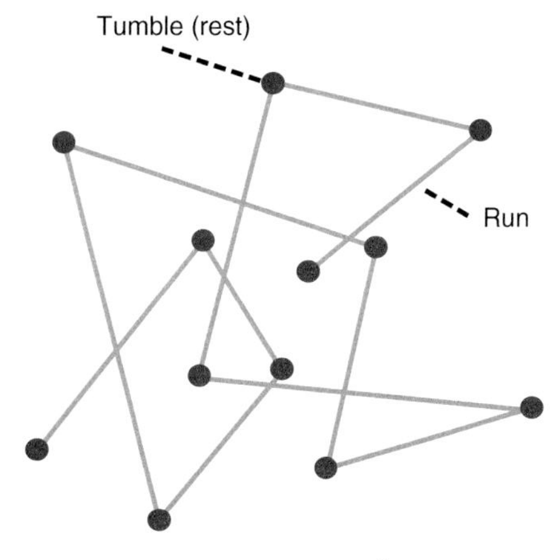

Figure 2.2 The *run-rest model* for bacterial motility due to Lovely and Dahlquist.[7] A hypothetical track for a single bacterium is shown in which exponentially distributed runs (straight lines) are interspersed with rests (circles, tumbles).

The subsequent calculations to describe bacterial motility based on the Lovely model borrow from the Flory model of a freely rotating polymer chain, where space is replaced by time in the derivation.[9] It is found that the run orientations ($\vec{p}(t)$, the unit vector along the swim direction) decay exponentially with time,

$$\left\langle \vec{p}(0).\vec{p}(t) \right\rangle = e^{-t/\tau_c}, \tag{2.9}$$

$$\text{and } \tau_c = \frac{T}{1-\alpha}, \tag{2.10}$$

where τ_c is the correlation time for the run orientation, T is the mean duration of a run and α is the mean value of the cosine angle between successive runs (this allows for a bias of the bacterial motility towards certain directions to be included in the calculations). The long time limit of the model tends towards Brownian motion (i.e. a linear scaling of the MSD with time interval following Equation (2.1)) with an effective diffusion coefficient[8] (D) given by

$$D = \frac{v^2 T}{3(1-\alpha)}. \tag{2.11}$$

Thus, in this model at short times, the bacterial motility is ballistic and the transition to diffusive motion occurs at longer time scales.

Although the Lovely model has some successes, application of this model to real bacteria poses some problems. Runs are rarely completely straight in experiments (implying a reduced persistence of the motion), the angular tumbles are not completely random, and the run and tumble times are modulated by sensory circuits in the bacteria, which can change over time (they have a memory). Run and tumble times also tend to be modulated in concentrated solutions due to interbacterial collisions, which again means the model needs to be recalibrated. Thus, Poisson models are reasonably successful in describing the low concentration motility of *E. coli* but often fail at high concentrations or when they are compared with high-resolution tracking measurements e.g. the run times do not follow perfect Poisson distributions. More recent work has tried to remedy some of these issues.

Often bacterial transport has some tortuosity (the tracks followed by bacteria between tumbles are not perfectly straight), which is quantified via the persistence of the bacterial motion (e.g. how the angular correlations of displacements decorrelate with time, Equation (1.5)). Some work has considered fractional Brownian motion (fBm) as a model for the persistence of bacterial motility.[10] This has the advantage of a simple scaling interpretation that can be directly related to the geometry of tracks, so it provides a robust description of changes in persistence of motion.

Formally, fBm is a continuous Gaussian process that is self-similar and has stationary increments.[11] The self-similarity is motivated by searching for a stochastic process $X_{\alpha t}$, $t \in [0,1]$, which has the same probability distribution as $\alpha^H X_{\alpha t}$, $t \in [0,1]$. Technically, fBm is defined through its covariance function (E),[11] which defines correlations as a function of time (t and s),

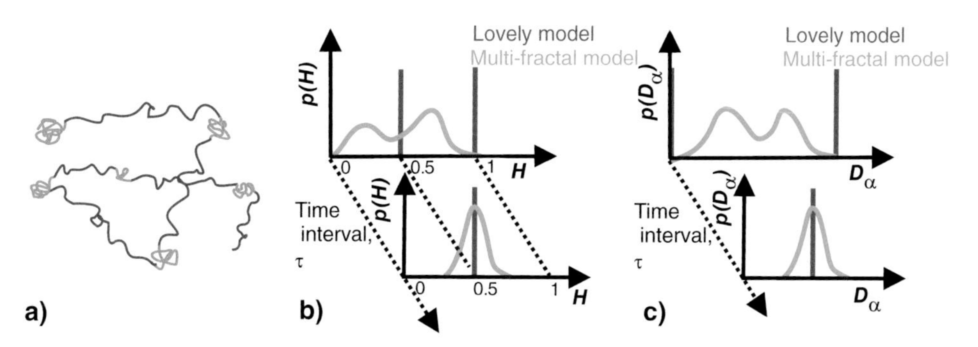

Figure 2.3 *Multi-fractal analysis* of *E. coli* motion. (a) Segmented tracks in terms of a multi-fractal model. The runs (purple) and rests (green) correspond to high H and high D_α, and low H and low D_α fractional Brownian motion (fBm), respectively, where H is the Hurst exponent and D_α is the generalised diffusion coefficient. (b) Predicted multi-fractal probability distribution ($p(H)$) of the Hurst exponent (H) for a prototypical bacterium swimming (following the Lovely model, red, and the multi-fractal model, blue) as a function of time interval (τ). The long time limit tends to diffusive transport ($H = 0.5$). (c) Probability distribution ($p(D_\alpha)$) of the generalised diffusion coefficient (D_α) for a prototypical bacterium swimming (following the Lovely model, red, and the multi-fractal model, blue) as a function of time interval (τ). More detailed multi-fractal analysis with neural networks can determine additional factors in the probability distributions of $p(H)$ and $p(D_\alpha)$ e.g. due to surface interactions. Note that D_α needs to be rescaled in practice to plot it on a single axis due to the fractional units.

$$E\left[B_H(t)B_H(s)\right] = \frac{1}{2}\left(|t|^{2H} + |s|^{2H} - |t-s|^{2H}\right), \tag{2.12}$$

where $B_H(t)$ is a continuous time Gaussian process and H is a bounded real number $[0,1]$. The persistence of fBm is quantified by the Hurst (H) fractal exponent. $H > 1/2$ describes *persistent motion*, $H = 1/2$ is a *random walk* and $H < 1/2$ is *anti-persistent motion*. Thus, fBm can conveniently describe motion over the range of interest in low Reynolds number motility, $0 < H < 1$ i.e. from stationary to ballistic motion. For $H > 1/2$, the motion is a long memory process, whereas for $H < 1/2$, it is a short memory process. fBm is only Markovian (memoryless) for $H = 1/2$. In multi-fractal fBm, the Hurst exponent is allowed to vary in space and time. A simple intuitive application of multi-fractal fBm is shown for *E. coli* in Figure 2.3 (blue). At short time intervals, there is a peak for low Hurst exponents corresponding to tumbling (anti-persistent motion) and a peak for high Hurst exponents corresponding to running motion. The two peaks merge together at longer time intervals as the motion tends to the diffusive type scaling and similar behaviour is expected for the generalised diffusion coefficients (Figure 2.3c). This is following the paradigm of *heterogeneous anomalous transport* (HAT) i.e. assuming anomalous transport, Equation (1.8), which varies in time and space.[12] New neural network methods can extract additional information from particle tracks and help differentiate between alternative models for multi-fractional Brownian motion.

In addition to issues associated with the directional persistence of runs, variations in the distributions of both the run and tumble times require a more flexible

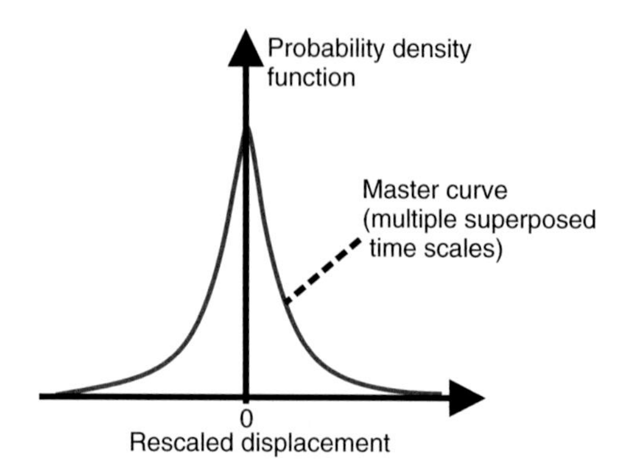

Figure 2.4 *E. coli* motion can be described as a Lévy walk.[14] Schematic probability distribution of the scaled displacements as a function of different time intervals (Δt). The displacements have been scaled according to predictions from the Lévy walk model and superposed on a master curve.

framework to perform calculations. For analytic calculations, this is provided by Lévy processes and more specifically Lévy walk models.[11,13] For example, such models can allow for memory effects to be included e.g. the adaptation of sensory circuits during motility. Formally, Lévy processes are defined as stochastic processes that have both stationary and independent increments.[11] Examples of Lévy processes include Brownian motion, Brownian motion with drift, Poisson processes and compound Poisson processes.

Lévy walks have been used to model the motility of *E. coli* as an extension of the traditional Poisson model presented by Lovely and others[15,16] (Figure 2.4). A particle moves with a constant velocity for a random time (this is known to be an approximation with *E. coli*[17]) and then at a turning point instantaneously switches to a new direction and then moves again. Periods of rest can also be added to the model and non-constant velocities can be included to make it more realistic. Fat-tailed distributions of run times can then be conveniently included in this mathematical framework, which is harder to do on an *ad hoc* basis.

The Lévy walk model also allows the experimentally observed power law statistics of run times seen with high-concentration suspensions of bacteria to be more accurately described. Thus, Lévy walks have been successful in describing the motion of swarming bacteria[18,19] (Chapter 21). Dilute bacterial suspensions show some additional evidence for Lévy walks, since power law runs are observed with individual tethered flagellar motors,[20] and the displacements of swimming *E. coli* collapse on a master curve scaled using Levy predictions[14] (Figure 2.4b). However, currently, Lévy walk models are unable to account for the varying degree of persistence observed experimentally for swimming bacteria, which are well described by multi-fractal fBm.

Further evidence for Lévy walk models is found for the upstream motion of *E. coli* in pressure-driven pipe flow. A very broad range of run times (power law distributed) for bacterial motility was required to explain the long-term upstream motility of *E. coli* using a biased random walk model i.e. random walks under flow.[21] A power law probability distribution of run times was superior to a model of exponential run times to describe this behaviour i.e. the power law,

$$\Psi(t) = \frac{\gamma}{\tau_0\left(1 + t/\tau_0\right)^{\gamma+1}}, \tag{2.13}$$

where $\Psi(t)$ is the distribution of run times (a survival time probability, Equation (1.5)), t is the run time, τ_0 is a characteristic time and γ is the scaling exponent. This expression can be compared with the run times for the standard Poisson case, which is exponentially distributed,

$$\Psi(t) = \frac{e^{-t/\tau_0}}{\tau_0}. \tag{2.14}$$

The statistics of motion of chemotactic bacteria are considered in more detail in Chapter 18.

Attempts have been made to model bacterial motility to include the dynamics of the sensory gene networks used in chemotaxis i.e. a hybrid systems biology approach. For example, the *behavioural variability* (BV) model quantifies the role of fluctuations of the phosphorylation of the protein CheY-P that regulates the statistics of motor switching in *E. coli*[22] (Chapter 18). A memory time is the central parameter in the BV model and describes the evolution of the persistence times of the running motion. The BV model is thus non-Markovian i.e. the statistical processes are time dependent.

Lévy walks and fBm models both describe non-Markovian processes and in general, such models are considered in the field of anomalous transport.[13,23] Bacteria are found to act as non-Markovian walkers in experiments with *E. coli* in fractal mazes.[24] Lévy walks are a possible description for such non-Markovian behaviour in fractal mazes, but there are others e.g. diffusion on fractal lattices or fractional Brownian motion.[23] Models (subordinated random walks) that include simultaneous mixtures of fBm and continuous time random walks were developed to combine the advantages of fBm to describe the persistence of motion and continuous time random walks (CTRWs) to describe broad dwell time distributions.[25] They are yet to be applied to bacterial motility.

Numerical data-driven approaches are expected to be most flexible in describing the motility of the huge range of bacteria that naturally occur e.g. to experimentally measure the run and rest survival times, Hurst exponents, generalised diffusion coefficients, angular correlations, pdfs and so on and insert them into stochastic agent-based models that experience anomalous dynamics (Chapter 27). Numerically calculated conditional probabilities (e.g. the probability of a run time given a previous run time $\left(P\left(t_1 / t_2\right)\right)$ and ergodicity parameters (Section 1.5)) are important to quantify memory effects over a range of time and length scales. However, analytic models do provide important lessons to understand the behaviour of particles experiencing anomalous motility and can be used to direct high-resolution machine learning approaches.[26]

Suggested Reading

Berg, H., *Random Walks in Biology*. Princeton: 1993. Classic elementary introduction to diffusive processes.

Ibe, O. C., *Elements of Random Walk and Diffusion Processes*. Wiley: 2013. Good introduction to the mathematics of stochastic processes for non-specialists.

Waigh, T. A.; Korabel, N., Heterogeneous anomalous transport in molecular and cellular biology. *Reports on Progress in Physics* **2023**, *86* (12), 126601. Considers the field of heterogeneous anomalous transport with molecules and cells.

References

1. Einstein, A., Uber die von der molekularkinetschen Theorie der Warme geforderte Bewegung von in ruhenden Flussigkeiten suspendierten Teilchen. *Annalen der Physik* **1905**, *322* (8), 549–560.
2. Chaikin, P. M.; Lubensky, T. C., *Principles of Condensed Matter Physics*. Cambridge University Press: 1995.
3. Berne, B. J., *Dynamic Light Scattering: With Applications to Chemistry, Biology and Physics*. Dover: 2003.
4. Hunter, G. L.; Edmond, K. V.; Elsesser, M. T.; Weeks, E. R., Tracking rotational diffusion of colloidal clusters. *Optics Press* **2011**, *19* (18), 17189–17202.
5. Zhang, Z. T.; Zhao, X.; Cao, B. Y., Diffusion tensors of arbitrary-shaped nanoparticles in fluid by molecular dynamics simulations. *Scientific Reports* **2019**, *9* (1), 18943.
6. Goldstein, H., *Classical Mechanics*, 3rd ed. Pearson: 2013.
7. Lovely, P. S.; Dahlquist, F. W., Statistical measures of bacterial motility and chemotaxis. *Journal of Theoretical Biology* **1975**, *50* (2), 477–496.
8. Berg, H. C., *Random Walks in Biology*. Princeton University Press: 1993.
9. Lauga, E., *The Fluid Dynamics of Cell Motility*. Cambridge University Press: 2020.
10. Zonia, L.; Bray, D., Swimming patterns and dynamics of simulated *Escherichia coli* bacteria. *Journal of the Royal Society – Interface* **2009**, *6* (40), 0397.
11. Ibe, O. C., *Elements of Random Walks and Diffusion Processes*. Wiley: 2013.
12. Waigh, T. A.; Korabel, N., Heterogeneous anomalous transport in cellular and molecular biology. *Reports on Progress in Physics* **2023**, *86* (12), 126601.
13. Klafter, J.; Sokolov, I. M., *First Steps in Random Walks: From Tools to Applications*. Oxford University Press: 2011.
14. Huo, H.; He, R.; Zhang, R.; Yuan, J., Swimming *Escherichia coli* cells explore the environment by Levy walk. *Applied and Environmental Microbiology* **2021**, *87* (6), e02429-20.

15. Matthaus, F.; Jagodic, M.; Donikar, J., *E. coli* superdiffusion and chemotaxis – search strategy, precision and motility. *Biophysical Journal* **2009**, *97* (4), 946–957.
16. Zaburdaev, V.; Denisov, S.; Klafter, J., Levy walks. *Review of Modern Physics* **2015**, *87*, 483.
17. Berg, H. C., *E. coli in Motion*. Springer: 2004.
18. Ariel, G.; Be'er, A.; Reynolds, A., Chaotic model for Levy walks in swarming bacteria. *Physical Review Letters* **2017**, *118* (22), 228102.
19. Ariel, G.; Rabani, A.; Benisty, S.; Partridge, J. D.; Harshey, R. M.; Be'er, A., Swarming bacteria migrate by Levy walk. *Nature Communications* **2015**, *6* (1), 8396.
20. Korobkova, E.; Emonet, T.; Vilar, J. M. G.; Shimizu, T. S.; Cluzel, P., From molecular noise to behavioural variability in a single bacterium. *Nature* **2004**, *428* (6982), 574–578.
21. Figueroa-Morales, N.; Rivera, A.; Soto, R.; Lindner, A.; Altshuler, E.; Clement, E., *E. coli* 'super-contaminates' narrow ducts fostered by broad run-time distribution. *Science Advances* **2020**, *6* (11), eoay0155.
22. Figueroa-Morales, N.; Soto, R.; Junat, G.; Darnige, T.; Douarche, C.; Martinez, V. A.; Lindner, A.; Clement, E., 3D spatial exploration by *E. coli* echoes motor temporal variability. *Physical Review X* **2020**, *10* (2), 021004.
23. Metzler, R.; Klafter, J., The restaurant at the end of the random walk. *Journal of Physics A: General Physics* **2004**, *37* (31), R161–R208.
24. Phan, T. V.; Morris, R.; Black, M. E.; Do, T. K.; Lin, K.; Nagy, K.; Sturm, J. C.; Bos, J., Bacterial route finding and collective escapes in mazes and fractals. *Physical Review X* **2020**, *10* (3), 0310107.
25. Tabei, S. M. A.; Burov, S.; Kim, H. Y.; Kuznetsov, A.; Huynh, T.; Jureller, J.; Philipson, L. H.; Dinner, A. R.; Scherer, N. F., Intracellular transport of insulin granules is a subordinated random walk. *PNAS* **2013**, *110* (13), 4911–4916.
26. Murphy, K. P., *Probabilistic Machine Learning: An Introduction*. MIT Press: 2022.

3 The Electrochemical Potential of a Cell

Electrochemical potentials are relatively simple for a biological system to develop.[1] All that is needed is a partition (e.g. a membrane) that has an unequal amount of charged ions on either side e.g. a cell that enriches its environment for a particular charged ion.[2] Electrochemical potentials will thus have evolved fairly early on for life on planet Earth, as soon as membranes were created by the self-assembly of surface-active molecules.[2] However, the active modulation of membrane potentials requires more sophisticated molecular machinery e.g. the evolution of voltage-gated ion channels was needed with neurons, muscle and sensory cells in humans. Another perspective is that all organisms, including microorganisms, are fundamentally electroactive due to their respiration processes, which shuttle around charged ions to make use of chemical stores of energy. Electrochemical potentials are thus closely connected with the ability of cells to use chemical energy, which also presents a key evolutionary step.

The voltage across cell membranes due to imbalances of charged molecules (including ions) between the intracellular and extracellular environments[2] can be quantified (Figure 3.1). The *Nernst equation* allows the potential (V) due to each type of ion to be calculated at equilibrium via the equation,

$$V = \frac{kT}{ze}\ln\left(\frac{c_{out}}{c_{in}}\right), \tag{3.1}$$

where z is the valence of the ion, kT is the thermal energy, e is the electronic charge, c_{out} is the ion concentration outside the cell and c_{in} is the ion concentration inside the cell. The Nernst equation is derived from simple Boltzmann statistics for the movement of charged ions across the energy barrier (the membrane) due to the voltage difference over the membrane. The total resting potential (V_{rest}) across a cell can then be calculated by adding up the individual Nernstian potentials of each ion (V_i) weighted by the conductances (g_i) of the ion channels that transmit them,

$$V_{rest} = \frac{\sum_i g_i V_i}{\sum_i g_i}. \tag{3.2}$$

Although all biological cells have potentials across their membranes, not all of them are *excitable*. Excitable cells are classified as those that can actively modulate the potential across their membranes and in humans, well-studied examples include neurons, sensory cells, endocrine cells and muscle cells e.g. cardiac cells in hearts. Due to their larger sizes and importance in medicine, excitable eukaryotic cells have been

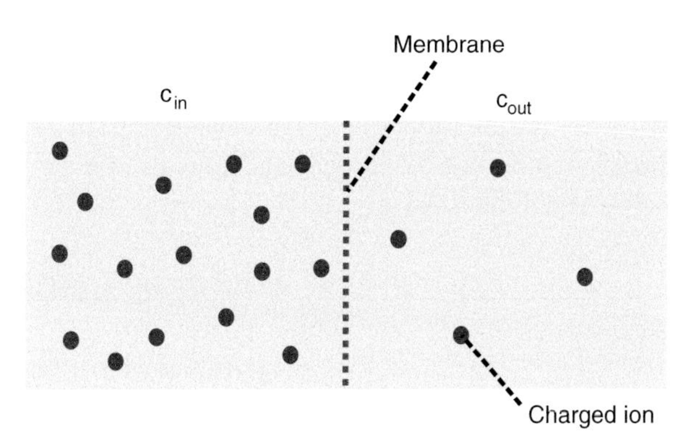

Figure 3.1 An imbalance of charged ions (circles) across a partition leads to a *Nernstian potential*. c_{in} is the concentration of ions on the left (e.g. inside a cell), whereas c_{out} is the concentration of ions on the right (e.g. outside the cell).

explored in detail and lots of quantitative analysis has been performed.[1,3] The study of the electrophysiology of bacterial cells is much less well established.

Action potentials are also observed in eukaryotic microorganisms, such as fungi, *Amoeba proteus* and paramecium.[4,5] Again, there are many gaps in our understanding of these organisms.

Voltage-gated ion channels are a signature of *excitable* cells, since they provide the cells with a direct mechanism to modulate their membrane potentials. Voltage-gated ion channels exist in a large number of bacteria[6] and it is thus possible that many of them are excitable.

The *Hodgkin–Huxley* (HH) equations provide a standard description of how membrane potentials in excitable cells are modulated with time and such models are based on the development of equivalent electrical circuits.[1,7,8] They were originally developed to describe the electrophysiology of neurons from giant squid but subsequently were widely used to describe all excitable eukaryotic cells. They can also be employed with bacterial cells. The main HH equations used consist of two types: a current balance for the transfer of ions through an equivalent circuit for the membrane (from Kirchhoff's law) and empirical equation(s) for the opening dynamics of the ion channels as a function of time (often simple first-order kinetics are assumed) and voltage (to allow for feedback from voltage gating). An example of the current balance equation for *Bacillus subtilis* cells using Kirchhoff's law and a single variety of ion channel is

$$C\frac{dV}{dt} = -g_I n^4 \left(V - V_I\right) - g_{leak}\left(V - V_{leak}\right), \tag{3.3}$$

where C is the membrane capacitance, n is a gating parameter, V is the membrane voltage, g_I and g_{leak} are the ion channel and leak conductances, respectively, and V_I and V_{leak} are the ion and leak Nernst potentials, respectively.[7] The leak currents are required to describe non-ion channel currents seen during measurements. This HH model is for a single ion channel e.g. the YugO potassium channel with *B. subtilis*. More sophisticated models can be easily constructed by adding further contributions to the current from different ion channels using Kirchhoff's law for current conservation e.g. models of human hearts can invoke >20 different varieties of ion channels and there are thus >20 separate contributions to the currents on the right-hand side of Equation (3.3). To solve Equation (3.3) numerically, an additional expression is needed to describe the kinetics of n. A possibility (originally based on data for squid neurons) is

$$\frac{dn}{dt} = \alpha(S)(1-n) - \beta(V)n, \tag{3.4}$$

where the opening rate $\alpha(S)$ of the ion channel is a function of the signal strength (S). The response of the opening rate of the ion channel on the signal is often sigmoidal (a Hill function),

$$\alpha(S) = \frac{\alpha_0 S^{\mu}}{S_{th} + S^{\mu}}, \tag{3.5}$$

where μ is the cooperativity parameter and S_{th} is the threshold voltage. The ion channel closing rate $\beta(V)$ is assumed to depend on the voltage (V, a voltage-gated ion channel),

$$\beta(V) = \beta_0 e^{-V/\sigma_h}, \tag{3.6}$$

where β_0 is the maximum closing rate and σ_h gauges the sensitivity to voltage.

Together, Equations (3.3) to (3.6) constitute a closed model that can be numerically solved to calculate the electrochemical response of a bacterial cell to an external signal based on the activity of K^+ ions.[9] For example, the HH model can describe the formation of action potentials in bacteria due to irradiation with blue light (Figure 3.2). Action potentials are inherently non-linear phenomena due to positive feedback (a cascading domino effect for the opening of ion channels in cell membranes during a spike) and require ideas from non-linear physics for a quantitative understanding (Chapter 12).[10] For example, hysteresis, bifurcations, bursting and chaos for the action potentials can all occur.

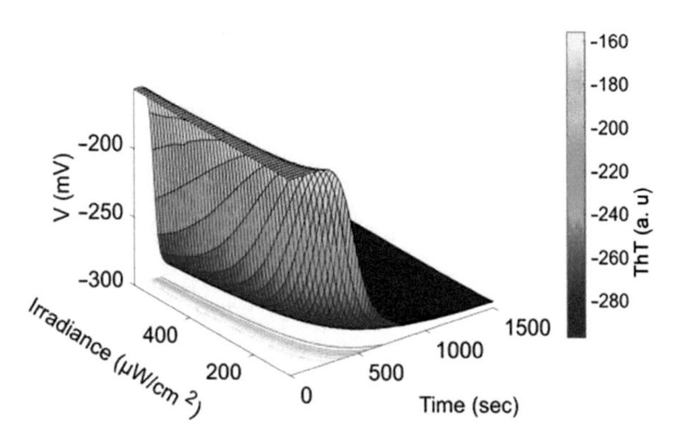

Figure 3.2 *Action potential* (V) for a bacterium in response to irradiation with blue light as a function of time calculated with a Hodgkin–Huxley model.[9]

In eukaryotic cells, the HH equations are the foundation for modelling neurons, sensory cells and cardiac cells.[11] They are the equivalent in electrophysiology of the Schrödinger equation in quantum mechanics. The HH model is able to describe action potentials in the cells (the form of spikes), refractoriness of the cells (gaps between spikes), the velocity of electrical spike transmission (via the cable equation) and how the populations of ion channels affect electrical behaviour e.g. how mutations affect neuronal or cardiac electrical activity.

Other sophisticated non-linear models have been developed to describe eukaryote electrophysiology e.g. for neuronal dynamics, and it is expected that they will be transferred to bacteria.[3] Thus, it is expected that more sophisticated models for ion channel conductivity and non-linear oscillators will be applied to bacteria as data sets improve e.g. to understand the response of bacterial cells to pulsed excitation, such as electrical fields or laser light.

Bacterial cell electrophysiology has some unique features compared to eukaryote electrophysiology due to the smaller size of bacteria.[12] Bacterial cells typically have volumes of $\sim 10^{-15}$ L and surface areas of $\sim 6\ \mu m^2$, whereas eukaryotic cells are much larger having $\sim 10^{-12}$ L and $\sim 1\,600\ \mu m^2$ respectively. There are thus three orders of magnitude differences between the volumes and areas of bacteria compared with eukaryotic cells. The capacitance of the membranes in bacteria is therefore much less and a thousandfold fewer ions are required to create an equivalent membrane potential. Lower capacitance implies the time for charging and discharging will be much smaller (RC is the characteristic time constant for a simple equivalent circuit, where R is the resistance and C is the capacitance). Due to the small number of ions, bacterial potentials are thus very sensitive to fluctuations in the number of ions and the opening of a single ion channel can deplete the potential difference across the entire membrane within a few seconds.

There are four general methods that have been used to measure the potentials inside bacteria (Figure 3.3):

1. *Fluorescent probes* can be used to measure Nernstian potentials (Figure 3.3(ai)) e.g. positively charged fluorophores enter cells at higher concentrations when the cells' voltage becomes more negative (during hyperpolarization). The amount of intracellular fluorescence thus correlates with the magnitude of the potential. Voltage-sensitive fluorescent proteins (Figure 3.3(aii)) can also be genetically expressed inside cells. These proteins are less invasive than extrinsic fluorophores and can be specific to a certain type of ion (e.g. Ca^{2+}), although fluorescent proteins tend to photobleach quickly. Synthetic fluorophores are often very bright, but they need to enter the cell (be membrane permeable), may perturb intracellular processes and careful control experiments are needed with all types of voltage-sensitive fluorophores to check the perturbations they cause.
2. *Flagellar rotation* can be used to measure the electrochemical potential (Figure 3.3b), since the potential provides the source of energy that drives the motility of bacteria i.e. the cells act as electrochemical batteries.[13] However, the models currently used to interpret data need to be revisited to describe the excitability

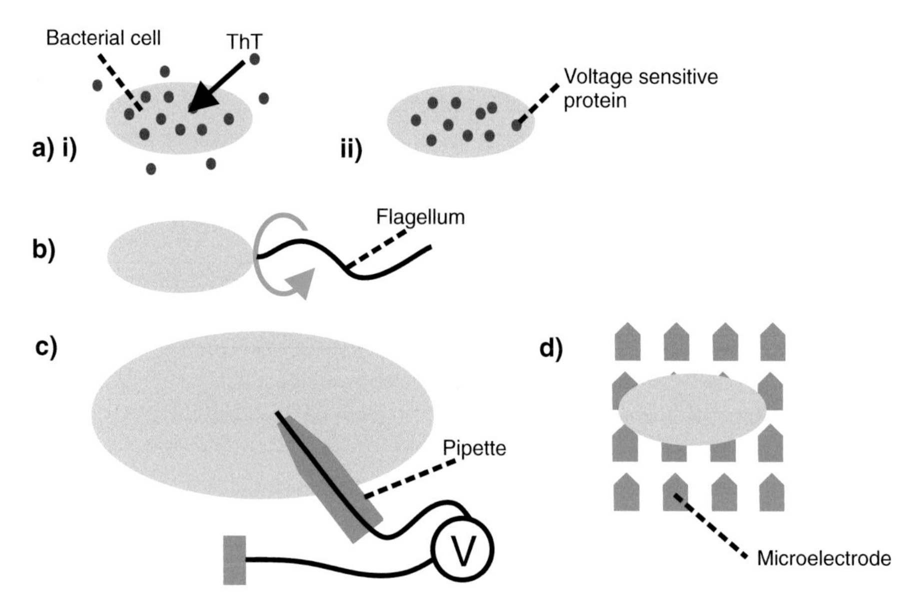

Figure 3.3 Methods to measure the electrochemical potentials inside bacteria. (a) *Fluorophores* are used with fluorescence microscopy: (i) membrane permeable *synthetic Nernstian dyes* (e.g. ThT) enter the cells when they hyperpolarize (become more negatively charged) and (ii) *optogenetic techniques* express voltage-sensitive fluorescent proteins inside cells; (b) the rate of *rotation of flagellar motors* can be used to gauge the membrane potential; (c) a *micropipette* can allow an electrode to be inserted in large mutant cells (spheroplasts are used, which are created from mutant cells treated with detergents) to measure the membrane potential; and (d) *microelectrode arrays* can be used to measure the potentials from many bacteria at the same time e.g. within biofilms.

of the cells e.g. they need to include voltage-gated ion channels via the HH Equations (3.3) to (3.6). Furthermore, the clutch that exists in some bacterial motors provides some experimental complications[14] i.e. the flagellar motor can be switched off, even at high membrane potentials. In larger eukaryotic micro-organisms, the intracellular electrochemical potentials play additional roles in their physiology, such as synchronising the movement of cilia in paramecium.[5] Whether this also occurs in bacteria via synchronisation of flagella has not yet been demonstrated.

3. *Patch clamps* allow direct measurements of the potential of a membrane held in a micropipette (Figure 3.3c). The use of traditional patch clamps requires giant mutant bacterial cells (spheroplasts formed by detergent treatment) to be made due to size constraints i.e. they require membranes that are greater than 1 – 2 μm in diameter.[15]
4. *Microelectrode arrays* (MEAs) were originally developed for spatially resolved measurements of neuronal tissue and have only recently been applied to bacteria. MEAs can provide spatially resolved measurements on surface adsorbed bacteria and inside biofilms[16] (Figure 3.3d). Other types of microelectrodes could also be used e.g. those based on AFM cantilevers.

The electrochemical potential across the cell membrane is used to drive a series of processes in bacterial physiology that would not occur otherwise e.g. ATP synthesis (a specific component of respiration), cell division and flagellar motors (predominantly via gradients in H^+ ions, although Na^+ ions are also possible). There is also evidence that the membrane potential regulates pH homeostasis, membrane transport, antibiotic resistance, electrical communication and environmental sensing. An ion motive force (IMF) is needed for active membrane transport i.e. to move cargoes against a concentration gradient. The uptake of positively charged antibiotics is also driven by the IMF and it thus modulates the susceptibility of bacteria to these antibiotics. Reducing the membrane potential of bacteria can increase the number of cells that persist in the presence of antibiotics by three orders of magnitude (persister cells are discussed in Chapter 24, where in addition to the reduced uptakes of antibiotics due to membrane potentials, this behaviour may be due to the reduction of their rate of metabolism).

There are a wide range of other roles of ion channels emerging in bacterial physiology. Bacterial chloride ion channels are important in the acid stress response. Deletion of the K^+ ion channel gene in *B. subtilis* prevents biofilm formation.[17] Ca^{2+} ion channels are used for sensing mechanical stress on membranes.[18]

Electrical effects in bacteria are considered in more detail in Chapter 29. A complex interplay is expected between metabolism, motility, signalling and many other physiological processes.

Suggested Reading

Benarroch, J. M.; Asally, M. The microbiologists guide to membrane potential dynamics. *Trends in Microbiology*, **2020**, *28* (4), 304. A very simple introduction to electrophysiological phenomena in bacteria.

Gerstner, W. *Neuronal Dynamics: From Single Neurons to Networks and Models of Cognition*. Cambridge University Press: 2014. Very clear theoretical development of models for neuronal dynamics. Many of the models could be applied to bacteria.

Keener, J.; Sneyd, J. *Mathematical Physiology*. Springer: 2009. Classic book on mathematical modelling of electrophysiology.

Smith, G. C. *Cellular Biophysics and Modelling: A Primer on the Computational Biology of Excitable Cells*. Cambridge University Press: 2019. Reasonably straightforward account of cellular excitability.

References

1. Keener, J.; Sneyd, J., *Mathematical Physiology*. Springer: 2009.
2. Waigh, T. A., *The Physics of Living Processes*. Wiley: 2014.

3. Gerstner, W., *Neuronal Dynamics: From Single Neurons to Networks and Models of Cognition*. Cambridge University Press: 2014.
4. Adamatzky, A., On spiking behaviour of oyster fungi *Pleurotus djamor*. *Scientific Reports* **2018**, *8* (1), 7873.
5. Greenspan, R. J., *An Introduction to Nervous Systems*. Cold Spring Harbor: 2007.
6. Ensembl. Database, G. bacteria.ensembl.org.
7. Prindle, A.; Liu, J.; Asally, M.; Ly, S.; Garcia-Ojalvo, J.; Sudel, G. M., Ion channels enable electrical communication in bacterial communities. *Nature* **2015**, *527* (7576), 59–63.
8. Smith, G. C., *Cellular Biophysics and Modelling: A Primer on the Computational Biology of Excitable Cells*. Cambridge University Press: 2019.
9. Blee, J. A.; Roberts, I. S.; Waigh, T. A., Membrane potentials, oxidative stress and the dispersal response of bacterial biofilms to 405 nm light. *Physical Biology* **2020**, *17* (3), 036001.
10. Izhikevich, E. M., *Dynamical Systems in Neuroscience: The Geometry of Excitability and Bursting*. MIT Press: 2010.
11. Hobbie, R. K.; Roth, B. J., *Intermediate Physics for Medicine and Biology*, 4th ed. Springer: 2007.
12. Benarroch, J. M.; Asally, M., The microbiologists guide to membrane potential dynamics. *Trends in Microbiology* **2020**, *28* (4), 304.
13. Mancini, L.; Tian, T.; Guillaume, T.; Pu, Y.; Li, Y.; Lo, C. J.; Bai, F.; Pilizota, T., A general workflow for characterization of Nernstian dyes and their effects on bacterial physiology. *Biophysical Journal* **2020**, *118* (1), 4–14.
14. Blair, K. M.; Turner, L.; Winkelman, J. T.; Berg, H. C.; Kearns, D. B., A molecular clutch disables flagella in the *Bacillus subtilis* biofilm. *Science* **2008**, *320* (5883), 1636–1638.
15. Martinac, B.; Rohde, P. R.; Cranfield, C. G.; Nomura, T., Patch clamp electrophysiology for the study of bacterial ion channels in giant spheroplasts of *E. coli*. *Methods in Molecular Biology* **2013**, *966*, 367–380.
16. Masi, E.; Ciszak, M.; Santopolo, L.; Frascella, A.; Giovannetti, L.; Marchi, E.; Viti, C.; Mancuso, S., Electrical spiking in bacterial biofilms. *Journal of the Royal Society – Interface* **2015**, *12* (102), 1036.
17. Liu, J.; Martinez-Corral, R.; Prindle, A.; Lee, D. Y. D.; Larkin, J.; Gabalda-Sagarra, M.; Garcia-Ojalvo, J.; Süel, G. M., Coupling between distant biofilms and emergence of nutrient time-sharing. *Science* **2017**, *356* (6338), 628–642.
18. Bruni, G. N.; Weekley, R. A.; Dodd, B. J. T.; Kralj, J. M., Voltage-gated calcium flux mediates *Escherichia coli* mechanosensation. *PNAS* **2017**, *114* (35), 9445–9450.

4 Mesoscopic Forces and Adhesion

Physicists have detailed models of the fundamental forces in nature, such as gravity, electromagnetism and the strong and weak nuclear forces. However, forces at the mesoscale (nm-micron length scales) are of primary interest to biophysicists and these involve complicated coarse-grained averages over very large numbers of molecules (e.g. often on the order of Avogadro's number, 6×10^{23}) interacting via fundamental interactions that are often very long-ranged.[1] Calculations with long-range interactions can be intractable without restrictive simplifying assumptions. Furthermore, subtle blends of fundamental interactions occur e.g. in steric forces, and they can take on well-defined characteristics on the mesoscale, not seen clearly at smaller length scales, that lead to distinct phenomena e.g. depletion interactions. The good news is that the soft matter community has developed a range of models for mesoscopic forces and tested them with simpler synthetic systems.[1] Quantitative agreement with models is often possible and studies have subsequently been extended to more complex biological systems.

4.1 Models for Mesoscale Interactions

In practice, a broad range of mesoscopic interactions can occur simultaneously with bacteria. However, it is useful to individually isolate some of the main contributors, such as *Derjaguin, Landau, Verwey and Overbeek (DLVO) forces*, *sterically stabilised polymer brushes*, *bridging interactions* and *hydrophobic interactions* to understand how they control the cumulative interaction potentials experienced by the bacteria.

4.1.1 DLVO Potentials

The dominant force of interaction between bacteria and their external environments is often electrostatic in nature, since charged groups occur on bacterial surfaces and charged interactions can be long-ranged. Bacterial surfaces are commonly negatively charged. Electrostatic interactions can be described using a *Yukawa potential* for colloidal spheres (bacteria are a type of colloidal matter, Chapter 7),

$$V = \frac{V_0}{r} e^{-\kappa_D r}, \qquad (4.1)$$

where κ_D is the inverse Debye length (κ_D^{-1} is the length scale over which the charge interaction decays), r is the distance from the charged colloidal sphere and V_0 is the interaction strength (a constant). The Yukawa potential can be compared with the Coulombic interaction in a dielectric material of relative permittivity ε,

$$V = \frac{q}{4\pi\varepsilon\varepsilon_0 r}, \tag{4.2}$$

where q is the charge on the point sphere and ε_0 is the permittivity of free space. Clearly, the potential described by Equation (4.1) decays much faster with separation distance (r) than Equation (4.2) due to the exponential dependence. The more rapid decay is due to the screening of the charges on the bacteria by ions in the solution e.g. salts in the growth media and counterions that dissociate from the bacterial surfaces.

The solution to the *Poisson–Boltzmann* (PB) equation that gives rise to Equation (4.1) is derived based on the linearisation of this non-linear differential equation. Equation (4.1) is thus only rigorously accurate at low charge fractions and small potentials that allow the process of linearisation. Furthermore, it neglects any charge fluctuations in the system. Better accuracy, avoiding the linearisation approximation, can be found from the numerical solution of the PB equation. For yet further improvements in accuracy, more sophisticated methods are needed to account for fluctuating charges[2] and have not yet found extensive use with bacteria. In general, it is harder to rigorously control the ionic environment with bacteria compared with analogous experiments on synthetic colloids, which in practice also makes quantitative analysis challenging (it can also be challenging in synthetic systems).

When charged or neutral surfaces approach very closely, the electrostatic forces can be dominated by forces due to spontaneously fluctuating dipoles i.e. *van der Waals* forces. Simply adding the van der Waals force to the screened electrostatic force (the spatial derivative of Equation (4.1), $F = -\frac{dV}{dr}$) leads to the successful *DLVO model.* Whether the two forces are simply additive is not obvious theoretically (it requires rigorous confirmation in a specific system), but the model has been reasonably successful in describing experimental data from a wide range of soft matter systems.[1]

DLVO is the most successful theory to describe the mesoscopic forces experienced by charged nanoparticles (Figure 4.1). To make it fully quantitative (including the case of multivalent counterions), the challenge is to measure the combined effects of surface charge densities, charge regulation phenomena and ion pairing equilibria[2] to nail down all the prefactors in the equations.

The *DLVO force* is given by

$$F = F_{vdw} + F_{dl}, \tag{4.3}$$

where F_{vdw} signifies the van der Waals force and F_{dl} is the screened electrostatic force due to double layer interactions. For the interactions of small spheres, the leading term in the van der Waals force between two spheres can be approximated by

$$\frac{F_{vdw}}{R_{eff}} = -\frac{H}{6r^2}, \tag{4.4}$$

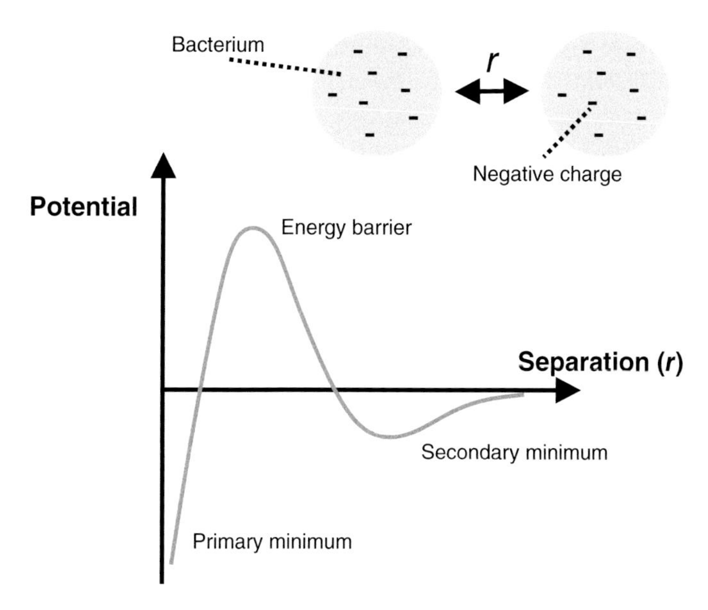

Figure 4.1 The *DLVO potential* between two negatively charged bacteria as a function of separation distance (r).

where r is the smallest surface separation, H is the Hamaker constant and R_{eff} is the effective radius of the sphere (Figure 4.1). A different expression is required for the van der Waals forces if the geometry is changed e.g. for the interaction of a bacterium with a flat surface (a sphere and a plane). However, it is relatively straightforward to look up alternative expressions to insert into Equation (4.3) for different simple geometries from the literature.[3] A specific form of the screened electrostatic force (the spatial derivative of Equation 4.1) is

$$\frac{F_{dl}}{R_{eff}} = 4\pi\varepsilon\varepsilon_0\kappa_{eff}\psi_{eff}^2 e^{-\kappa_{eff}r}, \tag{4.5}$$

where ψ_{eff} is the effective potential on the particle, κ_{eff} is the effective decay length, ε is the relative permittivity and ε_0 is the permittivity of free space. For weakly charged surfaces

$$\psi_{eff} = \frac{4kT}{Zq}\tanh\left(\frac{Zq\psi_{dl}}{4kT}\right), \tag{4.6}$$

where ψ_{dl} is the potential at the origin of the diffuse layer, kT is the thermal energy and Z is the valence of the electrolyte. More sophisticated approaches are possible with DLVO interactions (e.g. to describe charge reversal phenomena and charge regulation), but they have not yet been rigorously applied to bacteria. From the solution of the linearised PB equation, the Debye screening length $\left(\kappa_D^{-1}\right)$ is given by

$$\kappa_D^2 = \frac{2q^2 I}{\varepsilon\varepsilon_0 N_A kT}, \tag{4.7}$$

where I is the salt concentration, q is the charge on the salt ions, N_A is Avogadro's constant and kT is the thermal energy. It is found that $\kappa_D \approx \kappa_{eff}$ for 1:1 electrolytes e.g. sodium chloride (NaCl) in the solutions. There are important differences for the Debye screening length with multivalent electrolytes (e.g. the length reduces in proportional to Z^{-2} where Z is the ion valence and they can induce attractive interactions between nanoparticles[4]) and these cases are of practical importance for bacterial flocculation e.g. in water purification.[5]

4.1.2 Sterically Stabilised Surfaces

van der Waals forces are the default interaction between all matter and they cause nanoparticles that are made of similar materials to stick together when they approach each other closely (the primary minimum in Figure 4.1). Often bacteria need to control such aggregation phenomena and a common mechanism is to modulate their electrostatic interactions e.g. the expression of more negatively charged groups on their surfaces will cause the bacteria to disperse via the repulsive Yukawa potential (Equation (4.1)). However, such electrostatic interactions are very sensitive to the ionic environment (e.g. salt concentration), so alternative mechanisms are also useful. Synthetic chemists found they could graft flexible polymer chains to the surfaces of nanoparticles to provide strong repulsive interactions due to the entropy of the chains (compression of the chains causes a reduction in entropy and is resisted). These polymeric brushes are commonly used in industrial applications e.g. to improve the phase stability of colloidal particles used in paint. Naturally occurring bacterial cells often have polymeric brushes attached to their surfaces e.g. *rough/smooth lipopolysaccharides* (small/intermediate-sized brushes) and *capsules* (giant brushes). Models from polymer physics can be used to quantify the interbacterial forces from these brushes.[6,7] The polymers in the brushes are often charged and polymer models for charged polymers (i.e. polyelectrolytes) encounter some technical challenges due to the long range of the charge interactions, but practical solutions have been developed using scaling theories (based on the concept of electrostatic blobs, Chapter 6) in tandem with molecular dynamics simulations.[8]

The most successful theories for the *polyelectrolyte brushes* commonly found in capsules are based on the scaling approach (Chapter 6).[8] These models aim to capture the scaling behaviour of key parameters (brush sizes, energies, forces, etc.) on

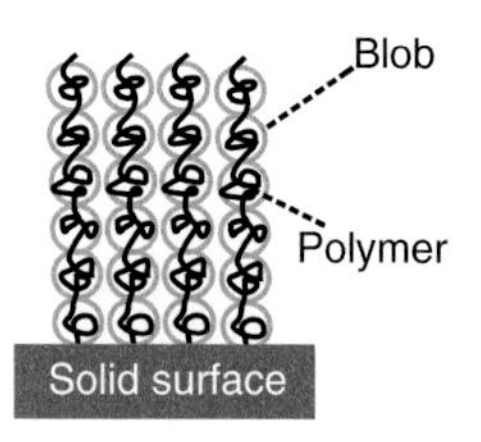

Figure 4.2 Flexible polymeric brush on a solid/liquid interface. A simple model for lipopolysaccharides and capsular polymers with bacteria can be made using the idea of *blobs*.

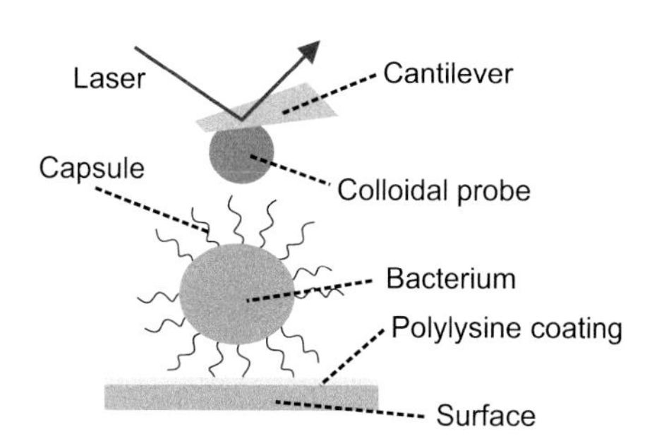

Figure 4.3 *Atomic force microscopy* (AFM) measurements of the forces due to bacterial capsules. A crucial experimental factor is to promote good adhesion of the bacteria with the surface using positively charged polylysine coatings (capsules are often negatively charged). The colloidal probe on the AFM cantilever tends to amplify the forces measured and provides a simple geometry to study the forces experienced by the capsular brush.

crucial control variables (e.g. the length of the chain or the grafting density), but are agnostic with respect to the prefactors in the resultant equations and as such are often semi-quantitative. Blob models can be used to predict the structure of the brush (Figure 4.2). Blobs are hypothetical substructures in the polymer chains, which facilitate scaling calculations e.g. in charged polymers, the blob size is set by equating the energy from the charge interaction with the thermal energy. Blob models are thus a form of multi-fractal in which the chains' fractal exponent (determining the chain size as a function of the number of monomers) has one value below the blob size, which switches to another value above the blob size.

The grafting density of the chains in the brush affects the forces experienced by the bacteria and the spatial extent of the brush e.g. increasing the grafting density (placing chains in the brush closer to one another) causes the polymer chains in the brushes to extend. Corrections to the steric forces exerted by the brushes due to the curvature of the surfaces to which they are attached are expected if the brushes are tethered to nanoparticles, although such corrections were considered to be relatively small for capsular *Escherichia coli*.[6] Different regimes for the behaviour of the brushes are defined in the framework of polyelectrolyte scaling theories e.g. neutral, low charged and osmotic brush regimes.[9–11]

The steric forces for capsular brushes on bacteria were measured by the Melbourne (*Klebsiella pneumoniae*) and Manchester (*E. coli*) groups[6,7] and a rich range of physical phenomena are predicted theoretically. Atomic force microscopy (AFM) experiments were performed to quantify the forces from single bacterial capsules (Figure 4.3). Bacterial forces can be directly measured in AFM experiments (Section 13.2.1). A colloidal sphere attached to the AFM cantilever can simplify calculations and provide a larger area of contact (it magnifies the force experienced by the cantilever compared with a bare pyramidal cantilever). Multivalent counterions (e.g. Ca^{2+}) can collapse capsular brushes reducing the forces measured in AFM experiments and similar behaviour is observed with synthetic polyelectrolyte brushes.[7]

Polyelectrolyte scaling theories provide predictions for many physical properties of the charged polymer brushes in capsules. In the osmotic regime (used to describe *E. coli* capsules[6]), the interbacterial force (f_{osm}) due to the polymeric brushes is given by

$$\frac{f_{osm}}{kT} \simeq \frac{\alpha N}{H}, \tag{4.8}$$

where N is the degree of polymerisation of the polymer chains (i.e. the number of monomers in a chain), k is Boltzmann's constant, T is the temperature, H is the thickness of the capsular brush and α is the charge fraction on the polymer chains. The thickness (H) of the capsular brush can also be calculated as

$$H \cong Nb\alpha^{1/2}. \tag{4.9}$$

where b is the size of a monomer. Due to the complexity of bacterial surfaces, many corrections might be expected to the forces on polyelectrolyte brushes attached to bacteria e.g. competition with pili or the effects of patchy membrane microheterogeneity (*E. coli* brushes are longer on their poles[6]). However, the scaling theories provide useful first-order approximations to approach quantitative modelling.

Figure 4.4 shows the different force regimes observed in AFM experiments with *E. coli* capsular brushes as a function of the cantilever/bacterium separation distance. At larger separations, the forces are dominated by double layer (screened electrostatic) interactions, followed by osmotic brush and then Hookean (compression of the bacterial cell walls) as the distance of separation decreases.

Pili can extend beyond the capsules providing a long-range (up to 10 μm)[12] attractive interaction with high specificity in some standard pathogenic bacteria that balances the repulsive capsular brush potential. Flagella also extend beyond the capsular brushes and are expected to contribute a force of active steric repulsion between neighbouring bacteria and surfaces.

Capsules can occur simultaneously with biofilms and slimes. How the repulsive steric potential of the bacterial capsule modifies the formation of the biofilm gel formed from the extracellular polymeric substance is unknown (Chapter 23). Capsules could have an anti-gelation effect and they will affect the viscoelastic properties of the composite gels in biofilms (via nanoparticle filler effects), such as strain hardening.[13]

4.1.3 Depletion and Bridging Interactions

During water purification, it is common to encourage the flocculation of bacteria to remove them from suspension (Figure 4.5b). This is often done by the addition of polyvalent ions to induce attractive *bridging interactions* e.g. Ca^{2+} or cationic polymers.[14] Similarly, in immunology, bridging interactions are caused by polyvalent proteins e.g. antibodies, such as immunoglobulin A. These encourage the bacteria to stick together and help with the clearance of the infections they cause.

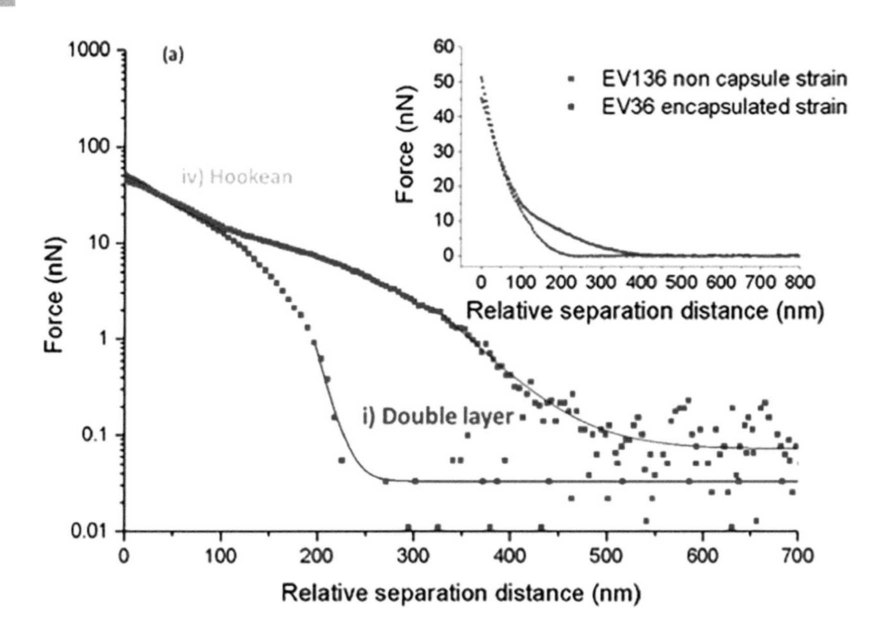

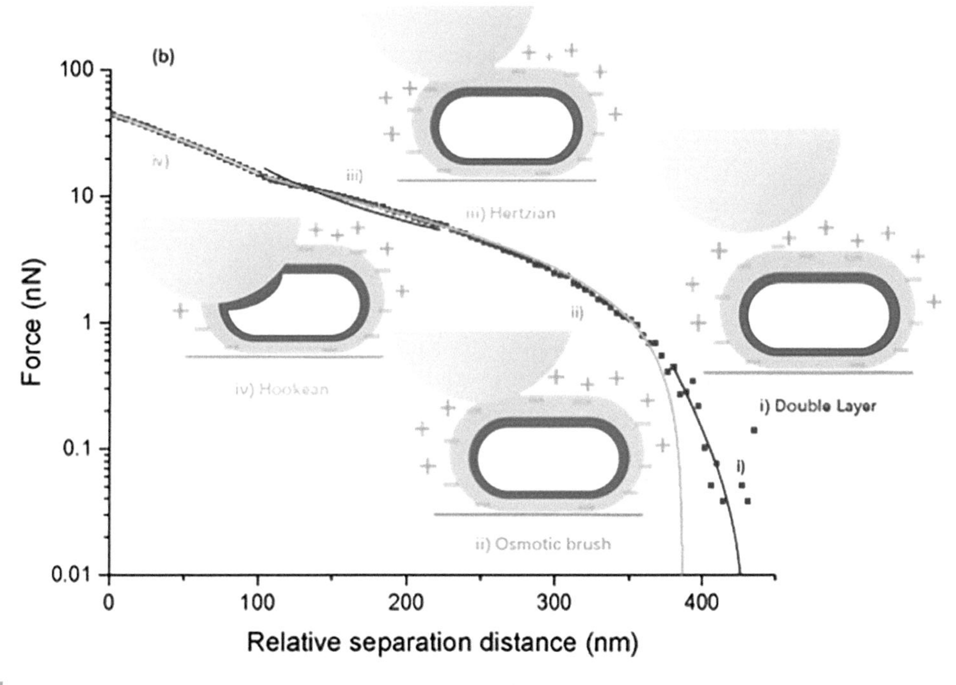

Figure 4.4 Force as a function of separation distance for capsular brushes from *atomic force microscopy* experiments on single bacteria.[6] (a) The force experienced by an *E. coli* cell with (blue dots) and without (red dots) the capsular brush. (b) Schematic diagram of the different regimes during the compression of *E. coli* capsular brushes.

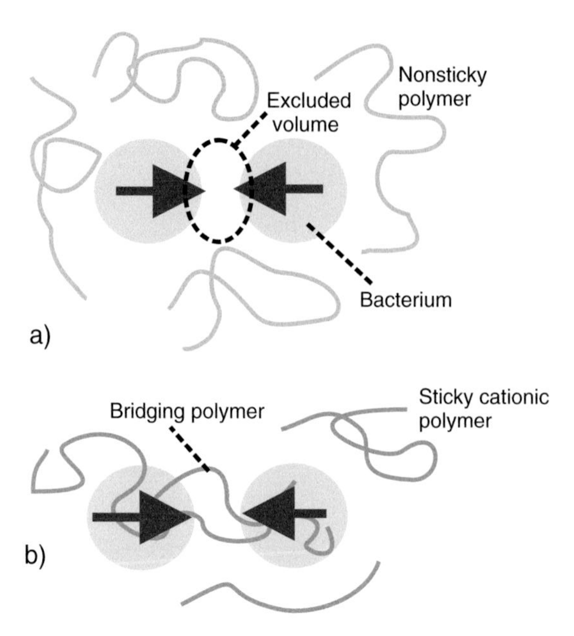

Figure 4.5 (a) *Depletion interactions* in which non-adsorbing non-sticky polymers are added to a bacterial suspension and (b) *bridging interactions* in which adsorbing sticky polymers are added to a bacterial suspension can both induce attractive forces between colloidal particles. The arrows indicate the direction of the interparticle forces (attractive in both cases).

In addition to bridging interactions, which are expected to be more common, the *depletion interaction* (Figure 4.5a) due to exopolysaccharide production in *Sinorhizobium meliloti* is found to produce increased aggregation of the bacteria.[15] Depletion interactions have been well studied with synthetic colloidal systems e.g. polyethylene glycol (PEG) is added to protein solutions to increase crystallisation, which is required in X-ray structural studies.[16] Depletion is an entirely repulsive interaction driven by entropic effects i.e. the entropy of the polymer chains drives the colloids together.

4.1.4 Hydrophobic Interactions

Hydrogen bonding is a strong polar interaction due to the high charge density of hydrogen nuclei (the highest possible, since the nuclei of hydrogen contain a single proton).[1] Hydrogen bonding helps determine the conformation of biomolecules, since all life exists in an aqueous environment and biomolecules constantly interact with water. Inconveniently for would-be modellers, hydrogen bonding is a complex many-body interaction and accurate *ab initio* calculations for the anomalous properties of water still do not exist[1] e.g. why ice floats on liquid water. Furthermore, the introduction of biomolecules in simulations only tends to complicate matters. Hydrogen bonds can bifurcate or trifurcate and complex many-body interactions occur with charged ions (including those with hydrogen ions due to acid/base equilibria).[1]

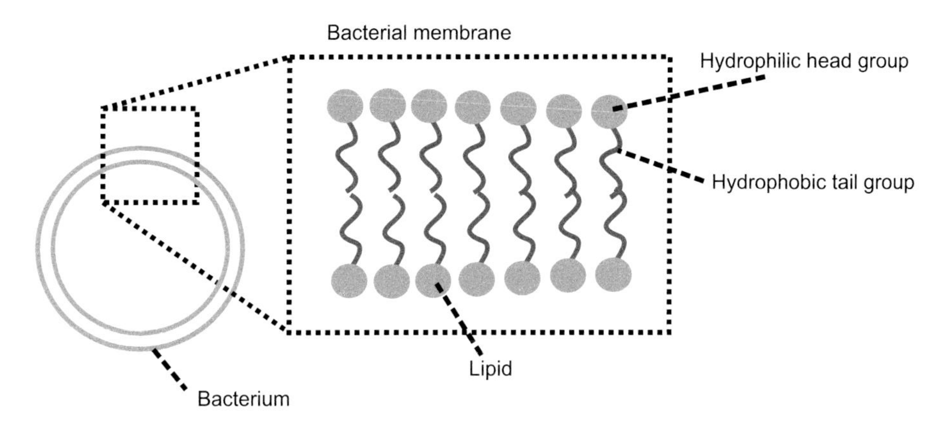

Figure 4.6 The inner and outer membranes of Gram-negative bacteria self-assemble due to the *hydrophobic effect*. Only a rough zoomed-in schematic of the outer membrane is shown.

The inability of a molecule to form hydrogen bonds can be thought of as a driver of another mesoscopic interaction; the *hydrophobic effect*.[1] Such hydrophobic moieties on molecules can be an important determinant of physical behaviour e.g. they direct the self-assembly of membranes due to the hydrophobicity of lipid tails (Figure 4.6) and the globular structure of proteins due to hydrophobic amino acids inside the globules.

4.2 Adhesion

Adhesion of synthetic materials has been considered in detail by the soft matter community.[1] With bacteria or biofilms attached to surfaces, adhesion describes the force per unit length (or equivalently the surface-free energy) to pull them off the surface (Figure 4.7). Equilibrium surface-free energies are used to motivate adhesive energies. A positive difference (E) in surface-free energies is required for an adhesion process to occur spontaneously,

$$E = \gamma_{12} - \gamma_1 - \gamma_2, \tag{4.10}$$

where γ_{12} is the surface-free energy of surfaces 1 and 2 in contact, and γ_1 and γ_2 are the surface-free energies of surfaces 1 and 2 respectively exposed to the surrounding solution.

In practice, the application of the theory of adhesion with bacteria is hampered by long-range charge effects on bacterial surfaces (e.g. Lewis acid/base phenomena that modulate the effective surface-free energies) and dissipative phenomena e.g. viscous effects. Furthermore, at the nanoscale, thermodynamic processes are often non-equilibrium. For example, this is observed in *Bell's equation* in which the forces observed depend on the time scales over which they are measured,[1]

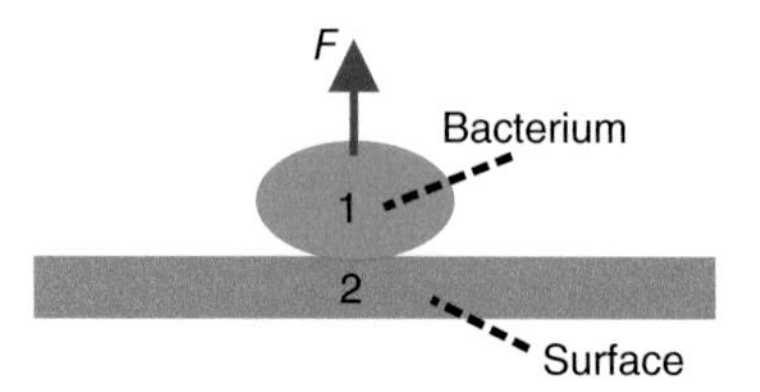

Figure 4.7 Schematic diagram of the *adhesive force* (F) applied to a bacterium (1) attached to a surface (2).

$$F(t) = \frac{kT}{r} \ln\left(\frac{\tau}{t}\right), \tag{4.11}$$

where τ is the characteristic time scale of the interaction, kT is the thermal energy and r is the length of the interaction e.g. a bond length. Such time-dependent forces will be commonly observed in measurements on bacterial adhesion e.g. those measured via AFM, optical or magnetic tweezers.

In general, bacteria are very good at sticking to a huge number of surface chemistries and the creation of effective anti-adhesion coatings to disrupt infections is thus challenging. The adhesion of bacteria to a surface is one of the earliest events in the formation of a biofilm and it is crucial for the development of viable biofilm structures[17] (Chapter 23). Bacteria *in vivo* use monomeric adhesins/invasins or sophisticated machines, such as type III secretion systems or type IV pili (Chapter 19), to adhere to their hosts and for information transfer.[18]

Some pathogenic bacteria have adhesive structures adapted for human cells e.g. tip adhesins on the end of pili that bind to receptors on mucosal surfaces. Specific examples of tip adhesins include: lectins on pili that bind to glycoproteins on cells, polysaccharides on capsules that bind to lectins on macrophages, fibronectin-binding proteins that attach to fibronectin-presenting cells and lipid-binding proteins that attach to lipid membranes.[19] One bacterium may express several types of adhesive moiety simultaneously and each one can be targeted at a separate receptor. Clones of bacterial cells exhibit phenotypic variability for the adhesion complexes they express, which seems to be driven by an evolutionary pressure to adhere to the maximum number of host cells (bet hedging). Further adhesive interactions occur between bacteria and extracellular matrix components (e.g. collagens) in addition to those directly between bacterial and eukaryotic cells.

A challenge to blocking the adhesion of bacterial cells to surfaces and the creation of biofilms is thus the large diversity of adhesive complexes e.g. mannose might block one variety of adhesive complex in a urinary tract infection, but other backup adhesion complexes will be used by the bacteria to anchor themselves on the wall of the bladder.

Detailed morphological studies have been performed on *E. coli* pili as adhesion complexes.[20] *Staphylococcal* adhesins, such as SdrG, ClfAp and ClfB, can resist substantial forces before failure (~2 nN) due to the cooperative effect of many hydrogen bonds in parallel between the adhesin and the surface. AFM experiments with the *E. coli* adhesin protein, Fim H, provide evidence for a catch bond model i.e. the adhesive bonds become stronger with more tensile force.[21] Similar behaviour has been observed with *Staphylococcus aureus* adhesins that connect to blood proteins (e.g. fibrinogen) and in fungal biofilms (e.g. *Candida albicans*), so it appears to be a relatively common mechanism for cellular adhesion.[17]

The aquatic bacterium *Caulobacter crescentus* has a specialised adhesive apparatus called a holdfast that it uses to strongly adhere to surfaces in water.[22] The holdfast is thought to cause this bacterium's unusual mechanism of asymmetric cell division and it is a promising source of inspiration for synthetic underwater adhesives.

Velocity gradients due to the interaction of curved surfaces with shear flows can change the regions of surface attachment chosen by *Pseudomonas aeruginosa* and *E. coli* cells in microfluidic geometries[23] (Section 17.3). *P. aeruginosa* cells increase the residence time of their adhesion events when they experience an increased shear flow in microfluidic experiments.[24] This can improve the chances that the bacteria will go on to create a biofilm. A steady flow of media is often associated with a plentiful supply of nutrients, which implies it is a favourable location for bacterial growth.

Scanning electron microscopy was combined with focused ion beam tomography to understand the formation of early-stage *S. aureus* biofilms.[25] Bacterial cells were modelled as elastic spheres that deformed due to adhesion with the substrates (a Hertzian model[1]) and adhesive forces of 1–6 nN were found per bacterium in the biofilm.

More specialised examples of adhesive complexes have been found in the Antarctic bacterium *Marinomonas primoryensis*.[26] A specific ice-binding protein helps the bacteria avoid damage due to freezing.

Some success has been found to block adhesion of bacteria by etching pores on surfaces e.g. 15/25 nm pores on aluminium surfaces were found to impede the adhesion of *E. coli*, *Listeria monocytogenes*, *S. aureus* and *Staphylococcus epidermidis*.[27] The impact of roughness on adhesion is expected to be a general phenomenon due to the modified surface-free energies, since very rough surfaces trap lots of air and can become superhydrophobic.

Micropillars in microfluidic growth chambers have been used to measure the pressures exerted by a range of biofilms that adhere to the substrates using the deflection of the pillars.[28] Differential pressures in the range 1.8–20.5 kPa were measured for *E. coli*, *S. epidermidis*, *S. aureus*, *P. aeruginosa* and *Shewanella oneidensis* adsorbed to the pillars.

Suggested Reading

Israelachvili, J. *Intermolecular and surface forces*, 3rd ed. Academic Press: 2011. Classic textbook on surface forces in soft matter physics with many biological examples.

Smith, A. M. et al., Forces between solid surfaces in aqueous electrolyte solutions. *Advances in Colloid and Interface Science* **2020**, *275*, 102078. Modern approaches to DLVO forces are discussed from an experimental physical chemistry perspective.

References

1. Israelachvili, J. N., *Intermolecular and Surface Forces*. Academic Press: 2011.
2. Smith, A. M.; Borkovec, M.; Trefalt, G., Forces between solid surfaces in aqueous electrolyte solutions. *Advances in Colloid and Interface Science* **2020**, *275*, 102078.
3. Parsegian, V. A., *Van der Waals Forces: A Handbook for Biologists, Chemists, Engineers and Physicists*. Cambridge University Press: 2005.
4. Muthukumar, M., *Physics of Charged Macromolecules: Synthetic and Biological Systems*. Cambridge University Press: 2023.
5. Duan, J. M.; Gregory, J., Coagulation by hydrolysing metal salts. *Advances in Colloid and Interface Science* **2003**, *100*, 475–502.
6. Phanphak, S.; Georgiades, P.; Li, R.; King, J.; Roberts, I. S.; Waigh, T. A., Super-resolution fluorescence microscopy study of the production of K1 capsules by *Escherichia coli:* Evidence for the differential distribution of the capsule at the poles and the equator of the cell. *Langmuir* **2019**, *35* (16), 5635–5646.
7. Wang, H.; Wilksch, J. J.; Lithgow, T.; Strugnell, R. A.; Gee, M. L., Nanomechanics measurements of live bacteria reveal a mechanism for bacterial cell protection. *Soft Matter* **2013**, *9* (31), 7560.
8. Dobrynin, A. V.; Rubinstein, M., Theory of polyelectrolytes in solutions and interfaces. *Progress in Polymer Science* **2005**, *30* (11), 1049–1118.
9. Zhulina, E. B.; Borisov, O. V., Polyelectrolytes grafted to curved surfaces. *Macromolecules* **1996**, *29* (7), 2618–2626.
10. Pincus, P., Colloid stabilization with grafted polyelectrolytes. *Macromolecules* **1991**, *24* (10), 2912–2919.
11. Zhulina, E. B.; Birshtein, T. M.; Borisov, O. V., Curved polymer and polyelectrolyte brushes beyond the Daoud-Cotton model. *European Physical Journal* **2006**, *20* (3), 243.
12. Telford, J. L.; Barocchi, M. A.; Margarit, I.; Rappuoli, R.; Grandi, G., Pili in Gram-positive pathogens. *Nature Reviews Microbiology* **2006**, *4* (7), 509–519.
13. Jana, S.; Charlton, S. G. V.; Eland, L. E.; Burgess, J. G.; Wipat, A.; Curtis, T. P.; Chen, J., Nonlinear rheological characterisation of single species bacterial biofilms. *npj Biofilms and Microbiomes* **2020**, *6* (1), 19.
14. Borkovec, M.; Papastavrou, G., Interactions between solid surfaces with adsorbed polyelectrolytes of opposite charge. *Current Opinion in Colloid and Interface Science* **2008**, *13* (6), 429–437.
15. Dorken, G.; Ferguson, G. P.; French, C. E.; Poon, W. C. K., Aggregation by depletion attractions in cultures of bacteria producing exopolysaccharides. *Journal of the Royal Society – Interface* **2012**, *9* (77), 3490–3502.
16. Foffi, G.; et al., Phase equilibria and glass transition in colloidal systems with short-ranged attractive interactions: Application to protein crystallization. *Physical Review E* **2002**, *65* (3), 031407.

17. El-Kirat-Chatel, S.; Beaussart, A.; Mathelie-Guinlet, M.; Dufrene, Y. F., The importance of force in microbial cell adhesion. *Current Opinion in Colloid and Interface Science* **2020**, *47*, 111–117.
18. Pizarro-Cerda, J.; Cossart, P., Bacterial adhesion and entry into host cells. *Cell* **2006**, *124* (4), 715–727.
19. Ofek, I.; Bayer, E. A.; Abraham, S. N., Bacterial adhesion. In *The Prokaryotes: Human Microbiology*, DeLong, E. F., Lory, S., Stackebrandt, E., Thompson, F., Eds.; Springer: 2013; pp. 107–123.
20. Bullitt, E.; Makowski, L., Bacterial adhesion pili are heterologous assemblies of similar subunits. *Biophysical Journal* **1998**, *74* (1), 623–632.
21. Thomas, W.; Forero, M.; Yakovenko, O.; Nilsson, L.; Vicini, P.; Sokurenko, E., Catch-bond model derived from allostery explains force-activated bacterial adhesion. *Biophysical Journal* **2006**, *90* (3), 753–764.
22. Nyarko, A.; Barton, H.; Dhinojwala, A., Scaling down for a broader understanding of underwater adhesion – a case for the *Caulobacter crescentus* holdfast. *Soft Matter* **2016**, *12* (45), 9132–9141.
23. Secchi, E.; Vitale, A.; Mino, G. L.; Kanstler, V.; Eberl, L.; Rusconi, R.; Stocker, R., The effect of flow on swimming bacteria controls the initial colonization of curved surfaces. *Nature Communications* **2020**, *11* (1), 2851.
24. Lecuyer, S.; Rusconi, R.; Shen, Y.; Forsyth, A.; Vlamakis, H.; Kolter, R.; Stone, H. A., Shear stress increases the residence time of adhesion of *Pseudomonas aeruginosa*. *Biophysical Journal* **2011**, *100* (2), 341–350.
25. Gu, J.; Valdevit, A.; Chou, T. M.; Libera, M., Substrate effects on cell-envelop deformation during early stage *Staphylococcus aureus* biofilm formation *Soft Matter* **2017**, *13* (16), 2967–2976.
26. Dolev, M. B.; Bernheim, R.; Guo, S.; Davies, P. L.; Braslavsky, I., Putting life on ice: Bacteria that bind to frozen water. *Journal of the Royal Society – Interface* **2016**, *13* (121), 20160210.
27. Feng, G.; Cheng, Y.; Wang, S. Y.; Borca-Tasciuc, D. A.; Worobo, R. W.; Moraru, C. I., Bacterial attachment and biofilm formation on surfaces are reduced by small-diameter nanoscale pores: How small is small enough?. *npj Biofilms and Microbiomes* **2015**, *1*, 15022.
28. Chew, S. C.; Kundukad, B.; Teh, W. K.; Doyle, P. S.; Yang, L.; Rice, S. A.; Kjelleberg, S., Mechanical signatures of microbial biofilms in micropillar-embedded growth chambers. *Soft Matter* **2016**, *12* (23), 5224.

5 Reaction–diffusion Equations

Reaction–diffusion (RD) equations are useful mathematical tools to describe pattern formation[1] e.g. during the morphogenesis of organisms, such as how the leopard gets its spots. A prototypical RD equation is exactly what the name implies. Fick's second law is used to describe diffusion (Equation 2.4) and a term $(f(c))$ is added to describe how the concentration of molecules (c) is modified by reaction,

$$\frac{\partial c}{\partial t} = D\frac{\partial^2 c}{\partial x^2} + f(c), \tag{5.1}$$

where t is the time and x is the displacement in one dimension (straightforward 2D and 3D extensions are also possible using a Laplacian with Fick's second law). A specific example is the diffusion of a morphogen across an array of cells in which reaction with the morphogen determines how the cells differentiate into specialised cell types.

For rigorous solutions to such RD equations as Equation (5.1), reference should be made to the applied mathematics literature on *parabolic partial differential equations*.[2] Different specific choices of $f(c)$ have been studied in detail in the literature e.g. $f(c) = \alpha c(1-c)$ is called *Fisher's equation* (α is a positive constant). RD equations are a key model used to understand the interactions of bacteria in concentrated suspensions and biofilms e.g. in quorum sensing and electrical communication. Such coarse-grained analytic RD approaches can give valuable insights with bacteria, but they have some weaknesses e.g. they neglect the discrete nature of cells and molecules, fluctuations in the averaged quantities may not be accurately handled and they are unable to accurately describe rough wavefronts during signalling events.[3]

RD equations can support solutions in which the concentration profiles travel at constant velocities.[4] These can be wavefronts (e.g. step-like profiles), wave pulses, oscillatory waves (Figure 5.1) or more complex profiles.[4] For wavefronts, analytic solutions of the RD equations (Equation (5.1)) are explored for functions of the form

$$c(x,t) = U(x + st), \tag{5.2}$$

where s is the wavefront speed and U is the concentration profile of the wavefront travelling with constant velocity. In biological experiments, often propagating wavefronts do not follow this constant velocity assumption e.g. there are multiple forms of dynamic heterogeneity that perturb the wavefronts and anomalous dynamics occur, such as super-diffusive behaviour. Historically, analytic methods have been introduced to extend the applicability of solutions following Equation (5.2), using the ideas of pushed and pulled wavefronts.[5] However, such analytic solutions have only

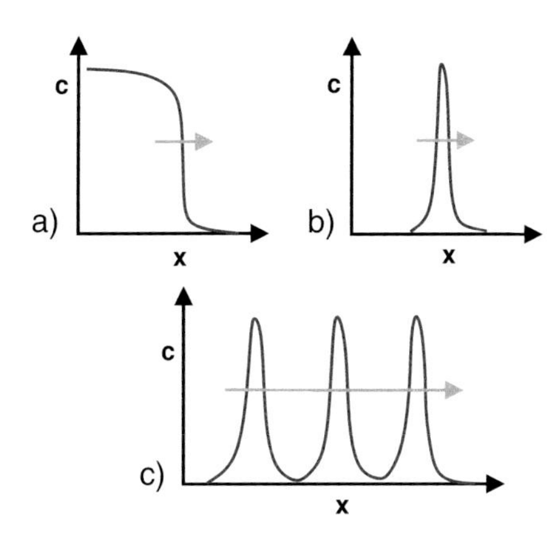

Figure 5.1 Schematic diagram of different types of *propagating waves*. (a) *Wave front* (e.g. with electrical signalling in *B. subtilis*[6] or *E. coli*[7]), (b) *wave pulse* (e.g. electrical signalling with *Neisseria gonorrhoeae*[8]) or (c) *oscillatory wave*. The arrows show the direction of propagation.

had modest success in quantitatively describing biophysics experiments and numerical solutions of the RD equations in complex biologically relevant geometries tend to be the more realistic approach e.g. using agent-based modelling (Chapter 27).

Practically, the square of the displacement of the wavefront position $\left(\left\langle R^2\right\rangle\right)$ can be considered as a function of time in analogy with the mean square displacement (MSD) for the anomalous transport of particles (Equation (1.8)) and can be used operationally as a definition of a wavefront with anomalous dynamics,

$$\left\langle R^2\right\rangle = R_c^2 + A\tau^{\alpha}, \tag{5.3}$$

where $\langle\ \rangle$ denotes averaging over the wavefront (which is straightforward for spherical wavefronts with a constant radius), τ is the time, α is the scaling exponent and R_c is the critical radius for nucleation of the wavefront (Figure 5.2). Different types of wavefronts can be defined; $\alpha = 1$ for *diffusive* wavefronts, $\alpha = 2$ for *ballistic* wavefronts, $2 > \alpha > 1$ for *super-diffusive sub-ballistic* wavefronts, $\alpha > 2$ for *super-ballistic* wavefronts and $\alpha < 1$ for *sub-diffusive* wavefronts.

Valuable analytic insights into the relationship between the wavefront velocity and its curvature can be described by RD equations and are provided by the *Eikonal approximation*, which holds for a reasonably wide range of RD equations of the form Equation (5.1),[9]

$$c_n(v) = c(v) - \frac{D}{R}, \tag{5.4}$$

where $c_n(v)$ is the normal speed of the curved wavefronts, $c(v)$ is the plane wavefront speed, D is the diffusion coefficient of the chemical in the RD equation and $1/R$ is the curvature of the wavefront. Thus, corrections are expected to the wavefront velocity (both positive and negative, depending on the sign of the curvature) and such phenomena are found in signalling waves that propagate through microbial communities[10] and electrical waves in cardiac tissue.[4] A detailed quantitative comparison of the

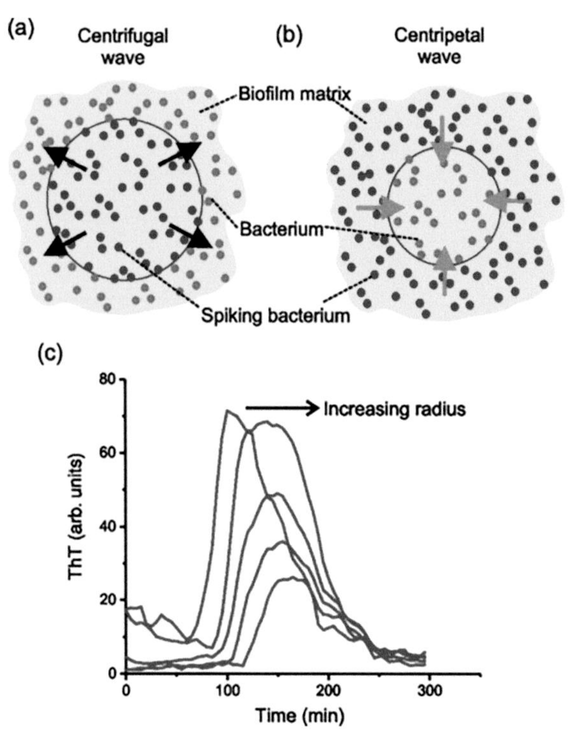

Figure 5.2 Circular wave fronts of electrochemical excitation propagating across two-dimensional biofilms of *B. subtilis*. (a) Schematic diagram of a *centrifugal wavefront* (outward moving), (b) a *centripetal wavefront* (inward moving) and (c) a plot of ThT concentration (a Nernstian voltage sensor) as a function of time indicating a potassium wavefront travelling through a biofilm from fluorescence microscopy.[6] ThT intensity is shown as a function of time for five different biofilm radii: 2, 10, 15, 100 and 150 μm.

effects of curvature on wavefront velocities measured in experiments tends to require numerical solutions to RD equations on realistic geometries that would be challenging to describe analytically.

Coupled RD equations can give rise to more complex patterning of multi-cellular communities e.g. *Turing patterns* that invoke the interaction of diffusing short-ranged activators and long-ranged inhibitors. An example of a simple Turing model is

$$\frac{\partial u_1}{\partial t} = f_1(u_1, u_2) + D_1 \frac{\partial^2 u_1}{\partial x^2}, \tag{5.5}$$

$$\frac{\partial u_2}{\partial t} = f_2(u_1, u_2) + D_2 \frac{\partial^2 u_2}{\partial x^2}, \tag{5.6}$$

where u_1 and u_2 are the activator and inhibitor concentrations, respectively. f_1 and f_2 describe the interaction of the activators and inhibitors, where D_1 and D_2 are their respective diffusion coefficients.[9] Turing patterns have been observed in genetically modified bacteria e.g. using synthetic biology techniques with *E. coli*[11,12] (Figure 5.3).

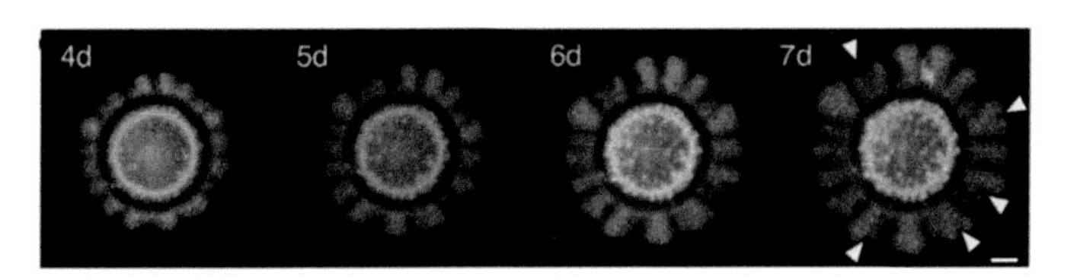

Figure 5.3 Pattern formation in colonies of *E. coli* modified using synthetic biology techniques can be described with Turing patterns.[11]

Wolpert's *French flag model* is another classic model for morphogenesis described by RD equations. The diffusion of a morphogen across the cells in the organism is thought to drive pattern formation. The morphogen gradients drive pattern development in combination with thresholds to determine cell fates and the patterns can be described by a single simple RD equation of the form Equation (5.1), where $f(c) = -\alpha c$,

$$\frac{\partial c}{\partial t} = D\frac{\partial^2 c}{\partial x^2} - \alpha c. \tag{5.7}$$

At long times, for stationary solutions $\left(\frac{\partial c}{\partial t} = 0\right)$, the concentration of the morphogens develops into an exponential decay,

$$c = c_0 e^{-x/\lambda}, \tag{5.8}$$

where $\lambda = \sqrt{\frac{D}{\alpha}}$ is a characteristic length scale of the decay and c_0 is a constant.

Thresholds on the morphogen concentration can lead to pattern formation in an organism (Figure 5.4). The French flag model is simple and has analytic solutions, but it is not robust to small changes in morphogen concentration and thus probably isn't very realistic. Non-linear expressions for $f(c)$ can help with the issue of robustness.[13]

Both Turing and French flag models can successfully describe some aspects of morphogenesis *in vivo*. More complex models are invoked to describe eukaryotic morphogenesis based on wavefront propagation (the clock and wavefront model in dynamic patterning), mechanotransduction and remodelling via cell migration.[14] These types of models may also be relevant to bacterial biophysics e.g. programmed cell death contributes to the morphogenesis of *Bacillus subtilis* biofilms.[15]

The Keller-Segel RD model is discussed in Section 18.2, where it is used to describe quorum sensing. The model is formed from two coupled RD equations and one equation that describes advection kinetics.

The specialist mathematical literature considers extensions of RD equations. A variety of non-linear expressions can be chosen for $f(c)$,[13] small number fluctuations can contribute[16] and the stochastic transport processes can be generalised beyond those of pure diffusion i.e. the effects of anomalous transport can be invoked.[17] There is good experimental data that suggests anomalous diffusion commonly occurs in biological systems due to congested environments and viscoelastic effects, particularly for the motility of larger particles.[18] Non-linear forms of $f(c)$ are thought to increase the

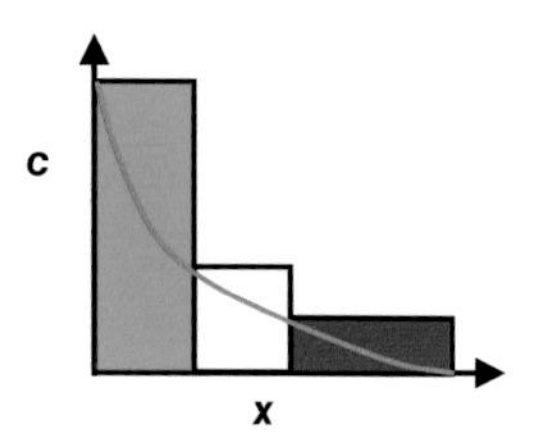

Figure 5.4 The *French-flag model* for the morphogenesis of an organism in one dimension. The exponential decaying morphogen concentration (c, orange line, Equation (5.8)) is plotted as a function of distance (x) in one dimension across an organism. The blue, white and red stripes of the tricolour correspond to three different fates for the cells (a striped pattern) e.g. three differentiated cell types.

robustness of pattern formation and this is often required biologically e.g. to create viable organisms during morphogenesis.[13]

Agent-based models (Chapter 27) allow the convenient inclusion of realistic geometries for comparison with the predictions of RD equations with real biological systems and provide a useful intermediate to bridge the gap between analytic solutions to RD equations and reality. Thus, agent-based models provide a platform for more accurate prediction of RD phenomena in a range of real bacterial systems.

Suggested Reading

Edelstein-Keshet, L., *Mathematical Models in Biology*. SIAM Classics: 2005. Excellent pedagogic account of reaction-diffusion equations.

Hillen, T.; Leonard, I. E.; van Rossel, H., *Partial Differential Equations: Theory and Completely Solved Problems*. FriesenPress: 2009. A good place to learn some of the mathematics relevant to reaction-diffusion equations.

Mendez, V.; Fedotov, S., *Reaction-transport Systems: Mesoscopic Foundations, Fronts and Spatial Instabilities*. Springer: 2010. A challenging mathematical book that considers how to generalize reaction-diffusion equations to include the anomalous transport kinetics that are common in biology.

Pismen, L., *Active Matter Within and Around Us*. Springer: 2021. Contains an interesting account of reaction-diffusion equations relevant to biophysics.

References

1. Fall, C. P.; Marland, E. S.; Wagner, J. M.; Tyson, J. T., *Computational Cell Biology*. Springer: 2003.
2. Hillen, T.; Leonard, I. E.; van Rossel, H., *Partial Differential Equations: Theory and Completely Solved Problems*. Friesen Press: 2019.

3. Ben-Jacob, E.; Cohen, I.; Levine, H., Cooperative self-organization of microorganisms. *Advances in Physics* **2010**, *49* (4), 395–554.
4. Keener, J.; Sneyd, J., *Mathematical Physiology*. Springer: 2009.
5. Dieterle, P. B.; Amir, A., Diffusive wave dynamics beyond the continuum limit. *Physical Review E* **2021**, *104* (1), 014406.
6. Blee, J. A.; Roberts, I. S.; Waigh, T. A., Spatial propagation of electrical signals in circular biofilms. *Physical Review E* **2019**, *100* (5-1), 052401.
7. Akabuogu, E. U.; Martorelli, V.; Krasovec, R.; Roberts, I. S.; Waigh, T. A., Emergence of ion-channel mediated electrical oscillations in *E. coli* biofilms. *eLife* **2023**, to appear.
8. Hennes, M.; Bender, N.; Cronenberg, T.; Welker, A.; Maier, B., Collective polarization dynamics in bacterial colonies signify the occurrence of distinct subpopulations. *PLOS Biology* **2023**, *21* (1), e3001960.
9. Cross, M.; Greenside, H., *Pattern Formation Dynamics in Nonequilibrium Systems*. Cambridge University Press: 2009.
10. Palsson, E.; Lee, K. J.; Goldstein, R. E.; Franke, J.; Kessin, R. H.; Cox, E. C., Selection for spiral waves in the social amoebae Dictyostelium. *Proceedings of the National Academy of Sciences of the United States of America* **1997**, *94* (25), 13719–13723.
11. Duran-Nebreda, S.; Pla, J.; Vidiella, B.; Pinero, J.; Conde-Pueyo, N.; Sole, R., Synthetic lateral inhibition in periodic pattern forming microbial colonies. *ACS Synthetic Biology* **2021**, *10* (2), 277–285.
12. Karig, D.; Martini, M.; Weiss, R., Stochastic Turing patterns in a synthetic bacterial population. *Proceedings of the National Academy of Sciences of the United States of America* **2018**, *115* (26), 6572–6577.
13. Alon, U., *An Introduction to Systems Biology: Design Principles of Biological Circuits*. 2nd ed. CRC Press: 2020.
14. Pismen, L., *Active Matter Within and Around Us: From Self-propelled Particles to Flocks and Living Forms*. Springer: 2021.
15. Asally, M.; Kittisopikul, M.; Rue, P.; Du, Y.; Hu, Z.; Cagatay, T.; Robinson, A. B.; Lu, H.; Garcia-Ojalvo, J.; Suel, G. M., Localised cell death focuses mechanical forces during 3D patterning in a biofilm. *Proceedings of the National Academy of Sciences of the United States of America* **2012**, *109* (46), 18891–18896.
16. Erban, R.; Chapman, S. J., *Stochastic Modelling of Reaction-diffuion Processes*. Cambridge University Press: 2020.
17. Mendez, V.; Fedotov, S.; Horsthemke, W., *Reaction-transport Systems: Mesoscopic Foundations, Fronts and Spatial Instabilities*. Springer: 2012.
18. Waigh, T. A.; Korabel, N., Heterogeneous anomalous transport in cellular and molecular biology. *Reports on Progress in Physics* **2023**, *86* (12), 126601.

6 Polymer Structure and Dynamics

Large molecules often play dominant roles in determining the mechanics and dynamics of biological materials. Large molecules in solution tend to move slowly and thus cause large increases in the viscosity of samples. Furthermore, large molecules in elastic materials are often the main contributors to the materials' elasticity. Large linear molecules and aggregates are commonly polymeric in origin.

Many of the successful physical theories originally developed to describe synthetic polymers ignore most of the molecular details and can thus be applied generally to any linear entity at the nanoscale. Such elongated molecules commonly occur within bacteria and bacterial capsules, in the extracellular polymeric substance (EPS) of bacterial biofilms and the slimes that bacteria excrete. The four most common classes of biomolecules in biology are proteins, nucleic acids, carbohydrates and lipids and all of them can have a polymeric character.

Polymers have some advantages over colloids (Chapter 7) in terms of the development of fundamental physical insights. With polymers, the physical properties often depend sensitively on the contour length of the chain ($L = Na$, where N is the number of monomers and a is the monomer size), which can be carefully controlled in reactions and provides an important control parameter to investigate the physical properties e.g. to develop scaling laws, such as how the size of the chain depends on the contour length.[1] The scaling approach has been very successful to quantify the structure and dynamics of polymeric solutions and gels.[1,2]

6.1 Structure of Polymers

A reductionist approach can be successful when considering polymeric materials. The aim is thus to first classify the behaviour of single polymer molecules and then try to explain their interactions with one another to understand more concentrated multi-chain systems.

The contour length (L) describes the length of the backbone of a polymer chain. It only equals the end-to-end distance of a polymer in the extreme limit of a very rigid chain (Figure 6.1d). After the contour length, perhaps the most important descriptor of polymers is the *persistence length*.[3] The persistence length of a polymer is the average length scale over which the tangent vectors to the chain decorrelate. The persistence length (l_p) is linearly related to the Young's modulus of the chain (E),

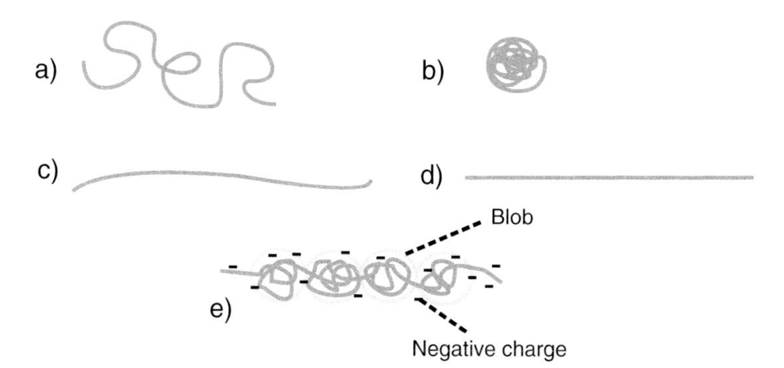

Figure 6.1 The structure of single polymers: (a) flexible, (b) globular, (c) semi-flexible, (d) rigid and (e) charged flexible chains (a bayonet of charged blobs).

since thermal forces tend to drive the directional decorrelation of the chain and are resisted in proportion to the modulus,

$$l_p = \frac{EI}{kT}, \tag{6.1}$$

where kT is the thermal energy and I is the geometrical factor (related to the cross-section of the chain). High persistence lengths imply the polymer could play an important role in the elasticities of the polymeric material (e.g. in the EPS of a biofilm), whereas for very low persistence lengths, lower elasticities are expected, as observed in flexible entropic polymer networks e.g. an elastomer. Single polymers are classified according to their flexibility, by comparison of the persistence length with the contour length, as *flexible* ($l_p << L$), *semi-flexible* $\left(l_p \sim L\right)$ (Figure 6.1c) or *rigid* ($l_p >> L$). Single flexible chains can be extended or globular (Figures 6.1a and b). Many biopolymers are polyelectrolytes and the long-range electrostatic interactions make calculations trickier. However, some reasonably robust scaling models are available to describe their behaviour e.g. bayonet models for charged flexible polymers that consist of rods of charged blobs[4] (Figure 6.1e). Another dominant interaction is of the polymers with water, which also presents further challenging many-body problems to describe the effects of hydrogen bonds on their conformations.

Statistical models can be used to calculate many of the physical properties of polymer chains.[1,5] For example, the mean square end-to-end distance $\left(\left\langle R^2\right\rangle\right)$ for a semi-flexible chain can be calculated in terms of its persistence length and contour length,

$$\left\langle R^2\right\rangle^{1/2} = \sqrt{2Ll_p}. \tag{6.2}$$

Thus, a DNA chain of contour length 1 μm and persistence length 50 nm will have a mean square end-to-end distance of 316 nm. The square root scaling in Equation (6.2) is for a random walk in space (classical diffusion is a random walk in time, compare with Equation (2.1)).

As the concentration of a polymeric solution is increased, there is an overlap concentration of the chains, c^* called the *semi-dilute overlap concentration* (Figure 6.2).

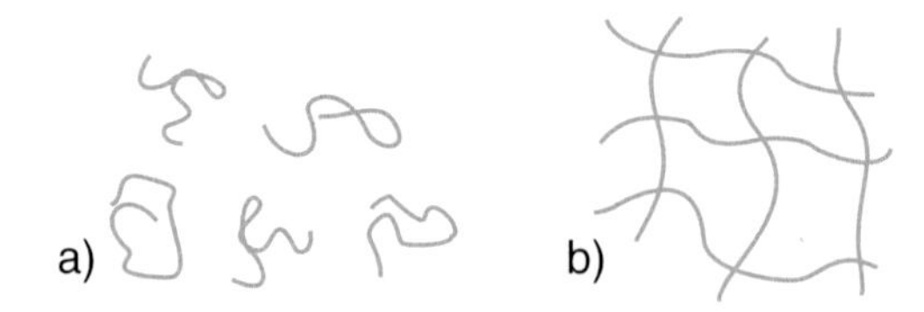

Figure 6.2 (a) A *dilute* polymeric solution occurs for polymer concentrations below the semi-dilute overlap concentration $c < c^*$ and (b) a *semi-dilute* polymeric solution occurs when $c > c^*$, where c is the polymer concentration and c^* is the semi-dilute overlap concentration.

c^* can be simply calculated as the concentration where one chain on average occupies a volume equal to the chain's unperturbed volume (V) i.e.

$$c^* = \frac{M}{V}, \tag{6.3}$$

where M is the mass of a chain and V can be calculated as

$$V = \frac{4}{3}\pi R_g^3, \tag{6.4}$$

where R_g is the radius of gyration of a chain (a measure of the chain's size).

Sticky interactions between polymer chains with concentrations above c^* will cause the semi-dilute solutions of the polymer to gel. The material becomes much more elastic, but the large amount of solvent bound to the polymer chain provides the gel with unique properties beyond those of typical glassy or rubbery amorphous solids e.g. high diffusion coefficients occur for permeants across the gel, syneresis is possible (solvent is exuded from the gel's surface) and liquid-like dynamics of the solvent are observed at small length scales. The process of percolation as the sticky aggregates bridge the sample volume during gelation has the form of a phase transition.[2]

Many of the polymeric components expressed in bacterial biofilms and slimes (e.g. gellan or xylan) form weak physical gels. Weak gels are formed when neighbouring chains cross-link due to non-permanent mesoscopic interactions e.g. charge (such as multivalent counterions), van der Waals or hydrogen bonding. Strong chemical gels in contrast form when the cross-links are chemical in nature (e.g. they are covalently bound) and require much more energy to disrupt. Glutaraldehyde has been applied to synthesise chemical cross-linked biofilms *in vitro* (e.g. for sample preparation with electron microscopy), but no *in vivo* examples are yet known of dedicated EPS chemical cross-linkers. They may be rare in naturally occurring biofilms because it is difficult to reverse the process of chemical cross-linking when bacteria need to return to their planktonic form.

Self-assembly can direct the creation of linear polymeric aggregates.[3] The aggregates spontaneously assemble from solutions containing monomers, which is driven by the minimisation of free energy (a thermodynamic phenomenon). This occurs in cytoskeletal polymers of bacteria and bacteria have homologues of the three main types of cytoskeletal proteins found in eukaryotes: *tubulins* (FtsZ, Tub Z, Phu Z, etc.), *actins* (MareB, FtsA, ParM, AlfA, MamK, etc.) and *intermediate filaments* (crescentin, FilP, Scy, etc.).[6] FtsZ filaments can experience treadmilling, which is important during cell division i.e. monomers are added to one end of the filament and removed from the other end, such that the filament length is constant. There are other bacterial cytoskeletal proteins that have no homologues in eukaryotic cells and they are not well understood e.g. bactofilins and septins.

In the EPS of biofilms, amyloid aggregates of proteins occur due to the self-assembly of misfolded proteins side by side into giant linear aggregates. Such aggregates are physically similar to those found in human disease e.g. Alzheimer's, although different protein sequences are involved. Due to their large size, amyloids are expected to have large contributions to the mechanical properties of biofilms and they can gel in the EPS due to attractive interaggregate interactions. An example of a gel formed from *de novo* peptide aggregates is shown in Figure 6.4, imaged with fluorescence microscopy, where single peptide fibres are visible and similar gelled behaviour is expected within biofilms.[7–9]

Detailed fundamental work on polymer structure has been done on single DNA chains, which is facilitated by the chains' large size. A rich range of phenomena are observed e.g. torsional elasticity, counterion condensation and topological interactions. Such phenomena are often important for the activity of machines that manipulate DNA *in vivo*[10] (Chapter 19). DNA chains are found in biofilms and due to their large sizes, they have important contributions to biofilm elasticity (DNases suspended in aerosols are used as clinical treatments to fluidise biofilms in the lungs of cystic fibrosis patients).

6.2 Dynamics of Polymers

There are three prototypical models for the dynamics of single polymer chains. Two of them are based around bead spring models: the *Rouse model* (describing flexible chains with screened hydrodynamics – good for molten polymers with no solvent) and the *Zimm model* (describing flexible chains with unscreened hydrodynamics – good for flexible polymers in solution). The third prototypical model is the *semi-flexible model* based on semi-flexible chains and considers the effect of bending rigidity on the dynamics (including unscreened hydrodynamics – good for semi-flexible polymers in solution) (Figure 6.3).[1] All three models for single polymer chain dynamics have been extensively tested with both synthetic and naturally occurring polymers e.g. using quasi-elastic scattering, fluorescence correlation spectroscopy, nuclear magnetic resonance and single-molecule fluorescence microscopy. Extensions to the models for single-chain dynamics have considered branching, helicity and charge effects (e.g. polyelectrolytes).[4]

Models for *polymer gel dynamics* need to consider how the dynamics of overlapping chains in semi-dilute solutions is affected by cross-links e.g. transverse shear waves

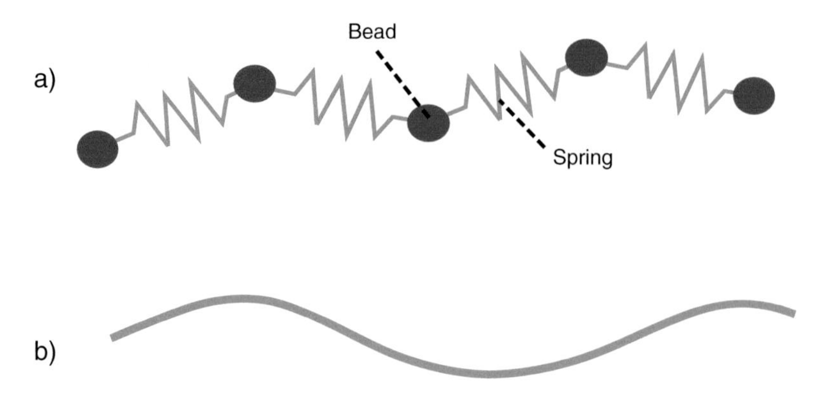

Figure 6.3 Models for the dynamics of polymers include (a) *Rouse* or *Zimm* (flexible) bead-spring chains and (b) *semi-flexible chains.*

can propagate across the gels, but not the precursor fluids. A rich range of dynamic behaviour of the polymers is modified by the molecular structure of the cross-links e.g. the effects of transient entanglements (reptation can contribute, which is the snake-like motion of polymers in tubes created by their neighbours), the lifetime of cross-links (sticker dynamics), the flexibility of the chains and the coordination number of the cross-links. Gels are typically non-equilibrium structures in terms of their thermodynamics (a rare possible exception is synthetic tetragels) and quenched disorder adds additional complications for their description e.g. their structures and dynamics sensitively depend on how they are made. Dynamic light scattering (DLS) measurements from a wide range of polymeric gels show that the correlation functions of scattered coherent light do not decay at long time scales.[11] The dynamics are thus non-ergodic and glassy phenomena occur. Single-molecule experiments also have evidence for quenched disorder for single chains inside gels in terms of states of prestress i.e. the polymer chains are not mechanically relaxed when they are cross-linked (Figure 6.4[7–9]). The quenched disorder again implies that the properties of the gel depend on how it was formed e.g. the kinetics of gel formation become important, since the gel has a memory.

There is a close connection between *polymer dynamics* and the *viscoelasticity* of polymer suspensions, including polymer gels. The viscoelasticity of polymeric solutions and gels is directly related to the dynamics of the molecules they contain. This is particularly noticeable with high-frequency rheology measurements, which are sensitive to smaller length scales in the samples and thus signatures of single chain dynamics can be observed in high-concentration solutions e.g. the complex shear moduli (G' and G'') scale as $\omega^{3/4}$ at high frequencies (ω) for semi-flexible polymers in solutions and gels, which is characteristic of the dynamics of single chains (Chapter 10). Weak polymeric gels in bacterial EPS often have power law rheology over a wide range of frequencies i.e. $G', G'' \sim \omega^{\alpha}$, where α has a constant low value. Operationally, the *gelation transition* can be followed as a function of polymer concentration via rheology measurements,[12] although fundamental questions on the universality of the phenomenon still exist. Experimentally, gel rheology is well described by a scaling relationship

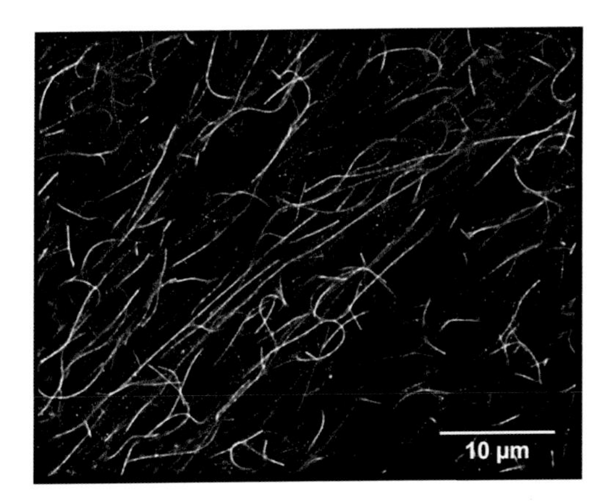

Figure 6.4 *Super-resolution fluorescence microscopy* images (STORM, stochastic optical reconstruction microscopy) of synthetic *de novo* amyloid peptides (with a lateral resolution of 20 nm).[8] Bayesian techniques demonstrate that states of self-stress exist in these peptide gels (quenched disorder, Figure 1.4c).[7,9] Similar tools could be used to quantify amyloids in biofilms.

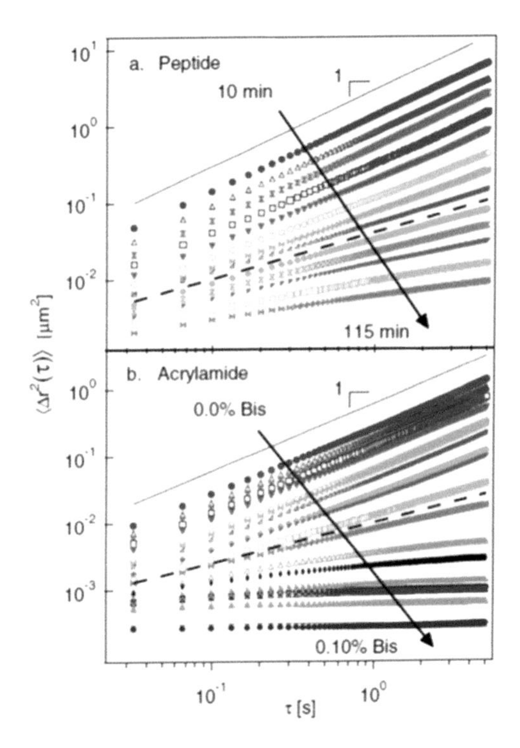

Figure 6.5 The mean square displacements $\left(\left\langle \Delta r^2(\tau)\right\rangle\right)$ of probes embedded in gelling polymeric solutions (peptides and polyacrylamides) as a function of time interval (τ) and gelation time.[13] The dashed line shows the point of gelation. Reprinted figure with permission from Larsen, T. H.; Furst, E. M., Microrheology of the liquid-solid transition during gelation. *Physical Review Letters* 2008, *100*, 146001. Copyright (2008) by the American Physical Society.

of the viscosity from the gel point[12] at which load-bearing structures percolate the sample. This has been carefully examined with particle tracking microrheology using synthetic polymers and peptides (Figure 6.5).

6.3 Bacterial Polymers

The dynamics of bacterial biopolymers play important roles in secreted slimes, capsules, biofilms, cell walls (e.g. peptidoglycans) and inside bacterial cells (e.g. during DNA replication and cell division). Cytoskeletal proteins play key roles in bacteria, since they determine their morphology and control the division of their cells. Extended aggregates of bacterial cells can also appear polymeric (on much larger length scales) e.g. sausage-like aggregates of *Bacillus subtilis* cells.

Both pure colloidal and polymeric systems (Figure 6.6) can form gels and these represent the two extremes of bacterial biofilm gel behaviour. If the bacteria stick together and produce little EPS, they form a *colloidal gel* (Figure 6.6a), whereas in the opposite limit in which the bacteria produce lots of polymeric EPS and the bacteria are well separated, they are better approximated by a *polymeric gel* (Figure 6.6b, a polymer gel with a small number of colloidal fillers). Analogous systems are observed with eukaryotic tissue in which the larger eukaryotic cells are combined with a fibrous polymeric extracellular matrix (ECM). This combination of polymers and colloids can lead to emergent properties that are not simply a weighted sum of the two volume fractions.[14] Synthetic polymer composites that include colloidal nanoparticles are commonly used as high performance materials e.g. rubber toughened glass.[15] Analogous high-performance polymeric composite gels also occur.[16]

The properties of *nanocomposite gels* in the context of biofilm research have been briefly reviewed.[17] Many bacterial cells act as relatively rigid inserts into the EPS (the bacteria are highly pressurised containers). Composite biofilm gels containing

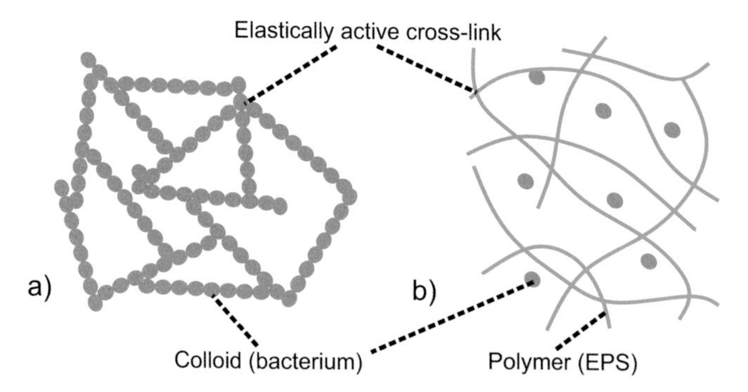

Figure 6.6 Bacterial biofilms can be considered as colloidal or polymeric gels. (a) A *colloidal gel* of aggregated bacterial cells (green). (b) A *polymeric gel* of extracellular polymeric substance (EPS, blue) components with occasional bacterial fillers (green).

bacteria are expected to have unique strain hardening phenomena compared to a reference system that contains EPS alone.[18]

The mechanics of *Pseudomonas aeruginosa* biofilms appear to be very robust to perturbation by a wide variety of chemicals in bulk rheology experiments.[19] This would appear to be regulated by cellular processes, since it is not expected for synthetic colloidal/polymeric gels or even single-component biopolymers, which tend to be sensitive to a wide variety of factors e.g. pH and salt concentration.

Bacteria have a wide range of specialised adhesive structures that can connect to the EPS, but non-specific physical connections are also possible. Both types of interaction can lead to additional cross-links in biofilms that will increase their elasticity.

The viscoelasticity of *E. coli* biofilms was studied as a function of the genetic expression of three EPS components (three knockdown mutants were available): curli, colonic acid and polyacetylglucosomae.[20] The curli protein which forms amyloid fibres dominated the mechanical response and the highest cohesive strength was observed in the wild-type film.

The classical theory of polymer gels has relatively limited quantitative use in biofilms (e.g. it assumes the polymer chains are all completely flexible and most biopolymers are not), but it is useful for a qualitative understanding. For a flexible polymeric gel where the elasticity is predominantly entropic, the shear modulus (G) is given by

$$G = \nu kT, \tag{6.5}$$

where ν is the density of cross-links and kT is the thermal energy. Thus, to a first approximation, increases of the density of cross-links will cause a linear increase in the elasticity of the gels. This is thought to be, in general, a reasonably robust result for gel mechanics, so a wide range of cross-linkers (e.g. bacteria themselves, multivalent counterions, adhesive molecules, etc.) are expected to increase the elasticity of biofilm gels.

Modern work has attempted to generalise results for the mechanics of flexible chains to *semi-flexible polymer gels* (e.g. Equation (6.5)) and a rich range of additional phenomena are observed. EPS chains are often semi-flexible and the geometry of cross-links can play an important role in such systems.[21] Semi-flexible networks can also demonstrate prestresses and quenched disorder (Figure 6.4). Biofilms demonstrate strain stiffening in a deformable microfluidic device and under shear flow.[22] This is a non-linear viscoelastic phenomenon and is typical of semi-flexible polymer gels.[23] To quantify the elasticity of biofilms in more detail, another challenge is their mixed nature e.g. colloidal gel phenomena can exist simultaneously with multi-component polymer gels that contain a broad mixture of types of cross-linker.

To reduce the viscoelasticity of a gel (e.g. during cleaning), both the number of effective cross-links (ν) as per Equation (6.5) and the length of the polymers need to be reduced (which reduces the viscosity in proportion to a power of the chain contour length e.g. $\eta \sim L^3$, and reduces the number of elastic entanglements in concentrated gels). Once the dominant polymer chemistry of the EPS is understood, enzymes can be chosen to disrupt the film e.g. proteinase to cut up protein chains or DNase for DNA chains. The enzymes can reduce the effective number of cross-links, decrease the modulus of the gels and make them easier to mobilise and thus remove. Mechanical forces, ultrasound, ionising radiation (for *ex vivo* applications only) and localised heating may also be successful to disrupt EPS gels. Alternatively, the adhesive interactions between the bacterial cells could be disrupted (for colloidal gels) or those between the bacterial cells and the EPS (for polymeric gels).

EPS biofilms often contain charged polymers i.e. polyelectrolytes. Polyelectrolyte gels have some unusual phenomena associated with charge effects, such as Donnan equilibrium, which explains how nappies hold on to water and the shape of human eyeballs.[24] The highly hydrophilic nature of biofilms combined with the high counterion pressure leads to the ability of polyelectrolyte gels to swell by orders of magnitude to hold on to lots of water and is clearly advantageous for bacterial communities that are vulnerable to desiccation.

The *swelling behaviour* of synthetic polyelectrolyte gels has been carefully studied.[25] Hydraulic permeability plays a role and Darcy's law explains how the transport of water can occur through the gel. Such models for synthetic systems have been successfully applied to biological tissues, such as the kinetics of swelling in a bruise, the shape of eyeballs and muscle.[26] The swelling behaviour of gels will help to keep bacteria in biofilms hydrated and thus alive. There are lots of hydrophilic groups along the polymeric backbone that allow gels to hold on to water, but the gelation mechanism via osmotic swelling will also play a role. Polyelectrolyte gels have a high swelling behaviour due to the dissociation of counterions, which increases the osmotic pressure. Syneresis (weeping of water at the surface of a gel) will also be important to maintain the hydration of bacterial cells and is commonly observed with synthetic gels.

Suggested Reading

Broedersz, C. P.; MacKintosh, F. C., Modelling semi-flexible polymer networks. *Reviews of Modern Physics* **2014**, *86* (3), 995. Overview of theoretical models for semi-flexible polymers that typically describe biopolymer gels reasonable well.

Djabourov, M.; Nishinari, K.; Ross-Murphy, S., *Physical Gels from Biological and Synthetic Polymers*. Cambridge University Press: 2013. Slightly old-fashioned approach to the polymer theory, but lots of interesting examples of gel systems are included.

Gong, J. P., Why are double network hydrogels so tough? *Soft Matter* **2010**, *6* (12), 2583–2590. Explores some unusual phenomena found in composite polymeric hydrogels. There are many others.

Muthukumar, M., *Physics of Charged Macromolecules: Synthetic and Biological Systems*. Cambridge University Press: 2023. Describes some modern developments in the physics of charged polymers.

Rubinstein, M.; Colby, R. H., *Polymer Physics*. Oxford University Press: 2003. Simple overview of polymer physics applied to synthetic polymers including theories of gelation.

References

1. Rubinstein, M.; Colby, R. H., *Polymer Physics*. Oxford University Press: 2003.
2. de Gennes, P. G., *Scaling Concepts in Polymer Physics*. Cornell: 1979.
3. Waigh, T. A., *The Physics of Living Processes*. Wiley: 2014.
4. Dobrynin, A. V.; Rubinstein, M., Theory of polyelectrolytes in solutions and interfaces. *Progress in Polymer Science* **2005**, *30* (11), 1049–1118.
5. Philips, R.; Kondev, J.; Theriot, J.; Garcia, H.; Kondev, J., *Physical Biology of the Cell*. Garland Science: 2012.
6. Ramos-Leon, F.; Ramamurthi, K. S., Cytoskeletal proteins: Lessons from bacteria. *Physical Biology* **2022**, *19*, 021005.
7. Cox, H.; Cao, M.; Xu, H.; Waigh, T. A.; Lu, J. R., Active modulation of states of prestress in self-assembled short peptide gels. *Biomacromolecules* **2019**, *20* (4), 1719–1730.
8. Cox, H.; Georgiades, P.; Xu, H.; Waigh, T. A.; Lu, J. R., Self-assembly of mesoscopic peptide surfactant fibrils investigated by STORM super-resolution fluorescence microscopy. *Biomacromolecules* **2017**, *18* (11), 3481–3491.
9. Cox, H.; Xu, H.; Waigh, T. A.; Lu, J. R., Single-molecule study of peptide gel dynamics reveals states of prestress. *Langmuir* **2018**, *34* (48), 14678–14689.
10. Bensimon, D.; Croquette, V.; Allemand, J. F., *Single-molecular Studies of Nucleic Acids and Their Proteins*. Oxford University Press: 2019.
11. Pusey, P. N.; Van Megen, W., Dynamic light scattering by non-ergodic media. *Physica A: Statistical Mechanics and Its Applications* **1989**, *157* (2), 705–741.
12. Winter, H. H.; Chambon, F., Analysis of the linear viscoelasticity of a cross-linking polymer at the gel point. *Journal of Rheology* **1986**, *30* (2), 367–382.
13. Larsen, T. H.; Furst, E. M., Microrheology of the liquid-solid transition during gelation. *Physical Review Letters* **2008**, *100* (14), 146001.
14. van Oosten, A. S. G.; Chen, X.; Chin, L.; Cruz, K.; Patteson, A. E.; Pogoda, K.; Shenoy, V. B.; Janmey, P. A., Emergence of tissue-like mechanics from fibrous networks confined by close-packed cells. *Nature* **2019**, *573* (7772), 96–101.
15. Gersappe, D., Molecular mechanisms of failure in polymer nanocomposites. *Physical Review Letters* **2002**, *89* (5), 058301.
16. Dannert, C.; Stokke, B. T.; Dias, R. S., Nanoparticle-hydrogel composites. *Polymer* **2019**, *11* (2), 275.
17. Even, C.; Marliere, C.; Ghigo, J. M.; Allain, J. M.; Marcellan, A.; Raspaud, E., Recent advances in studying single bacteria and biofilm mechanics. *Advances in Colloid and Interface Science* **2017**, *247*, 573–588.
18. Jana, S.; Charlton, S. G. V.; Eland, L. E.; Burgess, J. G.; Wipat, A.; Curtis, T. P.; Chen, J., Nonlinear rheological characterisation of single species bacterial biofilms. *npj Biofilms and Microbiomes* **2020**, *6* (1), 19.
19. Lieleg, O.; Caldara, M.; Baumgartel, R.; Ribbeck, K., Mechanical robustness of *Pseudomonas aeruginosa* biofilms. *Soft Matter* **2011**, *7* (7), 3307–3314.

20. Horvat, M.; Pannuri, A.; Romero, T.; Dogsa, I.; Stopar, D., Viscoelastic response of *E. coli* biofilms to genetically altered expression of extracellular matrix components. *Soft Matter* **2019**, *15* (25), 5042.
21. Broedersz, C. P.; MacKintosh, F. C., Modelling semiflexible polymer networks. *Review of Modern Physics* **2014**, *88*, 039903.
22. Wilking, J. N.; Angelini, T. E.; Seminara, A.; Brenner, M. P.; Weitz, D. A., Biofilms as complex fluids. *MRS Bulletin* **2011**, *36* (5), 385.
23. Wen, Q.; Basu, A.; Janmey, P. A.; Yodh, A. G., Non-affine deformations in polymer hydrogels. *Soft Matter* **2012**, *8*, 8039–8049.
24. Muthukumar, M., *Physics of Charged Macromolecules: Synthetic and Biological Systems*. Cambridge University Press: 2023.
25. Rubinstein, M.; Colby, R. H.; Dobrynin, A. V.; Joanny, J. F., Elastic modulus and equilibrium swelling of polyelectrolyte gels. *Macromolecules* **1996**, *29* (1), 398–406.
26. Ethier, C. R.; Simmons, C. A., *Introductory Biomechanics: From Cells to Organisms*. Cambridge University Press: 2008.

7 Colloidal Structure and Dynamics

Particulate matter on micron to nanometre length scales is *colloidal* (from the Greek for glue, kolla) and this includes a very wide range of everyday materials, including proteins, cells (e.g. tomato ketchup or planktonic bacteria), aerosols (smoke, fog, etc.) and surfactants (e.g. soaps and membranes). Intensive research has been performed on colloids due to the preponderance of their industrial applications e.g. in paints, detergents, opals and photonic crystals. The range of phenomena demonstrated by colloidal materials is surprisingly rich even for very simple synthetic systems e.g. monodisperse polymethylmethacrylate (PMMA) spheres in organic solvents (e.g. cis-decalin) demonstrate a wide range of phase behaviours. Similar to other related areas of soft matter (liquid crystals, polymers, etc.), theories developed for simple well-controlled inanimate systems can often be applied to the coarse-grained behaviour of living matter, which circumvents some of the problems associated with the extreme molecular complexity of living systems.

Bacterial cells, eukaryotic cells and viruses are all specific examples of *biological colloids*, albeit fairly complex ones. Standard colloidal theories predict the properties of interacting colloidal suspensions where most of the internal features of the colloids are neglected. Thus, the generic behaviour of particles of fixed size and buoyancy with simplified mesoscopic interparticle interactions can be predicted for bacterial colloids. The mesoscopic interactions experienced by biological colloids were described in Chapter 4.

7.1 Structure of Colloids

Simple synthetic colloids (e.g. neutrally buoyant monodisperse polystyrene spheres) provide a minimal system to understand the phase behaviour of colloidal suspensions.[2] The mesoscopic forces sensitively affect phase behaviour. Figure 7.1 shows how the strength of the sticky potential (ε) changes the phase behaviour of synthetic colloids as a function of the colloidal volume fraction (ϕ_e).[1,3] The colloids gel above a threshold value of the sticky potential. Glassy behaviour is also possible at high volume fractions of the colloids.

Colloids can be further sub-classified depending on their size. There are density and size limits that separate Brownian colloids from non-Brownian colloids (the limits can be deduced from a comparison of the buoyancy force to the thermal force). Non-Brownian colloids are so large and dense that they tend to sediment rapidly in

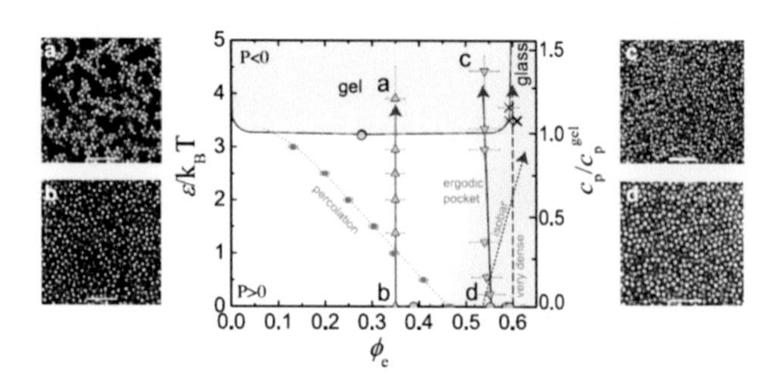

Figure 7.1 The *phase diagram* of sticky polystyrene colloidal spheres. The interparticle potential (ε) can be adjusted by adding more polymer and the phase diagram (centre) is shown as a function of colloid volume fraction (ϕ_e). k_BT is the thermal energy and is used as units for ε. The samples gel at higher interparticle potentials. Glassy non-ergodic behaviour is also observed at high volume fractions.[1] a, b, c and d show optical images from confocal microscopy corresponding to different parts of the phase diagram. Reprinted from [Royall C.P., Williams S.R., Tanaka H., Vitrification and gelation in sticky spheres. *The Journal of Chemical Physics* 2018, 148, 044501], with the permission of AIP Publishing.

solution. Furthermore, elongated cells of both Brownian and non-Brownian colloids can lead to liquid-crystalline colloidal suspensions i.e. phase transitions occur in suspensions due to the relative orientation of the cells (Chapter 8).

7.2 Dynamics of Colloids

At first sight due to their simple geometry and the neglect of internal dynamics, it might be thought that colloidal dynamics should be very simple. However, as the concentration of colloids in a suspension is increased, interparticle interactions cause large modifications to the motion of individual colloids. Cage hopping (similar to that seen with molecular fluids), glassification, crystallisation and gelation can all occur. The introduction of particle anisotropy can make the range of phases even richer and dynamics typical of lyotropic liquid crystalline phases can be observed, as well as anisotropic glasses, crystals and gels.

Two dimensionless ratios, the *Péclet number* (Pe) and the *Deborah number* (De), are very useful for gauging the effects of flow on colloidal suspensions.[3,4] The Péclet number allows the relative strength of convective flows to be compared with diffusive transport,

$$Pe = \frac{6\pi\eta a^3\dot{\gamma}}{kT}, \tag{7.1}$$

which is the characteristic time for diffusion of a sphere of radius a divided by the characteristic time for the flow ($\frac{1}{\dot{\gamma}}$, where $\dot{\gamma}$ is the shear rate), where η is the viscosity and kT is the thermal energy. For high Péclet numbers, convective transport dominates and diffusive transport can be neglected (e.g. bacteria in a strong flow in a pipe) and vice versa for low Péclet numbers (e.g. bacteria in a weak flow in a pipe). The Deborah number is defined as

$$De = \frac{\tau}{t}, \tag{7.2}$$

where τ is the time for stress relaxation in a material and t is the time scale of the measurement. When $De \sim 1$, the material is viscoelastic, whereas $De >> 1$ is solid-like and $De << 1$ is liquid-like. With viscoelastic materials, their flow behaviour becomes more complex (the field of rheology, Chapter 10).

7.3 Bacterial Colloids

Most bacteria are *non-Brownian colloids*,[3] since bacteria are denser than water and invariably sediment without active transport by motors (some marine bacteria adjust their buoyancy, but this is also an active process, since work is done in the thermodynamic sense, which introduces other complications). Brownian motion does lead to measurable fluctuations in the positions of the non-Brownian particles as they sediment, but its amplitude is insufficient to provide long-term stability of such sedimenting colloids.

Bacteria can demonstrate both attractive colloidal interactions (since they have a wide range of adhesive interactions that can be actively modulated by the bacteria)[5] and repulsive colloidal interactions (e.g. due to membrane-bound lipopolysaccharides, capsular polysaccharides or negative charges).[6] Evidence for *in vivo* colloidal gels of bacteria due to adhesive interactions has been presented in the guts of Zebra fish.[7]

The rigidity of colloids can affect their rheology.[3] Many medically relevant bacteria (e.g. *Staphylococcus aureus*, *Escherichia coli* and *Pseudomonas aeruginosa*) are expected to be relatively rigid due to their high intracellular osmotic pressures, but capsular bacteria will have softer interparticle potentials[8] (Section 4.1.2) and spirochete bacteria actively change the shape of their cells to move.

As the concentration of a colloidal system increases, the dynamics slows down and there can be a dramatic *jamming phase transition*.[10,11] Furthermore, a two-step shear yielding phenomenon is observed for synthetic colloidal gels and attractive glasses in large-amplitude rheology experiments (Chapter 10). It has been recently observed in non-linear oscillatory rheology experiments with the biofilms of *Comamonas denitrificans*, *Bacillus subtilis*, *Pseudomonas fluorescens* and *P. aeruginosa*[9,12,13] (Figure 7.2).

Studies of the intermediate concentrations of model synthetic PMMA colloids (volume fraction, $\phi = 0.4$) combined with gelling polymers (polystyrene) showed signatures of non-ergodicity in the colloidal dynamics as the gelation point was approached i.e. stretched exponential correlation functions occur in dynamic light scattering measurements.[14] *Caging phenomena* contribute to this behaviour and the dynamics of colloids

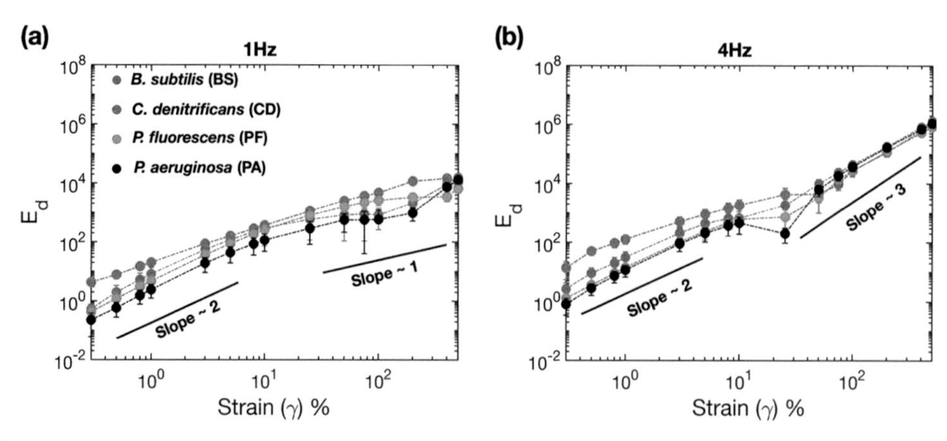

Figure 7.2 *Two-step shear yielding* is observed in bacterial suspensions during non-linear rheology experiments and it is characteristic of the glassy/gelled dynamics of synthetic colloids.[9] The elastic energy dissipated (E_d) is shown as a function of strain amplitude (γ) for (a) 1 Hz and (b) 4 Hz.

becomes split between intracage diffusion and cage hopping. Thus, non-ergodicity (including glassy behaviour) is expected in both biofilms that are predominantly polymeric (Chapter 6) and colloidal (and those in between), since it is observed in both types of matter.

When suspensions of colloidal spheres are sheared, there can be a mismatch in the velocity at their boundaries i.e. *wall slip*.[3] With standard molecular Newtonian fluids, the wall slip is very small, but in colloidal and polymeric systems, a giant wall slip is possible. Significant wall slip has not yet been observed with biofilms under physiologically relevant conditions due to their strong adhesive interactions with surfaces. However, the phenomena are possible in hard sphere colloidal systems and thus might be expected in high concentrations of capsular bacteria that have repulsive interparticle potentials (assuming adhesins on their pili are not strongly interacting with the surfaces).[15]

Rod-like colloids demonstrate other phenomena dependent on their aspect ratio[16] e.g. *liquid crystalline ordering*, which can cause a sudden decrease in viscosity as the concentration of rods is increased[17] (Chapter 8). Due to the preponderance of rod-like and filamentous bacteria, similar phenomena are expected with many bacterial colloids.

Anomalous transport is observed in many soft matter materials, including colloidal systems, where it is due to a combination of viscoelasticity, geometry, motor activity and glassy effects (Section 1.4). Both colloidal glasses and gels exist, so the expectation is that they may also occur in analogous bacterial systems, with similar-sized systems and similar interparticle potentials. Caging in high-concentration colloidal solutions has been much explored in synthetic systems. Cages lead to long-lived anomalous dynamics, kinetic jamming and a colloidal glass transition for colloids with hard sphere potentials[18] e.g. no attractive interactions are needed to drive jamming transitions.

Optical tweezers can be used to trap colloidal particles including bacteria (Chapter 13) to explore their interactions and phase behaviour. A challenge is to limit the radiation damage, so that realistic interparticle potentials can be measured for single colloids, but optical tweezers are very sensitive, allowing the measurement of piconewton forces. Atomic force microscopes (AFMs, Section 13.2.1) have fewer problems with radiation damage for measuring the forces of living organisms, but they are restricted to surface geometries and have a limited range of sensitivities (much lower sensitivities are possible with AFMs compared with optical tweezers e.g. nanonewtons compared with piconewtons). Magnetic tweezers can have a similar sensitivity to optical tweezers and can be made less invasive. However, magnetic tweezers can be more challenging to calibrate e.g. due to magnetic hysteresis (superparamagnetic beads are preferred over ferromagnets for this reason), the performance of the electromagnets and inductance effects.

Simple aggregation phenomena occur with synthetic colloidal systems that can lead to intricate fractal morphologies. One mechanism is *diffusion-limited aggregation* (DLA) in which colloids diffuse until they make contact with a neighbouring particle, which causes them to stick.[20] Some bacterial aggregates resemble DLA morphologies (Figure 7.3).[21]

Figure 7.3 Simulation of the *diffusion-limited aggregation* (DLA) of colloids. DLA-shaped aggregates are observed for aggregates of both synthetic and bacterial colloidal systems.[19] Reprinted figure with permission from Witten, T. A.; Sander, L. M., Diffusion-limited aggregates, a kinetic critical phenomenon. *Physical Review Letters* **1981,** *47* (19), 1400. Copyright (1981) by the American Physical Society.

The linear rheology of *active suspensions* of *E. coli* has a characteristic $\omega^{-1/2}$ scaling of the shear moduli (ω is the frequency), which can be explained by a colloidal model that couples stress, orientation and concentration fluctuations[22] (Chapter 21). The extension of thermodynamic techniques to understand active colloidal matter is still in development. For example, great care is needed when extending the concept of the state function (e.g. its dependence on pressure) for active colloidal fluids.[23]

Patchy colloidal models have been adapted to describe the adhesion of *E. coli*[24] (Section 4.2) i.e. colloids where the interparticle potential varies from point to point on their surfaces. The model was used to explain the high throughput data of *E. coli* on glass surfaces that exhibited large phenotypic variability for adhesion strength.

Suggested Reading

Berg, J. C., *An Introduction to Interfaces and Colloids: The Bridge to Nanoscience*. World Scientific: 2009. Well-explained pragmatic account of colloidal science.

Gazezelli, E., *A Physical Introduction to Suspension Dynamics*. Cambridge University Press: 2011. The majority of bacteria sediment as their default mode of transport (e.g. *S. aureus*) due to their relatively large colloidal size and their density compared with water (specifically they are non-Brownian colloids). Thus, active motility is required for a stable liquid colloidal phase of bacteria to be achieved and sedimentation is common.

Israelachvili, J. N., *Intermolecular Surfaces and Forces*, 3rd ed. Academic Press: 2011. Classic account of mesoscopic forces.

Mewis, J.; Wagner, N., *Colloidal Suspension Rheology*. Cambridge University Press: 2012. Excellent discussion of synthetic colloidal rheology.

References

1. Royall, C. P.; Williams, S. R.; Tanaka, H., Vitrification and gelation in sticky spheres. *The Journal of Chemical Physics* **2018**, *148*, 044501.
2. Berg, J. C., *An Introduction to Interface and Colloids: The Bridge to Nanoscience*. World Scientific: 2010.
3. Mewis, J.; Wagner, N. J., *Colloidal Suspension Rheology*. Cambridge University Press: 2011.
4. Goodwin, J. W.; Hughes, R. W., *Rheology for Chemists: An Introduction*. Royal Society of Chemistry: 2008.

5. Cates, M. E.; Fuchs, M.; Kroy, K.; Poon, W. C. K.; Puertas, A. M., Theory and simulation of gelation, arrest and yielding in attracting colloids. *Journal of Physics: Condensed Matter* **2004**, *16* (42), S4861.
6. Carrier, V.; Petekidis, G., Nonlinear rheology of colloidal glasses of soft thermo-sensitive microgel particles. *Journal of Rheology* **2009**, *53* (2), 245.
7. Schlomann, B. H.; Parthasarathy, R., Gut bacterial aggregates as living gels. *eLife* **2021**, *10*, 71105.
8. Vlassopoulos, D.; Cloitre, M., Tunable rheology of dense soft deformable colloids. *Current Opinion in Colloid and Interface Science* **2014**, *19* (6), 561–574.
9. Jana, S.; Charlton, S. G. V.; Eland, L. E.; Burgess, J. G.; Wipat, A.; Curtis, T. P.; Chen, J., Nonlinear rheological characterisation of single species bacterial biofilms. *npj Biofilms and Microbiomes* **2020**, *6*, 19.
10. Cipelletti, L.; Ramos, L., Slow dynamics in glasses, gels and foams. *Current Opinion in Colloid and Interface Science* **2002**, *7* (3–4), 228–234.
11. Ahuja, A.; Potanin, A.; Joshi, Y. M., 2 step yielding in soft materials. *Advances in Colloid and Interface Science* **2020**, *282*, 102179.
12. Kim, J.; Merger, D.; Wilhelm, M.; Helgeson, M. E., Microstructure and nonlinear signatures of yielding in a heterogeneous colloidal gel under large amplitude oscillatory shear. *Journal of Rheology* **2014**, *58* (5), 1359.
13. Koumakis, N.; Petekidis, G., Two step yielding in attractive colloids: Transition from gels to attractive glasses. *Soft Matter* **2011**, *7* (6), 2456–2470.
14. Laurati, M.; Petekidis, G.; Koumakis, N.; Cardinaux, F.; Schofield, A. B.; Brader, J. M.; Fuchs, M.; Egelhaaf, S. U., Structure, dynamics and rheology of colloid-polymer mixtures: From liquids to gels. *Journal of Chemical Physics* **2009**, *130* (13), 134907.
15. Ballesta, P., Wall slip and flow of concentrated hard-sphere colloidal suspensions. *Journal of Rheology* **2012**, *56* (5), 1005.
16. Solomon, M. J.; Spicer, P. T., Microstructural regimes of colloidal rod suspensions, gels and glasses. *Soft Matter* **2010**, *6* (7), 1391–1400.
17. Collings, P. J.; Goodby, J. W., *Introduction to Liquid Crystals: Chemistry and Physics*. CRC Press: 2019.
18. Pusey, P. N.; Van Megen, W., Phase behaviour of concentrated suspensions of nearly hard colloidal spheres. *Nature* **1986**, *320* (6060), 340–342.
19. Witten, T. A.; Sander, L. M., Diffusion-limited aggregates, a kinetic critical phenomenon. *Physical Review Letters* **1981**, *47* (19), 1400.
20. Lin, M. Y.; Lindsay, H. M.; Weitz, D. A.; Ball, R. C.; Klein, R.; Meakin, P., Universality in colloid aggregation. *Nature* **1989**, *339*, 360–362.
21. Vassallo, L.; Hansmann, D.; Braunstein, L. A., On the growth of non-motile bacteria colonies. *The European Physical Journal B* **2019**, *92*, 216.
22. Chen, D. T. N.; Lau, A. W. C.; Hough, L. A.; Islam, M. F.; Goulian, M.; Lubensky, T. C.; Yodh, A. G., Fluctuations and rheology in active bacterial suspensions. *Physical Review Letters* **2007**, *99* (14), 148302.

23. Solon, A. P.; Fily, Y.; Baskaran, A.; Cates, M. E.; Kafri, Y.; Kardar, M.; Tailleur, J., Pressure is not a state function of generic active fluids. *Nature Physics* **2015**, *11* (8), 673–678.
24. Vissers, T.; Brown, A. T.; Koumakis, N.; Dawson, A.; Hermes, M.; Schwarz-Linek, J.; Schofield, A. B.; French, J. M.; Koustos, V.; Arlt, J.; Martinez, V. A.; Poon, W. C. K., Bacteria as living patchy colloids: Phenotypic heterogeneity in surface adhesion. *Science Advances* **2018**, *4* (4), eaao1170.

8 Liquid Crystals

Elongated molecules and elongated colloids (e.g. cells) can spontaneously align themselves in concentrated suspensions to form *liquid crystalline phases*. This is a purely thermodynamic phenomenon due to the minimisation of the free energy. In equilibrium systems, such alignment has the characteristics of a phase transition e.g. a sharp enthalpy change associated with latent heat can be observed.[1] Synthetic liquid crystals have found applications in detergents, bullet-proof jackets and electronic displays. Modern developments with synthetic liquid crystals include a range of new phases, such as hockey puck molecules that assume hexatic phases (crystalline in two-dimensional [2D] and liquid-like in one-dimensional [1D]), bent core nematics and oriented polymers containing both rigid and flexible moieties (e.g. dendritic and side-chain liquid crystals phases).[2]

A wide variety of biological molecules also have liquid crystalline phases e.g. DNA, proteins, lipids and carbohydrates.[2,3] The discovery of liquid crystalline phases of suspensions of bacterial cells is more recent.[4] Liquid crystalline models enable more nuanced descriptions of bacterial matter at high concentrations in which orientation is observed, since the collective phenomena of the interacting bacteria (the phase) are well-defined for a large number of organisms under well-defined physical conditions (temperature, volume, pressure, etc.). Bacterial motility also leads to completely new phases of matter extending the field of liquid crystallinity i.e. *active liquid crystalline phases*.

8.1 Passive Liquid Crystals

Liquid crystals are intermediate phases of matter between liquids and crystals. They have features of both phases, with regular structuration at the molecular scale representative of crystals, but they can flow like fluids. Liquid crystal-forming particles are called *mesogens*. Simple phases of small-molecule liquid crystals can be motivated by thermodynamic considerations via the minimisation of the equilibrium free energy (F, the internal energy (U) combined with an entropic component (S), $F = U - TS$, where T is the temperature).[2,5]

Examples of liquid crystalline phase are shown in Figure 8.1. Nematic phases are commonly observed in high concentration phases of elongated bacteria (Figure 8.1a). Other simple liquid crystalline phases of bacteria could possibly occur, such as smectics in which the bacteria form layers (Figure 8.1b). However, smectic phases can be

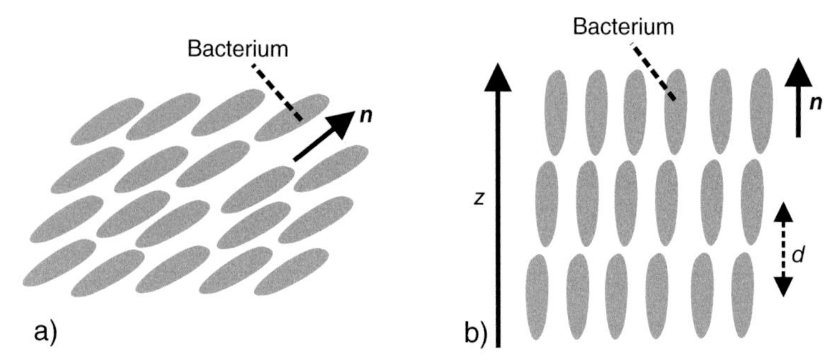

Figure 8.1 (a) Schematic diagram of rod-shaped bacteria demonstrating almost perfect *nematic* ordering along the director *n* in two dimensions. (b) *Smectic* ordering of bacteria (not yet observed in the literature for bacteria). *d* is the periodicity of the layered structure.

very sensitive to the polydispersity of the mesogen (bacterial) length and the processes of bacterial growth may disrupt their occurrence. The motility of bacteria in synthetic smectic liquid crystals has been considered.[6] Other synthetic liquid crystalline phases of interest include cholesterics (chiral nematics) and hexatics.

Order parameters are important mathematical tools to describe the progress of phase transitions and thus have been widely used in studies of soft condensed matter, including liquid crystals e.g. an orientational order parameter based on the *second Legendre polynomial* $\left(P_2\left(\cos\theta\right)\right)$ is used to describe nematic phase transitions of liquid crystals,[1,2]

$$P_2\left(\cos\theta\right)=\frac{3}{2}\left\langle\cos^2\theta\right\rangle-\frac{1}{2}, \tag{8.1}$$

where θ is the angle the long axis of a mesogen makes with the director (n, the average direction of alignment) and $\langle\ \rangle$ denotes averaging over all the molecules. $P_2\left(\cos\theta\right)$ will have values close to 1 in nematic phases and close to 0 in disordered phases.

Other standard order parameters for phase transitions are the *number density* of the mesogens (e.g. the bacteria) and the *lamellar order parameter* (ψ) for a smectic (Figure 8.1b),

$$\rho(z)=\rho_0\left[1+\psi\cos\left(\frac{2\pi z}{d}\right)\right], \tag{8.2}$$

where d is the periodicity of the smectic layers, $\rho(z)$ is the density perpendicular to the layers, ρ_0 is the average density and z is the coordinate perpendicular to the smectic layers. More sophisticated order parameters are required to fully describe the properties of anisotropic liquid crystalline phases and a second rank tensor, Q_{ij}, is typically used[1,2] e.g. to describe uniaxial and biaxial mesogens.

The *disclination strengths* (m) of orientational defects of liquid crystals (Figure 8.2) can be calculated using

$$m=\frac{1}{2\pi}\oint\frac{d\varphi(r)}{d\theta}d\theta=\frac{\varphi_{total}}{2\pi}, \tag{8.3}$$

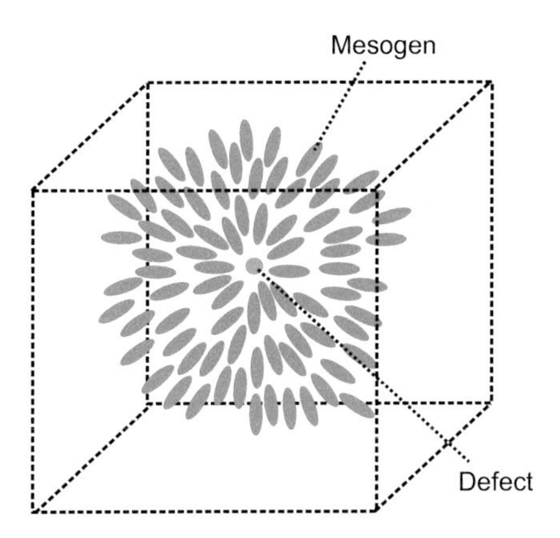

Figure 8.2 Hedgehog defect textures of a nematic phase ($m = 1$) in a liquid crystalline material.

where $\varphi(r)$ is the angle of the director at r and θ is the angle r makes with the horizontal axis. Many of the physical properties of liquid crystals are affected by such defects e.g. their flow behaviour.

Defect structures in the orientation are characteristic of liquid crystalline phases, so liquid crystalline textures measured with polarising microscopy are used to quantify the defects and thus identify the exact liquid crystalline phases that occur e.g. nematic versus smectic phases.[2,7] Defect textures have a rich range of possibilities in three dimensions e.g. hedgehog orientations from single point defects[2] (Figure 8.2). The defects are strongly affected by the anchoring of the mesogens at surfaces (similar to small molecule liquid crystals used in displays where the surface interactions are carefully controlled).

A simple successful continuum model to understand the free energies (F) of the elastic deformations of nematic liquid crystals is due to Franks,

$$F = \frac{1}{2}K_1(\nabla . n)^2 + \frac{1}{2}K_2(n.\nabla \times n)^2 + \frac{1}{2}K_3(n \times \nabla \times n)^2, \tag{8.4}$$

where the three elastic constants K_1, K_2 and K_3 correspond to *splay*, *twist* and *bend*, respectively.[2] Such theories allow a range of physical phenomena with liquid crystals to be understood e.g. elastic mechanical deformations associated with different types of defects.[5]

The concentration of mesogens at which lyotropic mesogens (e.g. rigid rods in suspension) experience liquid to nematic phase transitions can be calculated using the *Onsager theory*, which leads to

$$\frac{\phi L}{D} > 4.49 \tag{8.5}$$

for nematic phases, where ϕ is the volume fraction of the mesogens, L is the mesogen length and D is the mesogen diameter.[8] Purely hard sphere repulsive interactions are assumed with this Onsager theory and it would thus only be a first approximation for bacterial liquid crystals.

8.2 Active Liquid Crystals

The motility of mesogens (e.g. the motility of rod-like bacteria) extends the variety of phases expected with liquid crystals and they constitute a new type of matter, *active liquid crystals*.[9] Active liquid crystals provide challenges for modelling, since equilibrium thermodynamic tools are no longer suitable e.g. simple minimisation of the equilibrium free energies will not provide accurate results. Dense suspensions of elongated motile bacteria can be considered as active liquid crystals. Biofilms can also have some liquid crystalline character at high bacterial concentrations, although the gel-like extracellular polymeric substance (EPS) can lead to quenched disorder (Chapter 6), which also complicates simple thermodynamic descriptions e.g. biofilms are not in thermal equilibrium even when they contain immotile bacteria.

Active liquid crystalline phenomena are observed with elongated bacteria in concentrated suspensions, surface adsorbed swarms and biofilms. Such active liquid crystals are novel states of soft condensed matter that demonstrate some completely new behaviour compared with conventional liquid crystals.[10] An alternative analogy is also possible to treat dense bacterial suspensions as *living granular matter*, similar to vibrated granular rods.[11] The bacterial rearrangements are driven by cell growth or motility and they demonstrate non-equilibrium phenomena, similar to the granular rods, whereas conventional liquid crystals are often considered to be in thermal equilibrium. Both dry (granular matter) and wet (active liquid crystals) models can provide useful insights into the behaviour of bacterial suspensions.[12,13]

A detailed study of the *nematic liquid crystalline* phases formed by *Vibrio cholerae* biofilms was performed using confocal microscopy combined with image analysis, in which every cell in the biofilm could be located together with its orientation.[13] The results were then compared with a particle-based effective potential model (a model that assumes thermal equilibrium). The response of the nematic ordering of the bacteria to shear in the microfluidic device was also measured.

A *verticalisation transition* is found for the orientation of *V. cholerae* cells in biofilms as they mature[14] i.e. the cells predominantly adsorb horizontal to the surface and switch to the perpendicular orientation as they age. The transition sensitively depends on the cell length, since long cells are more likely to be peeled off the surface by nearby vertical cells. A fountain effect was also found to drive the three dimensional growth of *V. cholerae* biofilms using lattice sheet microscopy (Section 13.2.4).[15]

Under strong *shear flows*, biofilms can adopt aerofoil shapes.[16] With *V. cholerae*, the shear flow breaks the symmetry of the biofilms' hemispherical caps. Flow also altered the nematic alignment of the cells due to the drag-induced torque which caused the verticalisation of the cells at the front of the biofilms.

The rate of colony expansion was found to inversely correlate with agar concentration in the growth media of *V. cholerae* biofilms.[17] Addition of dextran to solutions containing submerged biofilms was found to cause them to shrink in volume. It was suggested that the osmotic pressure in the biofilm drives these changes. Osmotic pressure differences between the biofilm and the surroundings are also advantageous to increase food uptake by the bacteria, but they could also increase antibiotic uptake, a disadvantage for the bacteria (Chapter 23).

A *nematic liquid crystalline* phase occurs with *Escherichia coli* suspensions as the cell density is increased and it is a first-order phase transition.[18] Motility can be crucial for biofilm formation in *E. coli*, since paralysed flagellar mutants do not form biofilms, although non-motile *E. coli* biofilms from other strains do exist.[19] More sophisticated active nematic liquid crystal models have been applied to suspensions of *Serratia marcescens* that describe the dynamics of their defect structures.[20]

The *collective dynamics* of *Pseudomonas aeruginosa* across surfaces has been described in terms of active liquid crystal theory, focusing on the collisions of orientational defects.[21] The defect collision mechanism is an important factor that determines the speed of migration of the bacteria. A separate active liquid crystal study found that *Myxococcus xanthan* has topological defects (strength $+1/2, -1/2$) that promote layer formation in colonies.[22]

Small-angle X-ray scattering (SAXS) experiments show the carbohydrates in biofilm EPS from a marine bacterium look similar to standard heterogeneous physical polymer gels[23] and thus contain quenched disorder (Section 13.4.5). Biofilm EPS used in food (e.g. xanthan and gellan) are also weak physical gels with quenched disorder.[24–26] The appearance of such quenched disorder in biofilms obstructs more naïve thermodynamic approaches to model liquid crystalline biofilm behaviour e.g. calculations based on standard thermodynamic liquid crystalline approaches, such as those due to Landau de Gennes, Maier-Saupe or Onsager.[2,5] Chapters 6 and 7 consider the polymeric and colloidal gel aspects of biofilms. Both polymer and colloidal varieties of soft matter gel contain quenched disorder. Effective potential approaches have been introduced to describe the creation of biofilms,[13] but since they are based on equilibrium thermodynamics, they can only be considered approximations to non-equilibrium systems that contain quenched disorder. Active processes in terms of biofilm growth also raise numerous questions, since they require careful quantitative analysis e.g. the fluctuation-dissipation theorem is broken[27] (Section 2.1, Equation (2.2)). Thus, simple thermodynamic effective potential approaches might be a useful first approximation for biofilm liquid crystals and computationally efficient simulation tools, but they will always fail for high-resolution studies.

Suggested Reading

Collings, P. J.; Goodby, J. W. *Introduction to Liquid Crystals: Chemistry and Physics*, 2nd ed. CRC Press: 2019. Simple introduction to liquid crystalline materials.

Pismen, L., *Active Matter Within and Around Us*. Springer: 2021. Excellent pedagogic introduction to active matter.

References

1. Chaikin, P. M.; Lubensky, T. C., *Principles of Condensed Matter Physics*. Cambridge University Press: 1995.
2. Collings, P. J.; Goodby, J. W., *Introduction to Liquid Crystals: Chemistry and Physics*. CRC Press: 2019.
3. Neville, A. C., *Biology of Fibrous Composites*. Cambridge University Press: 1993.
4. Aranson, I. S., Bacterial active matter. *Reports on Progress in Physics* **2022**, *85* (7), 076601.
5. de Gennes, P. G.; Prost, J., *The Physics of Liquid Crystals*. Oxford University Press: 1995.
6. Lakey, C. C.; Turner, M. S., Emergent ordering of microswimmers in smectic liquid crystals. *Artificial Life and Robotics* **2022**, *27* (4), 218–225.
7. Dierking, I., *Textures of Liquid Crystals*. Wiley: 2003.
8. Vroege, G. J.; Lekkerkerker, H. N. W., Phase transitions in lyotropic colloidal and polymeric liquid crystals. *Reports on Progress in Physics* **1992**, *55* (8), 1241–1309.
9. Marchetti, M. C.; Joanny, J. F.; Ramaswamy, S.; Liverpool, T. B.; Prost, J.; Rao, M.; Simha, R. A., Hydrodynamics of soft active matter. *Review of Modern Physics* **2013**, *85* (3), 1143.
10. Tayar, A. M.; Hagan, M. F.; Dogic, Z., Active liquid crystals powered by force-sensing DNA-motor clusters. *PNAS* **2021**, *118* (30), 1–10.
11. Andreotti, B.; Forterre, Y.; Pouliquen, O., *Granular Media: Between Fluid and Solid*. Cambridge University Press: 2013.
12. Pismen, L., *Active Matter Within and Around Us: From Self-propelled Particles to Flocks and Living Forms*. Springer: 2021.
13. Hartmann, R.; Singh, P. K.; Pearce, P.; Mok, R.; Song, B.; Diaz-Pascual, F.; Dunkel, J.; Drescher, K., Emergence of three-dimensional order and structure in growing biofilms. *Nature Physics* **2019**, *15* (3), 251–256.
14. Beroz, F.; Yan, J.; Meir, Y.; Sabass, B.; Stone, H. A.; Bassler, B. L.; Wingreen, N. S., Verticalization of bacterial biofilms. *Nature Physics* **2018**, *14* (9), 954–960.
15. Qin, B.; Fei, C.; Bridges, A. A.; Mashruwala, A. A.; Stone, H. A.; Wingreen, N. S.; Bassler, B. L., Cell position fates and collective fountain flow in bacterial biofilms revealed by light-sheet microscopy. *Science* **2020**, *369* (6499), 71–77.
16. Pearce, P.; Song, B.; Skinner, D. J.; Mok, R.; Hartmann, R.; Singh, P. K.; Jeckel, H.; Oishi, J. S.; Drescher, K.; Dunkel, J., Flow-induced symmetry breaking in growing bacterial biofilms. *Physical Review Letters* **2019**, *123* (25), 258101.
17. Yan, J.; Nadell, C. D.; Stone, H. A.; Wingreen, N. S.; Bassler, B. L., Extracellular-matrix-mediated osmotic pressure drives *Vibrio cholerae* biofilm expansion and cheater exclusion. *Nature Communications* **2017**, *8* (1), 327.
18. Volfson, D.; Cookson, S.; Hasty, J.; Tsimring, L. S., Biomechanical ordering of dense cell populations. *PNAS* **2008**, *105* (40), 15346–15351.

19. van Houdt, R.; Michiels, C. W., Role of bacterial cell surface structures in *Escherichia coli* biofilm formation. *Research in Microbiology* **2005**, *156* (5–6), 626–633.
20. Li, H.; Shi, X. Q.; Huang, M.; Chen, X.; Xiao, M.; Liu, C.; Chate, H.; Zhang, H. P., Data-driven quantitative modeling of bacterial active nematics. *PNAS* **2019**, *116* (3), 777–785.
21. Meacock, O. J.; Doostmohammadi, A.; Foster, K. R.; Yeomans, J. M.; Durham, W. M., Bacteria solve the problem of crowding by moving slowly. *Nature Physics* **2020**, *17* (2), 205–210.
22. Copenhagen, K.; Alert, R.; Wingreen, N. S.; Shaevitz, J. W., Topological defects promote layer formation in *Myxococcus xanthus* colonies. *Nature Physics* **2020**, *17* (2), 211–215.
23. Dogsa, I.; Kriechbaum, M.; Stopar, D.; Laggner, P., Structure of bacterial extracellular polymeric substances at different pH values as determined by SAXS. *Biophysical Journal* **2005**, *89* (4), 2711–2720.
24. Morris, E. R.; Nishinari, K.; Rinaudo, M., Gelation of gellan – a review. *Food Hydrocolloids* **2012**, *28* (2), 373–411.
25. Cox, H.; Xu, H.; Waigh, T. A.; Lu, J. R., Single-molecule study of peptide gel dynamics reveals states of prestress. *Langmuir* **2018**, *34* (48), 14678–14689.
26. Cox, H.; Cao, M.; Xu, H.; Waigh, T. A.; Lu, J. R., Active modulation of states of prestress in self-assembled short peptide gels. *Biomacromolecules* **2019**, *20* (4), 1719–1730.
27. Mizuno, D.; Head, D. A.; MacKintosh, F. C.; Schmidt, C. F., Active and passive microrheology in equilibrium and nonequilibrium systems. *Macromolecules* **2008**, *41* (19), 7194–7202.

9 Low Reynolds Number Hydrodynamics

Biology needs to operate under strict physical restrictions (vitalism was rigorously disproved at the end of the 19th century) and many generic themes in bacterial motility can be understood from the physical perspective of fluid mechanics. A major issue with bacterial motility is the small size (~1 μm) of bacterial cells and how it relates to the hydrodynamics of fluids in which they move. Specifically, inertial effects are often negligible at the length scale of single bacteria, which means the swimming strategies used by larger organisms (e.g. the front crawl, butterfly or breast stroke for humans) become unviable.

At the length scale of bacterial molecules and cells, the equations for their hydrodynamics tend to simplify at time scales longer than about 10 μs (dependent on their exact size and density), since the inertial (mv, where m is the mass and v is the velocity) terms are negligible. Thus, once the force of propulsion is removed from bodies at low Reynolds number, they come to a halt almost immediately i.e. they do not glide, and they need the continuous application of a force to be motile. The development of low Reynolds number models provides useful descriptions of the overdamped motion of bacterial cells and the molecules they contain.

Fluid mechanics can often be reduced to the study of the *Navier–Stokes* (NS) *equation*,[1]

$$\rho\frac{\partial v}{\partial t}+\rho\left(v.\nabla\right)v=\rho g-\nabla p+\eta\nabla^{2}v, \tag{9.1}$$

where ρ is the fluid density, v is the fluid velocity (a function of both time and space), g is gravity, p is the local pressure in the fluid and η is the fluid viscosity (assumed constant). This is a continuum description of the fluid. The fluid is expected to have unique values for the velocity at all points and the molecular details have been averaged away. Equation (9.1) is often a good approximation for Newtonian fluids down to small length scales, ~10 times the molecular diameter.[2] Beyond this threshold, molecular details become important. The full NS Equation (9.1) is non-linear and can only be solved analytically under some restrictive sets of boundary conditions e.g. a spherical particle in an unbounded flow, which leads to the Stokes equation (Equation (2.3)), although numerical approximation schemes using modern computers can be very accurate under some conditions. A full analytic solution of Equation (9.1) has been defined as one of the top 10 unsolved problems in mathematics and a substantial prize has been offered.

The *Reynolds number* (Re) is a characteristic dimensionless ratio for flow geometries and is defined as

$$\mathrm{Re} = \frac{vL\rho}{\eta}, \tag{9.2}$$

where L is the characteristic length scale (e.g. the size of the microorganism considered or the diameter of the pipe that the fluid travels in) and the other symbols are as before. Fortunately for theoretical modellers, microorganisms are sufficiently small in size that the inertial effects are small over the time scales typically measured (≥ 10 μs). This means all the terms on the left-hand side of the NS Equation (9.1) can be dropped to a good approximation to calculate the motion of single microorganisms.[3] This creates a linear differential equation, from a non-linear one, that is much simpler to solve.

In large-scale flows of microorganisms in suspensions, the characteristic length scale (L) can be large, and so too is the Reynolds number calculated from Equation (9.2). Therefore, turbulence can be important, since it also depends on the inertial terms e.g. for marine bacteria in a tidal flow. Calculations can become very tricky as a result and turbulence is a challenging phenomenon that still lacks a completely quantitative framework for its analysis.[4]

The *Stokes equation* that results from the low Reynolds number approximation to Equation (9.1) is

$$\nabla p = \eta \nabla^2 v, \tag{9.3}$$

where gravity has been implicitly included in the pressure term.[1] This new equation is both linear and time-symmetric. Microorganisms need to break this time symmetry if they want to move at low Reynolds number. This is summarised in *Purcell's scallop theorem*[3] (Figure 9.1) in which a sufficient number of degrees of freedom are required for the organism at low Reynolds number to break the time symmetry and hence move e.g. a scallop with a single degree of freedom goes nowhere, no matter the manner in which it rotates its single hinge. However, the scallop theorem only rigorously holds for dilute swimmers in Newtonian fluids. It can be broken by reciprocal swimmers in viscoelastic fluids, swimmers near a deformable interface and via interactions with other neighbouring reciprocal swimmers.[5]

There have been some recent promising applications of artificial intelligence in solving fluid mechanics problems via the NS Equation (9.1) (Chapter 14)[6] that leverage automatic differentiation developed by computer scientists for fast numerical calculation of differentials.[7] Such a methodology can also be used for low Reynolds number

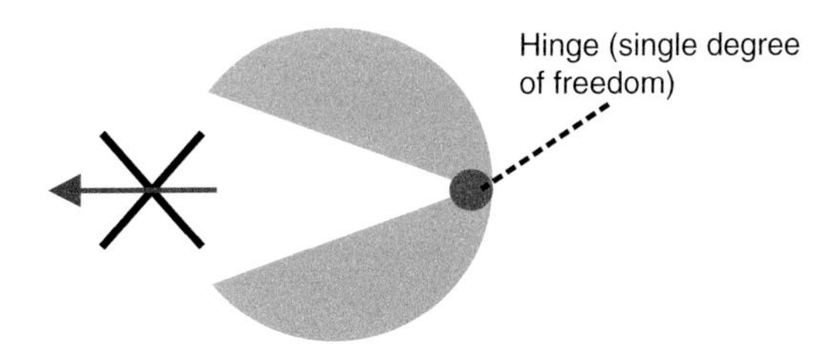

Figure 9.1 A scallop in dilute bulk suspensions cannot achieve motility at the microscale due to the lack of inertial effects and the time symmetry of its hinge motion (*Purcell's scallop theorem*[3]).

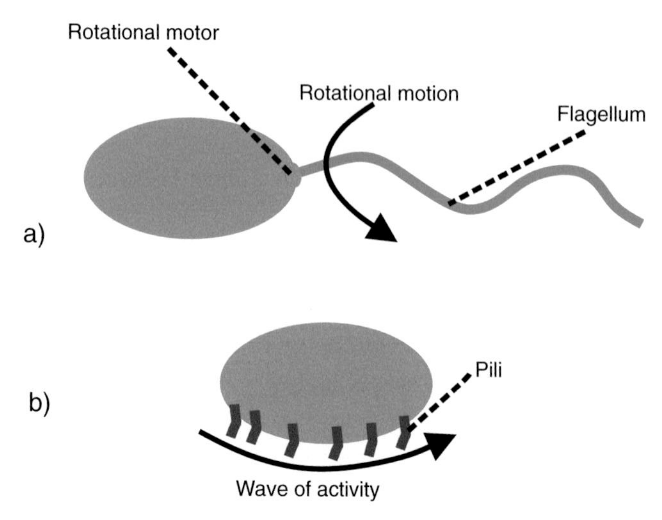

Figure 9.2 Bacteria often move using the time asymmetric motion of paddles e.g. (a) a flagellum or (b) pili.

dynamics and then extended to viscoelastic problems. Conventionally, numerical solutions of the NS equation are calculated using finite element models, but these only work well over a limited range of conditions.[8] AI tools could extend the range of applicability of these computational tools.

Given Purcell's scallop theorem, a question is how do bacteria actually achieve motility at such small length scales given the hydrodynamic constraints. Swimming of microorganisms at low Reynolds is often due to the time-asymmetric motion of a paddle e.g. a flagellum (from the Latin for whip, plural flagella), cilium (from the Latin for eyelash, plural cilia) or pilus (from the Latin for hair, plural pili, Figure 9.2). These paddles experience anisotropic resistance due to their elongated shapes e.g. pulling a rod perpendicular to its axis has much more resistance than along its axis and switching the rod's orientation provides the necessary drag anisotropy for motility. Calculations have been made for the slender body hydrodynamics of flagella, cilia and pili using approximations to Equation (9.3) with the appropriate boundary conditions.[9] Expressions can be calculated for the hydrodynamic friction coefficients of a slender filament.[9] Perpendicular to the filament, the force ($f_\perp$) is

$$f_\perp(s) \approx c_\perp u_\perp(s), \tag{9.4}$$

where $u_\perp$ is the perpendicular fluid velocity, s is the distance along the filament and $c_\perp$ is the friction coefficient. $c_\perp$ is calculated to be

$$c_\perp = \frac{4\pi\eta}{\ln\left(\frac{L}{a}\right)}, \tag{9.5}$$

where L is the length of the filament, a is the filament radius and η is the fluid viscosity. Parallel to the filament the force $\left(f_\parallel\right)$ is

$$f_{||}(s) \approx c_{||} u_{||}(s), \tag{9.6}$$

where $u_{||}$ is the velocity parallel to the filament and the parallel friction coefficient $(c_{||})$ is

$$c_{||} = \frac{2\pi\mu}{\ln\left(\frac{L}{a}\right)}. \tag{9.7}$$

Thus, $\frac{c_{\perp}}{c_{||}} = 2$, which provides the drag anisotropy that many microorganisms use for motility.

Flagella follow helical paths when attached to bacteria, which provide time asymmetry and allow them to move. Slightly less common is the motion of bacteria due to changes in the shape of their cellular bodies, but drag anisotropy still plays an important role (Chapter 17). Different motile strategies are often used when bacteria are near interfaces (Section 17.3) e.g. specialised walking, gliding or swarming can occur.

The flow field from many microorganisms swimming in the bulk resembles a point hydrodynamic dipole to a first approximation,[10]

$$\bar{u}(r) = \frac{\sigma \bar{r}}{8\pi r^3}\left(\frac{3x^2}{r^2} - 1\right), \tag{9.8}$$

where $\bar{u}(r)$ is the flow velocity at distance r from the dipole centre, x is the distance in the direction of the dipole and σ is the dipole strength. For negative σ, the particles are propelled from the rear (e.g. *Escherichia coli* with flagella), whereas for positive σ, they are propelled from the front (e.g. some unicellular algae). Motile microorganisms can therefore be defined in general terms as *pushers* versus *pullers*, dependent on how they perturb the flow and the sign of their hydrodynamic dipole (σ). The form of the hydrodynamic dipole then determines the first-order approximation required to describe large-scale flows of interacting organisms.

Collective hydrodynamic effects are challenging to understand with quantitative detail (Chapter 17). Under some circumstances, hydrodynamics can lead to the synchronisation of the motility of swimmers,[9,10] but effects need to be carefully considered in the context of chemotaxis, electrophysiology and intracellular communication to decide on the dominant contributions e.g. synchronisation in the motility of cilia on the surface of a single paramecium is electrophysiological rather than hydrodynamic in origin.[11]

Suggested Reading

Lauga, E., *The Fluid Dynamics of Cell Motility*. Cambridge University Press: 2020.
Excellent pedagogic account of the hydrodynamics of microorganisms.

References

1. Guyon, E.; Hulin, J. P.; Petit, L.; Mitescu, C. D., *Physical Hydrodynamics*, 2nd ed. Oxford University Press: 2015.
2. Israelachvili, J. N., *Intermolecular and Surface Forces*. Academic Press: 2011.
3. Purcell, E. M., Life at low Reynolds number. *American Journal of Physics* **1977**, *45* (1), 3–11.
4. Davidson, P., *Turbulence: An Introduction for Scientists and Engineers*. Oxford University Press: 2015.
5. Pismen, L., *Active Matter Within and Around Us: From Self-propelled Particles to Flocks and Living Forms*. Springer: 2021.
6. Raissi, M.; Yazdani, A.; Karniadakis, G. E., Hidden fluid mechanics: Learning velocity and pressure fields from flow visualizations. *Science* **2020**, *367* (6481), 1026–1030.
7. Baydin, A. G.; Pearlmutter, B. A.; Radul, A. A.; Siskind, J. M., Automatic differentiation in machine learning: A survey. *Journal of Machine Learning Research* **2018**, *18* (1), 1–43.
8. Pozrikidis, C., *Introduction to Theoretical and Computation Fluid Dynamics*. Oxford University Press: 2011.
9. Lauga, E., *The Fluid Dynamics of Cell Motility*. Cambridge University Press: 2020.
10. Aranson, I. S., Bacterial active matter. *Reports on Progress in Physics* **2022**, *85* (7), 076601.
11. Greenspan, R. J., *An Introduction to Nervous Systems*. Cold Harbor Spring: 2007.

10 Viscoelasticity

Materials that have intermediate responses to stress between those of perfect elastic solids and perfect viscous fluids are called *viscoelastic*. All materials made from condensed matter (e.g. liquids, solids and many other intermediate phases of soft condensed matter, such as gels, liquid crystals, glasses, etc.) are viscoelastic to some degree, with the possible exception of superfluid helium. Water is often considered as a prototypical fluid, but it acts as an elastic solid at very short time scales (picoseconds or less). Standard crystalline solids can flow over very long-time scales (greater than thousands of years) due to the motion of lattice defects, which explains the ductility of metals and challenges standard descriptions of solids.

Accurate measurement of the viscoelasticity of a sample is a key tool in biomaterials science and has important applications with bacteria e.g. to understand the mechanical cleaning of biofilms, the creation of food stuffs and marine ecology. The study of the flow behaviour of materials is called *rheology* and a device to measure the rheology of a material is a *rheometer*. The viscoelasticity of a sample is directly related to the dynamics of the molecules in the sample and thus rheological experiments can provide a window into the dynamics of bacterial cells and the molecules inside them. Materials respond to stresses as a function of time scale, so rheology experiments can be thought of as a form of mechanical spectroscopy.

Linear viscoelasticity is one of the simplest measurements to understand the flow behaviour of a material. Similar to Hooke's law, linear viscoelasticity can accurately describe the response of materials to very small stresses or strains, in which a linear approximation is valid. Measurements of linear viscoelasticity involve the measurement of strain as a function of stress and time (*stress-controlled* measurements) or stress as a function of strain and time (*strain-controlled* measurements).

Figure 10.1 shows a schematic diagram of a bulk fluids *rheometer*. Such devices are typically used to measure the viscoelasticity of soft materials, such as bacterial suspensions and biofilms. A crucial component of the instrumentation in a rheometer is the control loop, which adjusts the applied stress or strain in response to the non-linear flow behaviour of a sample e.g. it can hold the stress at a constant value.

For an oscillatory imposed strain (u) in a material that follows a sinusoidal pattern as a function of time (t) with a frequency (ω) and u_0 is the amplitude of the displacement,

$$u(t) = u_0 \sin \omega t, \tag{10.1}$$

the resultant stress ($\sigma(t)$) in the material in response to the stress can be calculated as

$$\sigma(t) = G'(\omega)u(t) + \frac{G''(\omega)}{\omega}\frac{du}{dt}, \tag{10.2}$$

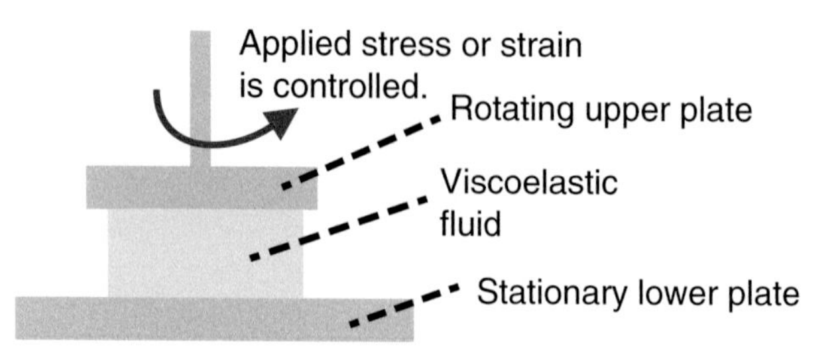

Figure 10.1 A *bulk fluids rheometer* applies a known stress or strain to a material and measures the corresponding strain or stress in response as a function of time or frequency. Sample volumes are typically on the order of ~1 mL.

where G' is the elastic shear modulus of the sample and G'' is the dissipative shear modulus ($\lim_{\omega\to 0}\frac{G''}{\omega}=\eta$, the viscosity). G' is in phase with the strain ($u(t)$), whereas G'' is $\frac{\pi}{2}$ rad out of phase, since it is proportional to the derivative of the strain ($\dot{u}(t)\sim\cos\omega t$, from Equation (10.1)). $G'(\omega)$ and $G''(\omega)$ can be used to construct the *complex shear modulus*, $G^*(\omega)=G'(\omega)+iG''(\omega)$, of the material as a function of frequency. $G^*(\omega)$ is one possible method to quantify the linear viscoelasticity of a specimen, which conveniently separates the viscous and elastic components as a function of frequency, but there are other measures, such as the *creep compliance* ($J(t)$), the *complex compliance* ($J^*(\omega)$), the *relaxation function* ($G(t)$), the *complex viscosity* ($\eta^*(\omega)$), the *retardation spectrum* and the *relaxation spectrum*.[1] The choice of the relevant parameter to quantify the linear viscoelasticity is often determined by experimental convenience. It is possible to transform between these measures of linear viscoelasticity without any significant loss of information.

Figure 10.2 shows *complex shear moduli* as a function of frequency for some standard polymeric and colloidal materials. Gelation of the fluids causes the magnitude of the moduli to increase. The fluid and gel states have different characteristic spectra e.g. fluids demonstrate flow behaviour at low frequencies ($G''>G'$ and the terminal relaxation regimes at long times are well described with a Maxwell model, Figure 10.3a) and gels often have weak power laws of the frequency with $G'>G''$.

Non-linear viscoelasticity is observed if the amplitude of u_0 in Equation (10.1) is increased beyond the linear limit. This is always the case if sample deformations occur continuously in one direction, since the threshold for linearity will inevitably be surpassed. Non-linear viscoelasticity has an even richer range of associated phenomena compared with linear viscoelasticity, such as *thixotropy*, *shear banding*, *wall slip*, *shear thinning*, *strain softening* and *shear thickening*.

Rheology can be considered as a generalised form of fluid mechanics and models to describe the phenomena are often mathematically sophisticated. An engineering approach is to construct simple equivalent mechanical models, such as those due to *Maxwell* and *Kelvin* (Figure 10.3). They consist of idealised elastic elements (springs) and idealised viscous elements (dashpots) arranged together in series or in parallel. Such constitutive models can be useful practically to describe the viscoelasticity of samples, but they lack a fundamental molecular interpretation. However, there is a

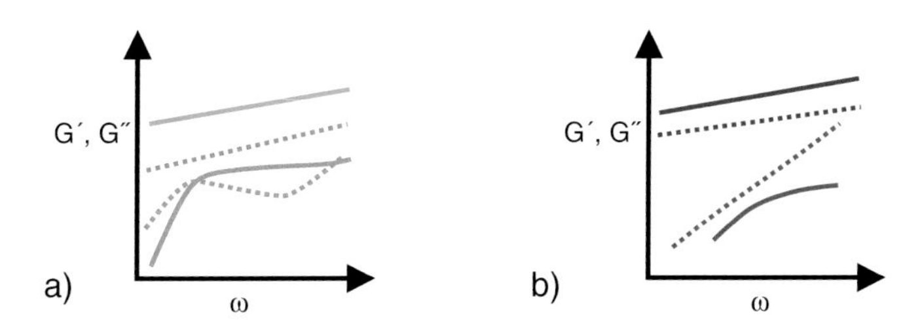

Figure 10.2 Schematic diagram of the *complex shear moduli* (G' continuous line and G'' dotted line) as a function of frequency (ω) for some typical materials relevant to bacterial suspensions and biofilms. (a) Polymeric fluid (green) and polymeric gel (blue), (b) colloidal fluid (red) and colloidal gel (navy blue). Gelation of the fluids causes the magnitude of the moduli to increase.

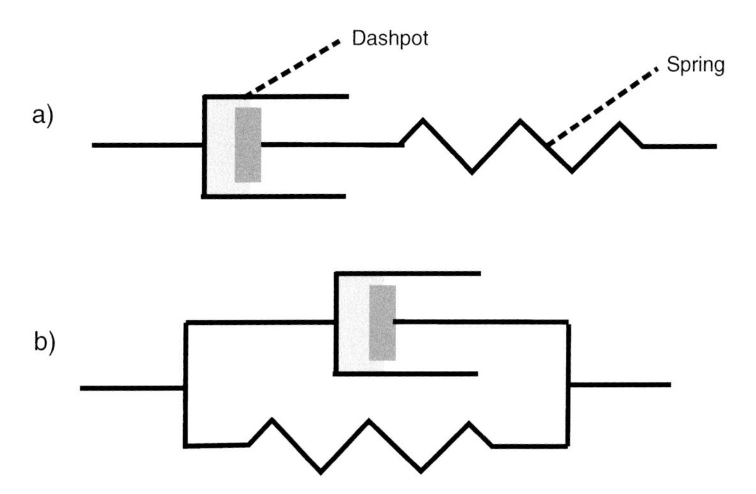

Figure 10.3 Simple constitutive models that are used to understand the rheology of viscoelastic materials. (a) *Maxwell model* and (b) *Kelvin model*.

large literature discussing more sophisticated models for soft matter physics that connect intrinsic molecular dynamics to viscoelasticity (e.g. the dynamics of polymers, colloids, liquid crystals, etc.) and reference should be made to the specialist literature for more details.[2]

In general, elastic materials are anisotropic and require additional terms to describe their response to a stress beyond just a single isotropic shear modulus (G). Furthermore, extensions can be made to describe viscoelasticity in terms of other moduli beyond those for shearing motions e.g. complex Young moduli can be defined for extensional deformations. With liquid crystalline materials, multiple viscoelastic constants are required to describe their anisotropic viscoelasticity (the Leslie–Ericksen theory is successful for purely viscous liquid crystals and has five independent viscosity components, but bacterial liquid crystals have substantial elasticity, complicating calculations).[2,3]

A wide range of experimental techniques can probe viscoelasticity. For bacterial suspensions and biofilms, predominantly their viscoelasticities are examined with

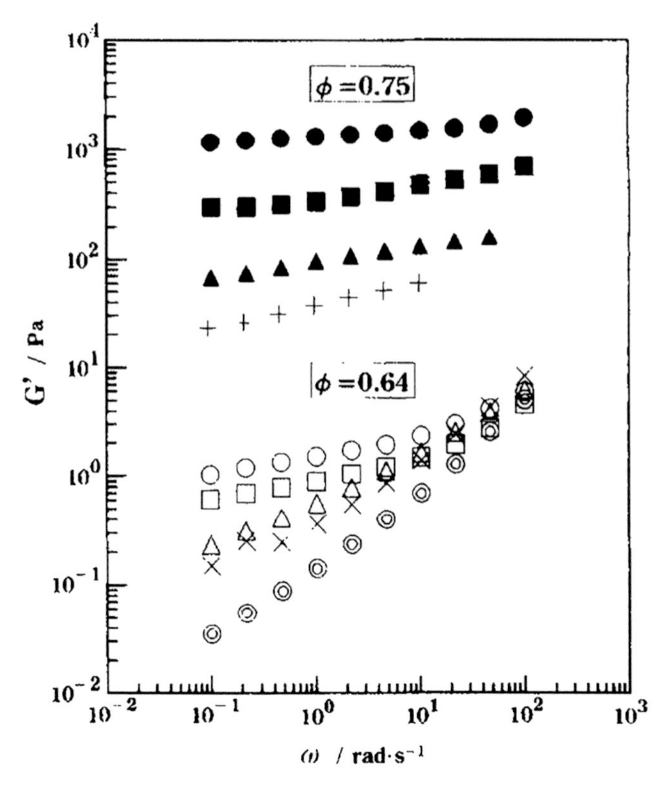

Figure 10.4 *Elastic shear modulus* (G') as a function of frequency (ω) for *Escherichia coli* suspensions with volume fractions of $\phi = 0.75$ and $\phi = 0.64$ from bulk rheology.[6] The different symbols indicate different strain amplitudes: circles (0.02), squares (0.1), triangles (0.5), + and × (1) and ellipsoid (3). Reprinted from Toda K., Furuse H., Amari T., Wei X., Cell concentration dependence of dynamic viscoelasticity of *E. coli* culture suspensions. *Journal of fermentation and bioengineering* 1997, 85, 4, 410–415 with permission from Elsevier.

solution state rheometers and *microrheometers* (particle tracking, magnetic microrheology and AFM), but many more methods are possible[4,5] e.g. ultrasound-based rheology. Microrheometers can measure microlitres of sample compared with millilitres for a standard fluids rheometer, which is a big advantage with scarce biological materials.[4,5] Microrheology thus requires much less sample and the lack of inertial effects at small length scales due to low Reynolds number flows (Equation (9.2)) means much faster time scales can be probed.[4] Most complex fluids have heterogeneous viscoelasticity as a function of length scale, so both averaged bulk and heterogeneous microrheology measurements provide useful complementary information to probe the response of samples.

The *elastic shear modulus* (G') as a function of frequency (ω) for *Escherichia coli* suspensions measured in bulk rheology experiments is shown in Figure 10.4. The elasticity decreases with strain amplitude, which is a non-linear phenomenon (strain softening).

Classical turbulence is not observed in the low Reynolds number flows that are normally measured in rheology experiments (viscosities of soft matter are typically much higher than water, causing the Reynolds number in Equation (9.2) to be small), but there are a wide range of other challenging nonlinear flow phenomena that are possible with bacterial suspensions, slimes and biofilms including *bacterial turbulence*[7] and *elastic turbulence*.[8]

Shear moduli observed for biofilms and suspensions tend to be in the range of $0.001–10^3$ Pa. Rheometers need to be chosen with transducers matched to these low values. Some recent work has explored the fungal biofilms of *Candida albicans*[9] and the shear moduli are of the same order of magnitude as bacterial biofilms.

Suggested Reading

Coussot, P., *Rheophysics: Matter in All its States*. Springer: 2016. A useful introduction to the field of rheology.

Goodwin, J. W.; Hughes, R. W., *Rheology for Chemists: An Introduction*, Royal Society of Chemistry: 2008. Excellent introduction to rheological phenomena with simple physical models. Since viscoelasticity is a generalized form of fluid mechanics, the literature can be formidably mathematical, but this is an approachable introduction.

Mewis, J.; Wagner, N. J. *Colloidal Suspension Rheology*. Cambridge University Press: 2011. Bacterial cells are examples of colloids. The book provides inspiration from studies of analogous synthetic colloidal materials.

References

1. Goodwin, J. W.; Hughes, R. W., *Rheology for Chemists: An Introduction*. Royal Society of Chemistry: 2008.
2. Larson, R. G., *The Structure and Rheology of Complex Fluids*. Oxford University Press: 1999.
3. de Gennes, P. G.; Prost, J., *The Physics of Liquid Crystals*. Oxford University Press: 1995.
4. Waigh, T. A., Microrheology of complex fluids. *Reports on Progress in Physics* **2005**, *68* (3), 685.
5. Waigh, T. A., Advances in the microrheology of complex fluids. *Reports on Progress in Physics* **2016**, *79* (7), 074601.
6. Toda, K.; Furuse, H.; Amari, T.; Wei, X., Cell concentration dependence of dynamic viscoelasticity of *Escherichia coli* culture suspensions. *Journal of Fermentation and Bioengineering* **1997**, *85* (4), 410–415.

7. Dunkel, J.; Heidenreich, S.; Drescher, K.; Wensick, H. H.; Bar, M.; Goldstein, R. E., Fluid mechanics of bacterial turbulence. *Physical Review Letters* **2013**, *110* (22), 228102.
8. Malm, A. V.; Waigh, T. A., Elastic turbulence in entangled semi-dilute DNA solutions measured with optical coherence tomography velocimetry. *Scientific Reports* **2017**, *7* (1), 1186.
9. Beckwith, J. K.; Ganesan, M.; VanEpps, J. S.; Kumar, A.; Solomon, M. J., Rheology of *Candida albicans* fungal biofilms. *Journal of Rheology* **2022**, *66* (4), 683–697.

11 Systems Biology

Systems biology and *synthetic biology* are two important modern paradigms in the life sciences.[1,2] However, the definition of *systems biology* is a little vague and depends on the context. For the purposes of this book, systems biology is framed in the context of the reverse engineering of the functioning of a cell. The field adopts ideas from chemical kinetics, network theory, non-linear physics and engineering control theory to understand the interaction of genetic circuits inside cells. It aims to answer a crucial question in post-genome research: we know the sequences of entire genomes, but how do sections of the genome interact with one another (through proteins expressed by genes, but also via regulatory RNA)? In contrast, *synthetic biology* considers the creation of new life forms (e.g. bacteria) and new biomolecules that have not been previously seen in nature (e.g. new types of peptide). Thus, it is possible to genetically modify bacteria so they will produce new pharmaceuticals, novel molecules in their biofilms[3,4] (e.g. for food applications) or provide fundamental insights into their behaviour.

Escherichia coli are often used as model systems for synthetic biology (the chassis).[5] *E. coli*-based technologies are sufficiently simple that they are used in standard undergraduate synthetic biology laboratory projects and there are international competitions for the bacteria created (iGEM). The availability of synthetic biology techniques provides a major motivation for systems biology, since gene circuits can be modified to test quantitative models e.g. via the construction of the *toggle switch*[6] or the *repressilator* in live bacteria[7] or in general using libraries of knock-down mutants that have been created for standard strains of bacteria.

Systems biology is a large field and references should be made to specialised texts for a full overview.[1,2] Here only a few simple useful tools for systems biology[1] with bacteria will be introduced.

The *central dogma of molecular biology* in its simplest form is shown in Figure 11.1. The information for all living cells is stored in DNA molecules. Transcription of this information creates RNA molecules and this information is then translated into proteins. A complication is that the proteins produced can then interact with other regions of the DNA chain modifying the expression of other proteins. Thus, chemical circuits are created by interacting proteins that can have similar functionalities to electronic circuits, such as negative feedback, amplification, logic gates, clocks and oscillators. The collection of all the interacting genes in an organism shown as a graph (with nodes for the genes and edges for the interactions) is called the *transcription network*.

Transcription networks of interacting genes are very useful for understanding organisms (called the *transcriptome* of an organism). All the genes in an organism can

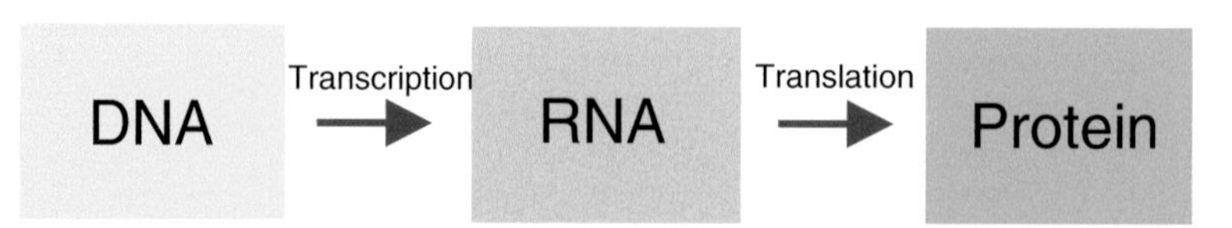

Figure 11.1 The *central dogma of molecular biology* in its simplest form (transcription of DNA to RNA and translation of RNA to protein).

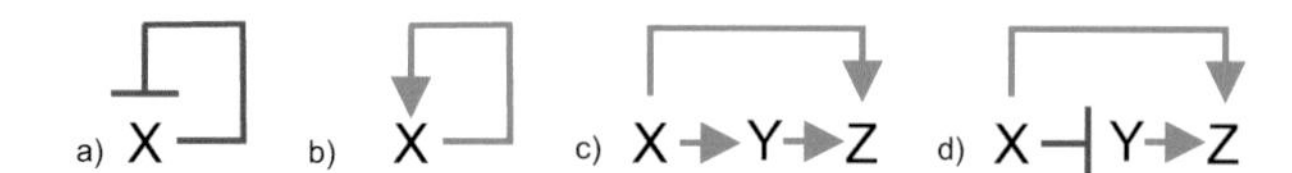

Figure 11.2 Some standard motifs for the interaction of genes in transcription networks.[1] (a) *Negative autoregulation*, (b) *positive autoregulation*, (c) *coherent feedforward* and (d) *incoherent feedforward* networks. X, Y and Z are genes. Motifs for interacting regulatory RNA have also been proposed.[8]

be mapped experimentally and their interactions quantified. This results in a complex network of interacting genes and specialist mathematical tools are needed to understand them. Most current models for gene networks neglect both the roles of regulatory RNA (motifs for RNA regulation are currently being discovered[8]) and the effects of spatial variation e.g. compartmentalisation of reaction kinetics. These are significant simplifications and can invalidate models when compared with results from real bacteria. The ingredients for standard systems biology models to understand the transcriptome use chemical kinetics, enzyme kinetics, network theory, control theory and the analysis of non-linear differential equations.

Some standard models for a quantitative understanding of the kinetics of gene transcription have been introduced, which can be handled with multiple levels of coarse graining.[1] The molecular machines involved in gene expression are very complex, leading to complex dynamic behaviour (Chapter 19). The statistical analysis of connectomes indicates certain gene motifs are much more common than expected at random e.g. autoregulation motifs or feedforward loops commonly occur. Some examples of gene motifs are given in Figure 11.2. Analysis of these gene motifs can then allow the quantitative analysis of the connectome dynamics to be developed, but it also provides inspiration for new bacterial genomes to be created using synthetic biology techniques. Motifs play a useful role during the reverse engineering of gene networks to interpret their functions.

As an example of modelling tools used in systems biology, consider a gene that represses itself (Figure 11.2a); a *negative autoregulation motif*. The dynamic behaviour can be modelled using differential equations (non-linear ordinary differential equations), but the details will be left to more specialist texts.[1] Such an arrangement makes the gene react faster and is more robust to fluctuations in its expression of genes (Figure 11.3). The negative autoregulation motif is very common in the transcription networks of bacteria as a result i.e. it often has an evolutionary advantage.

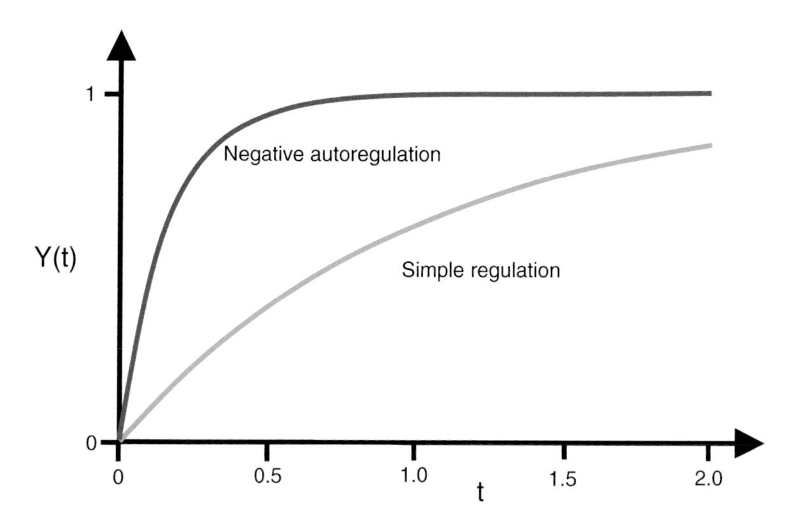

Figure 11.3 The *negative autoregulation motif* can speed up the response of a gene circuit compared with simple regulation. $Y(t)$ is the concentration of a regulated gene, which is shown as a function of time, t.

Robustness is a key theme in systems biology.[1] Robustness is crucial for determining the success of biological circuits and can be used to decide between competing alternative models to describe biological phenomena e.g. mechanisms of morphogenesis or chemical sensing should not depend on the exact number of molecules in the environment, otherwise they would experience fragility to minor perturbations (the systems often demonstrate adaptation of their sensitivity to provide the required robustness).

Suggested Reading

Alon, U.; *Introduction to Systems Biology*. CRC Press: 2020. Classic introduction to some of the phenomena observed in systems biology.

Ingalls, B. P. *Mathematical Modelling in Systems Biology: An Introduction*. MIT Press: 2013. A more mathematical introduction than Alon's, but the explanations are still clear.

References

1. Alon, U., *An Introduction to Systems Biology: Design Principles of Biological Circuits*, 2nd ed. CRC Press: 2020.
2. Ingalls, B. P., *Mathematical Modeling in Systems Biology: An Introduction*. MIT Press: 2013.

3. Chen, A. Y.; Deng, Z.; Billings, A. N.; Seker, U. O. S.; Lu, M. Y.; Citorik, R. J.; Zakeri, B.; Lu, T. K., Synthesis and patterning of tunable multiscale materials with engineered cells. *Nature Materials* **2014**, *13* (5), 515–523.
4. Fang, K.; Park, O. J.; Hong, S. H., Controlling biofilms using synthetic biology approaches. *Biotechnology Advances* **2020**, *40*, 107518.
5. Antony, C.; Liljeruhm, J.; Forster, E., *Synthetic Biology: A Lab Manual*. World Scientific: 2014.
6. Gardner, T. S.; Cantor, C. R.; Collins, J., Construction of a genetic toggle switch in *Escherichia coli*. *Nature* **2000**, *403* (6767), 339–342.
7. Elowitz, M. B.; Leibler, S., A synthetic oscillatory network of transcriptional regulators. *Nature* **2000**, *403* (6767), 335–338.
8. Storz, G.; Vogel, J.; Wassarman, K. M., Regulation by small RNAs in bacteria: Expanding frontiers. *Molecular Cell* **2011**, *43* (6), 880–891.

12 Non-linear Dynamics

Few biological phenomena are completely linear, and *non-linear processes* thus tend to be the rule rather than the exception in bacterial biophysics. Mathematical modelling becomes more challenging with non-linear equations and exact analytic solutions become much rarer. However, good robust mathematical tools exist to explore the qualitative behaviour of non-linear systems (e.g. the analysis of phase portraits) and numerical modelling with computers can produce quantitative results.[1–3]

Three prototypical models from non-linear dynamics that are commonly applied to bacteria are the *toggle switch* (systems biology, Chapter 11), the *Volterra–Lotka* model (population dynamics, Chapter 22) and the *Hodgkin–Huxley* model (electrochemical spiking of the membrane voltage, Chapter 3). Other important examples of non-linearity in bacterial biophysics include sigmoidal enzyme kinetics, non-linear viscoelasticities of suspensions, hysteresis during electroporation and synchronisation during signalling.

As an example of a non-linear system in systems biology, the *toggle switch* can be considered (Figure 12.1 and Chapter 11). This consists of two mutually repressing genes (a genetic network motif from systems biology, Figure 12.2). The kinetics of repression can be described by a non-linear Hill equation (sigmoidal kinetics, $\frac{dn}{dt} = \frac{\alpha}{1+n^{\kappa}}$). Thus, a mathematical description of the toggle switch is

$$\frac{du}{dt} = \frac{\alpha_1}{1+v^{\beta}} - u, \tag{12.1}$$

$$\frac{dv}{dt} = \frac{\alpha_2}{1+u^{\gamma}} - v, \tag{12.2}$$

where u is the concentration of repressor 1, v is the concentration of repressor 2, α_1 is the rate of synthesis of repressor 1, α_2 is the rate of synthesis of repressor 2, β is the cooperativity of repression of promoter 2 and γ is the cooperativity of repression of promoter 1. If the Hill equation is sufficiently cooperative (with both high β and γ) and thus has a high degree of non-linearity, two steady stables occur in the phase portraits (on and off, Figure 12.1a). For linear or low cooperativity kinetics, only a single unsteady state occurs (Figure 12.1b). Non-linearity is thus crucial for the function of the toggle switch gene motif and it would not function without it. The switch can be reset by pushing it through a bifurcation in the phase portrait.[3]

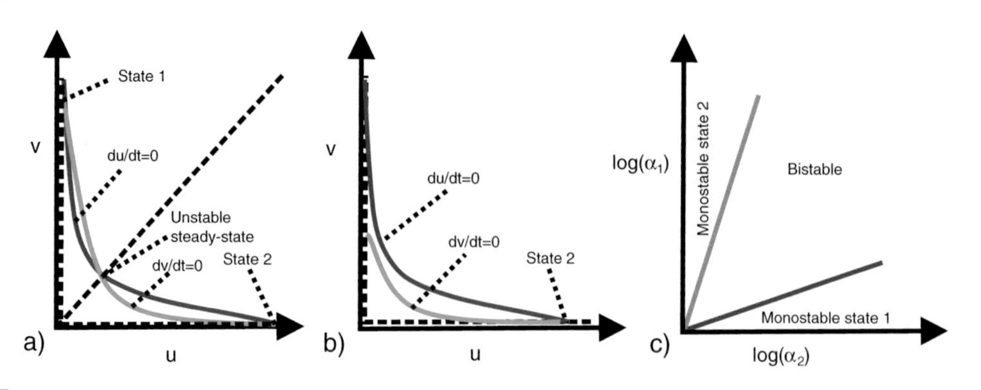

Figure 12.1 Dynamic phase portrait of the *toggle switch* created by the mutual inhibition of two genes[4] (protein concentrations *u* and *v*). (a) A *bistable toggle* switch that has balanced promoter strengths. It can switch between states 1 and 2. (b) A *monostable toggle* switch that has unbalanced promoter strengths. There is only state 2 that is stable. (c) The phase portrait in terms of the rate of creation of *u* and *v* (proportional to α_1 and α_2, respectively; the cooperativity parameters of the Hill kinetics of the enzyme).

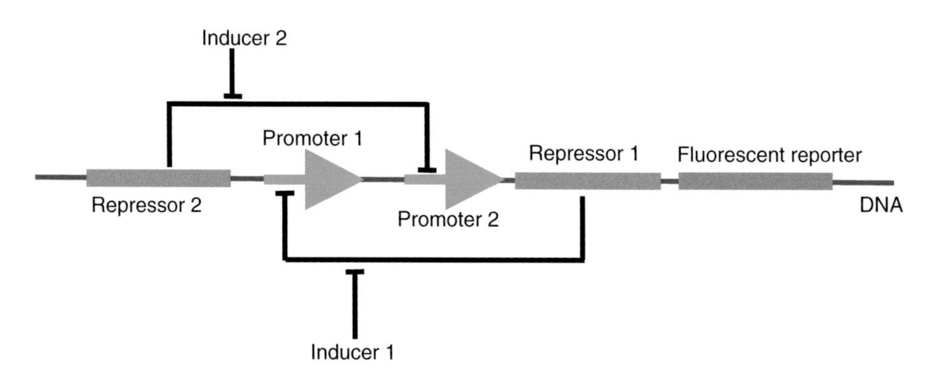

Figure 12.2 The design of a *toggle switch*. Repressor 1 inhibits the transcription by promoter 1 and is induced by inducer 1. In turn, repressor 2 inhibits the transcription by promoter 2 and is induced by inducer 2. The fluorescent reporter allows the state of the genes to be read out experimentally.

Suggested Reading

Nelson, P. *Physical Models of Living Systems*. Freeman: 2015. Pedagogic account of modelling techniques with an extensive discussion of the toggle switch.

Strogatz, S. *Non-linear Dynamics and Chaos*, 2nd ed. Westview Press: 2015. Classic overview of non-linear mathematics applied to some fascinating physical phenomena. The problem sets are extremely valuable.

References

1. Strogatz, S. H., *Nonlinear Dynamics and Chaos: With Applications to Physics, Biology, Chemistry and Engineering*, 2nd ed.; Westview Press: 2014.
2. Ingalls, B. P., *Mathematical Modeling in Systems Biology: An Introduction*. MIT Press: 2013.
3. Nelson, P., *Physical Models of Living Systems*. W.H. Freeman: 2015.
4. Gardner, T. S.; Cantor, C. R.; Collins, J., Construction of a genetic toggle switch in *Escherichia coli*. *Nature* **2000**, *403* (6767), 339–342.

13 Experimental Characterisation Techniques

A wide range of experimental techniques can be used to characterise bacteria and biofilms. A challenge is to provide high spatial resolution and high chemical specificity in these measurements while performing them in a non-invasive manner. Ideally, the measurements should be performed *in situ* and often medical applications require them to be *in vivo*. Bacteria and the biofilms they form are hierarchical and heterogeneous on nanometre to millimetre length scales, so spatially resolved methods are needed to characterise the structures. Thus, the challenge with measurements on biofilms is to perform them without perturbing either the delicate structures formed or the physiology of the integral bacteria. Reproducibility is crucial in all experiments, so random stochasticity must be carefully considered with both individual bacteria and biofilms. Every snowflake is unique due to a range of competing stochastic non-linear processes during its growth and so is every bacterium and every biofilm.[1]

There are many *non-linear stochastic processes* involved in biofilm formation. To make a controlled biofilm structure, it is possible to print individual bacteria onto a perfect lattice, but still the extracellular polymeric substance (EPS) would be different in every biofilm due to stochastic events at the molecular level (including quenched disorder during gelation of the EPS), every bacterium within the biofilm is a unique individual (even with perfect clones due to stochastic gene expression) and stochastic division events of growing bacteria would rapidly increase the variation of morphologies observed over time. Another heterogeneous biofilm would thus result from the printed lattice, albeit one with slightly better-defined initial conditions.

Many standard measurement techniques in physical chemistry and biological physics can be applied to bacteria, biofilms, slimes and capsules. Emergent techniques are stressed in the current approach, rather than motivating all the possibilities from first principles, which is better left to introductory biophysics texts.[2–4]

A key issue for bacterial biophysics is to successfully integrate incubators for bacteria and biofilm growth with the measurement techniques (Figure 13.1) to allow the growth processes to be followed in a reproducible manner when averaged over a large ensemble (individual bacteria and biofilms are highly variable). It is much harder to reliably interpret data from disrupted bacteria and biofilms, particularly with post-mortem measurements, such as electron microscopy.

Another challenge is to resolve key physical parameters at the level of single bacteria (or even single molecules inside live bacteria), since they are needed to inform more fundamentals models e.g. system biology models that describe interacting genes (Chapter 12). Techniques based on nanophotonics are in many cases rising to this challenge and can be used successfully on single bacteria. However, optical measurements *in vivo* inside biofilms still tend to be tricky, since thick mature biofilms are typically opaque, confounding the use of many photonics-based techniques due to multiple scattering. Thus, a goal is to measure the pH, mass, magnetic moment, temperature, cellular volume and so on at the level of single bacteria in both planktonic and biofilm states through all stages of their growth. In most cases, measurements are only currently possible with planktonic bacteria and at the early stages of biofilm growth. Measurements are also restricted to standard laboratory strains of bacteria in which culture conditions are well understood (Chapter 16). Otherwise, extensive fundamental microbiology work is required to establish the reproducibility of the cultures. A wide range of optical techniques can be used for the detection of pathogenic bacteria,[5] since mechanisms for their culture have already been well studied.

There are some large unfulfilled needs for bacterial biophysics instrumentation. For example, there are no standard clinical methods to directly diagnose biofilm growth *in vivo*.[6] It is possible to measure the presence of biomarkers of bacteria in mixed biofilms *in vivo* using indirect methods, such as DNA amplified via the polymerisation chain reaction (PCR), and antibodies can be used to detect other biofilm EPS components. Specific examples of biomarkers for biofilms include cellulose in urine from people with uropathogenic *Escherichia coli* and immunoglobulin G (IgG) antibodies raised against biofilm components in sputum from people with chronic *Pseudomonas aeruginosa* lung infections. Rapid robust *in vivo* diagnostics of bacteria and bacterial biofilms would be invaluable for targeted treatment of infections and it would decrease morbidity.

13.1 Sample Culture

Bacteria as they grow (either in the forms of a biofilm or planktonic states) in closed containers inevitably become pickled in their own juices. Although this scenario may be representative of their natural states in quiescent flows, it makes comparison of the results between different experiments difficult for fundamental studies of their biophysics. Slight changes in sample geometry, such as the size of the container, will make a big difference to bacterial growth (e.g. modifying the concentrations of the pickling juices) and affect the reproducibility of the experiments. To make experimental results more reproducible, chemostats tend to be used in which the growth medium is constantly flowed across the growing bacteria or biofilms, and this provides a better defined chemical environment.[9] More recently, chemostats have been shrunk to micron length scales in microfluidic geometries, where individual bacteria or emergent biofilms of a few thousand individuals can be grown and the fluid flow actively controlled (Figure 13.1). Microfluidic flows occur at low Reynolds number

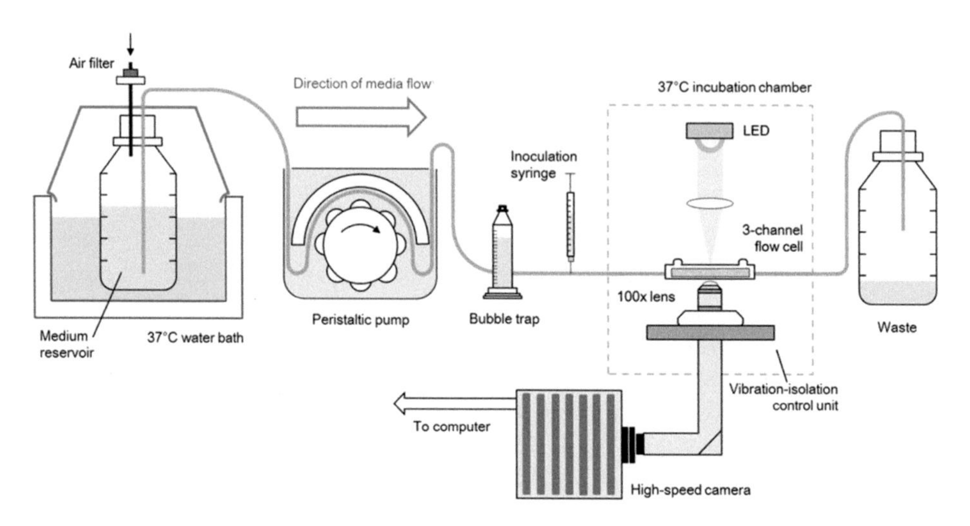

Figure 13.1 Example of a *microfluidic chemostat* used to create bacterial biofilms for measurements on single bacteria embedded in the biofilms.[7,8] The microscope was used to perform particle tracking microrheology experiments on early-stage biofilms of *Staphylococcus aureus* to study the viscoelasticity.

(Chapter 9) and they are thus laminar, so the effects of the fluid mechanics are more reproducible. Practically, care must be taken when using bacterial species that aggregate or form biofilms with microfluidics, since they tend to bung up all the pipes and it is a big nuisance. However, rigorous control of both the biochemical environment and the fluidics makes microfluidic chemostats the current gold standard for bacteria and biofilm experiments.

A variety of non-microfluidic devices have also been developed for *in vitro* biofilm culture[10] that include (Table 13.1): *closed systems*, such as agar plates and multi-well plates, and *open systems*, such as suspended substratum reactors, rotating reactors, Robbins devices, drip-fed biofilms and perforated biofilm fermenters. The open systems tend to be used more frequently in applied research due to the improved reproducibility when results from different geometries are compared (chemostats are open systems). Many of the open systems can also perform higher throughput screening when compared with microfluidic devices and they closely resemble industrially important applications.

The culture of bacteria is a technically accomplished area, where the exact species of bacteria, the physical environment (e.g. temperature, pressure and surface chemistry) and the nutrients the bacteria prefer to metabolise play key roles.[11] Standard tests and culture media to amplify bacterial samples are commonly available in hospitals for bacterial diagnostics.[12] Anaerobic bacteria have markedly different requirements compared with aerobic bacteria e.g. anaerobic bacteria need to be grown in a nitrogen filled environment (e.g. a glove box) with oxygen excluded. Similarly, extremophiles need carefully chosen growth conditions e.g. high/low temperatures, high pressures, high salt concentrations and low/high pH. This is one of the

Table 13.1 Standard culture techniques used to create biofilms *in vitro*.[10]

Method	Applications	Comments
Agar plates	Basic systems for biofilm experiments.	Accessible and simple to perform.
Multi-well plates	Widely used for replication and quantification e.g. for molecular genetics.	Simple and a high degree of reproducibility is possible.
Submerged substratum	Commonly used.	Representative of many real biofilms.
Robbins device	Biofilms in flowing systems. Commonly used.	Robust.
Flow cells/rotating reactors	Real-time visualisation in flowing systems.	
Constant-depth film fermenter	Dental biofilms and others.	Excellent for long-term continuous culture studies.
Drip-flow reactor	General biofilms.	Simple design.
Perfused membranes (1) Baby machine	Generation of thin uniform biofilms. Synchronous growth of daughter cells is possible.	Good control of the growth rate.
(2) Perfused biofilm fermenters.	Similar to (1).	Similar to (1).
(3) Swinnex fermenter.	Antimicrobial studies with physiological control.	Small, robust and cost effective.
(4) Sartorious fermenter.	For proteomic or transcriptomic analysis.	5 times the biomass is required compared with the Swinnex.
Sorbarod fermenter	General biofilms.	Substantial amounts of biomass.
Multiple Sorbarod fermenter.	As with Sorbarod, but allows for single unit replication.	Bespoke manufacturing required.

primary factors limiting bacterial research i.e. an inability to keep many bacteria alive in reproducible cultures.

Fortunately, single-cell genomics is now a standard technique in microbiology and it circumvents some of the issues involved with bacterial culture. This allows taxonomy studies to be performed on an unknown species.[13] The presence of specific types of bacteria can thus be routinely established, although often they cannot be reliably cultured.

A key measurement for bacterial physiology is that of population dynamics via *growth curves* (Chapter 16). Growth rates are a complex function of nutrient availability, the physiological state of the bacteria (often the growth rates are used as practical definitions of the physiological states), the interaction of neighbouring bacteria and so on. Frequently growth curves are measured off-line for bacteria and then bacteria are collected for measurement during distinct phases of growth to better define their physiologies e.g. often the exponential growth phase is chosen where the bacteria are

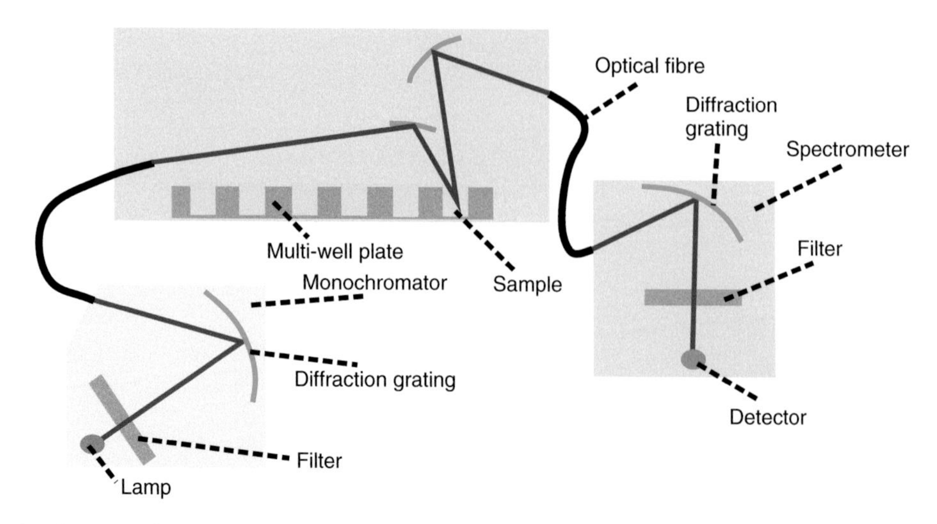

Figure 13.2 *Growth curves* can be measured using a plate reader. Light from the monochromator is directed onto the sample in one of the wells and it is subsequently detected using a spectrometer. Bacterial concentrations can be quantified based on the amount of light absorbed in the well using the Beer–Lambert law.

most physiologically active. Thus, an additional item on the wish list for live bacterial measurements is to measure growth curves *in situ* to avoid the perturbations otherwise involved. This can be possible with three-dimensional (3D) microscopy tools or scattering techniques, but in general, it can be challenging.

Growth curves are typically measured using light scattering and are thus constrained to lower concentrations of bacterial cells where multiple scattering does not complicate measurements e.g. using *plate readers* (Figure 13.2). Light absorption experiments due to scattering from the bacteria provide fairly coarse information on bacterial populations, but can be collected very quickly and can be robust.

13.1.1 Flow Cytometry

Flow cytometry offers a fast method to quantify the proportions of different species of bacterial cells in a suspension (Figure 13.3). In its simplest incarnation, flow cytometry uses light scattering to size individual bacterial cells under flow. Tandem use of spectroscopy (including fluorescence and impedance spectroscopy) with light scattering can make the discrimination between different bacterial strains and species more sensitive. Such techniques are very useful to quantify interacting populations of bacteria (Chapter 22).

13.1.2 Microfluidics and 3D Printing

3D printing techniques allow sterile bespoke microfluidic geometries to be rapidly created using synthetic polymers, such as polydimethylsiloxane (Figure 13.4). The

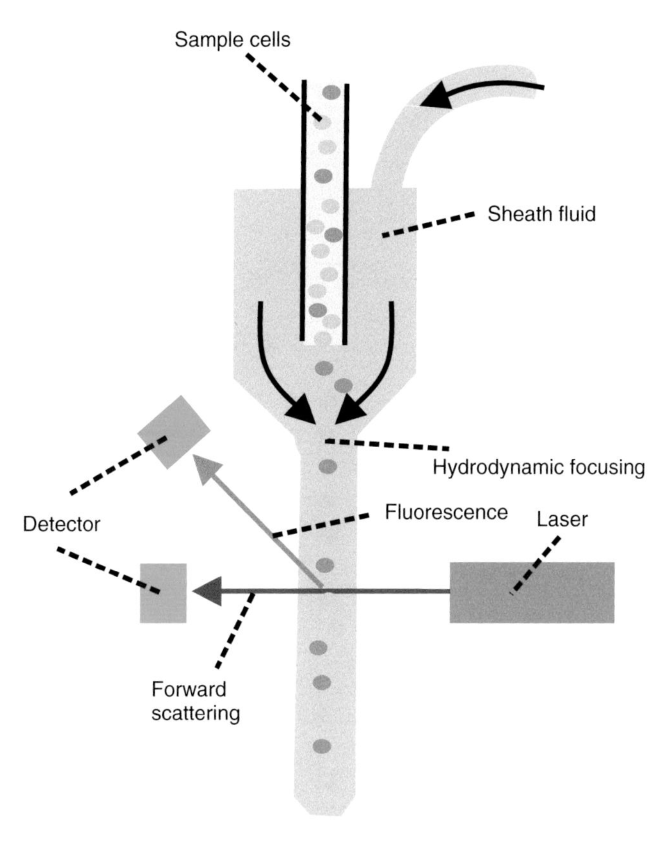

Figure 13.3 *Flow cytometers* can be used to measure the populations of different bacterial species in a mixture. Different bacterial species can be identified using the elastic scattering of laser light or the intrinsic fluorescence of the cells.

microfluidic geometries can then be used for cell culture (e.g. chemostats), to explore the hydrodynamics of bacteria or as part of an integrated characterisation platform. For smaller length scales, photolithography is a standard method for the creation of microfluidic geometries with > 20 μm features. Micromilling and ion beam lithography can be used for microfluidics that require yet smaller features. Many of the studies of bacterial hydrodynamics described in Chapter 17 were possible due to microfluidic geometries designed using such techniques.

Effective *microchemostats* for bacterial growth have been made with 16 nL volumes that allow long-term measurements. The smallest volumes that can be practically explored are mainly determined by issues of convenience, since the bacteria can easily clog up such devices.[14]

3D printing is possible using bacteria as the inks[15] (Figure 13.4). It is thus possible to make a model 3D-printed biofilm, although the method has not yet been extensively used. It would be a useful tool to observe the dependence of intercellular distance on emergent phenomena e.g. studies of signalling (Chapter 18).

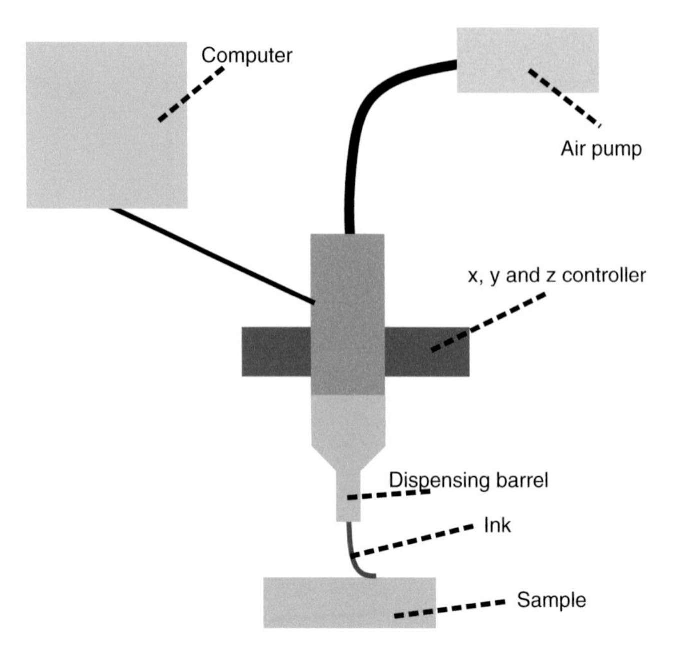

Figure 13.4 Schematic diagram of a *3D printer* that can be used to print bacteria onto surfaces or to create microfluidic geometries from synthetic polymers.

13.2 Microscopes

There are several new microscopy techniques that might be useful for imaging biofilms, such as *coherent X-ray imaging*[16] and *polarising optical coherence tomography* (OCT).[17] Here, the methods best established with bacteria are considered.

13.2.1 Atomic Force Microscopy

Atomic force microscopy (AFM) consists of a cantilever that is scanned across the surface of a sample by the movements of a piezo crystal under the sample (Figure 13.5). The position of the cantilever is measured using the reflection of a laser beam off the back of the cantilever onto a detector e.g. a quadrant photodiode. AFMs can be used for single-point measurements of the mechanical properties of a sample's surface (e.g. to map sample elasticities) by quantifying z displacements transverse to the surfaces in response to different applied forces via the piezo crystal. Alternatively, AFMs can be used to image surfaces, translating the surfaces in x, y and z using the piezo crystal.

AFM has found numerous applications to study the morphology of single bacterial cells. Molecular resolution images have been achieved with live bacteria e.g. the structure of peptidoglycan in the walls of bacteria has been imaged.[18] The forces experienced by a single bacterium have also been studied e.g. to develop Derjaguin–Landau–Verwey–Overbeek (DLVO) models for intercellular potentials or the steric forces due to capsules (Chapter 4).[19]

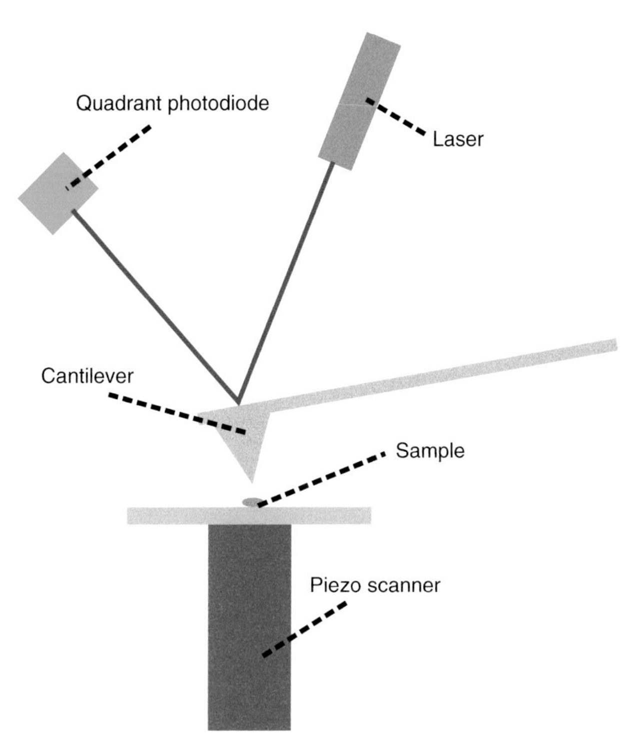

Figure 13.5 Schematic diagram of an *atomic force microscope* (AFM). A laser is reflected off the back of a cantilever and used to detect the position of the cantilever. Images of a surface can be created by scanning the piezo crystal in *x*, *y* and *z*.

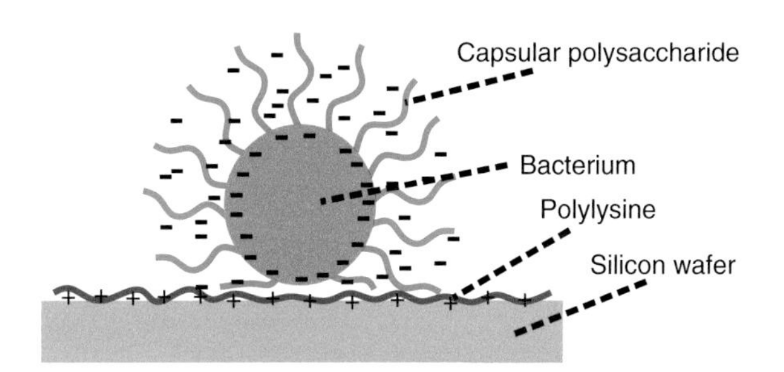

Figure 13.6 *Sample preparation* for bacterial imaging is crucial for microscopy experiments. Capsular bacteria with negative charges can be firmly attached to positively charged polylysine-coated silicon wafers.[19] The surface of silicon wafers is weakly negatively charged, which is why the polylysine coating is necessary.

Sample preparation is an issue in AFM studies, in common with other high-resolution microscopy techniques.[20] Care must be taken to ensure the cantilever does not damage the bacteria and non-contact tapping modes are often used. High-resolution imaging studies of bacteria made with AFM require strong adhesion of the bacteria to surfaces e.g. polylysine-coated surfaces (positively charged) can promote adhesion to negatively charged bacteria (most bacteria are negatively charged, Figure 13.6). Care must be taken that the physiology of the bacteria is not unduly perturbed by the method used to adhere the bacteria to the surfaces (cationic polyelectrolytes, such as polylysine, can kill bacteria, so control experiments are needed). Atomically flat surfaces tend to simplify data interpretation; therefore, silicon wafers or mica are typically used as the substrates on which bacteria are adhered.

Biofilm-coated beads have been used for studies of adhesion and viscoelasticity with AFM. The uniformity of biofilm formation during the growth process has also been studied with AFM.[21]

Electrostatic force microscopy is a modern extension of AFM that can spatially resolve the dielectric spectroscopy of surfaces as a function of the electric field frequency i.e. the current through a surface in response to an oscillatory voltage across the cantilever at a range of frequencies can be measured at different points across the surface.[22] The dielectric measurements can be performed in tandem with force or imaging studies. The dielectric properties of surfaces modulate the electrostatic forces experienced and these are often the dominant interactions felt by the bacteria (Chapter 4). This implies much better models of the local mesoscopic forces experienced by bacteria (and biomolecules in general) can be created e.g. electrostatics, van der Waals and hydrogen bonding.[22]

13.2.2 Confocal Microscopy

Confocal microscopy is normally based on visible wavelengths of light and can provide $\sim \frac{\lambda}{2\sqrt{2}}$ resolution for 3D images in plane and slightly worse axially, where λ is the wavelength of the light.[23] Images are typically constructed from naturally occurring absorption contrast or fluorescence labelling (Figure 13.7). A further advantage is that scanning pin holes across a sample tends to suppress out of focus light, allowing thicker samples to be imaged. Confocal microscopy has thus played a central role in monitoring the physiology of live bacterial biofilms. Every single cell within a biofilm can be located along with their orientation and the EPS components can be specifically labelled with fluorophores.[24] It is expected that modern updates of confocal microscopy methods, such as *lattice sheet microscopy*, *spatial illumination microscopy*, *selective plane illumination microscopy* and *two-photon microscopy* will allow imaging of biofilms at higher resolutions and with lower light dosages (reducing radiation damage).[23]

Fluorescence recovery after photobleaching measurements often are based on confocal microscopes and have been used to estimate the diffusion coefficients of fluorophores in biofilms.[25] The sample fluorescence is photobleached using high laser powers and the dynamics of fluorescence recovery is then measured at low laser powers.

Fluorescence correlation spectroscopy (FCS) is another method that can be used with a confocal microscope and it can allow the fast dynamics of biomolecules to be resolved. A laser illuminates a volume of sample using a pinhole and the fluctuations of fluorescent light emitted are measured with a point detector. The statistics of the light fluctuations can be related to the dynamics of the sample using correlation functions (Section 1.5), but some spatial information is lost in the process of Fourier averaging, so imaging techniques tend to be preferred if they are possible. However, FCS often can provide access to faster dynamics than direct imaging and can be performed with single molecules.

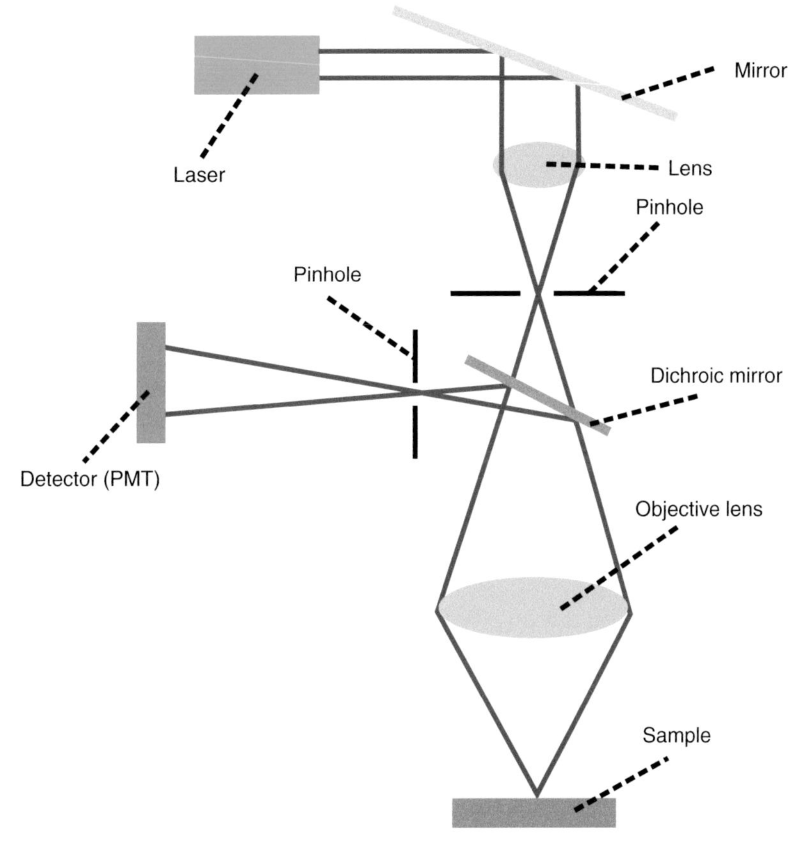

Figure 13.7 A schematic diagram of a *confocal microscope*. Pin holes after the light source and before the detector scan across the sample and allow the construction of three-dimensional images.

13.2.3 Fluorescence Microscopy

Fluorescence microscopy is a standard workhorse for molecular biology (Figure 13.8). It is exquisitely sensitive (single molecules can be detected) and exquisitely specific (e.g. fluorescent labels can be placed on specific molecules using genetic techniques). Reference should be made to standard molecular biology books for more details of labelling strategies with fluorophores.[3] Typically, genetically expressed fluorophores (e.g. green fluorescent proteins [GFPs]) or antibody-labelled synthetic fluorophores are used, but there are a wide range of other possibilities e.g. quantum dots, nanodiamonds and nucleic acids labelled with synthetic fluorophores. Functional fluorophores can be made e.g. that detect pH, voltage or oxidative stress and fluorescence used to read out the values. Fluorescence microscopy can allow the real-time kinetics of gene activity in individual bacteria to be followed.[26] Antibody-conjugated fluorescent nanoparticles can provide a very sensitive method to detect different types of bacteria.[27]

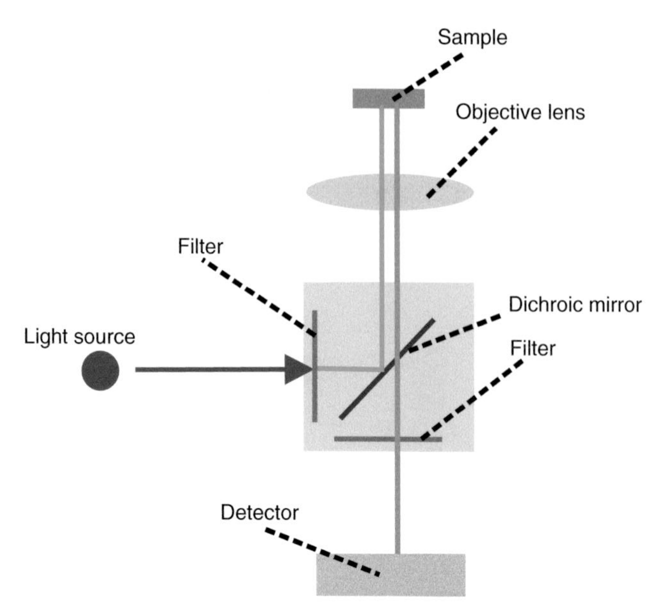

Figure 13.8 Schematic diagram of an epi illumination (the light enters and leaves the sample via the same lens) *fluorescence microscope* used to explore single molecules or bacterial cells.

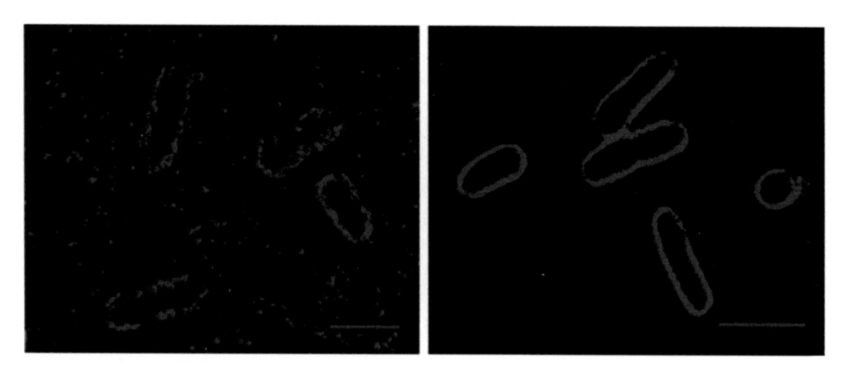

Figure 13.9 Capsular structure of live *E. coli* cells from *super-resolution fluorescence microscopy* (STORM), (a) without graphene oxide coverslips, and (b) with graphene oxide coverslips.[29] The graphene oxide increases the contrast by switching off non-specifically bound fluorophores (specifically, synthetic fluorophores attached to antibodies) via its large resonant energy transfer.

Evidence has been found for variations in pH across biofilms using fluorescent nanosensors.[28] The centres of *P. aeruginosa* and *Streptococcus mutans* biofilms were found to be more acidic (±2 pH).

Graphene oxide (GO) substrates are a convenient method to increase the contrast of fluorescence microscopy experiments (Figure 13.9).[29] They are particularly good for extinguishing the non-specifically bound background when antibody labelling is used.

High throughput molecular biology techniques often use fluorescence labelling. For example, HiP-FISH has high phylogenetic resolution and allows thousands of different species of bacteria to be identified simultaneously via coloured barcodes.[30]

Two photon fluorescence microscopy shows promise for the imaging of bacterial biofilms.[31] The technique reduces the effects of multiple scattering (via the non-linear effects of two low-energy photons exciting a single transition of twice the energy) and thus allows structures more deeply embedded in biofilms to be resolved.

13.2.4 Super-resolution Fluorescence Microscopy

The diffraction limit (d) of conventional optical microscopes is on the order of 200 nm (the Abbe limit is $d = \frac{\lambda}{2NA}$, where NA is the numerical aperture of the lenses, which is ~ 1 for standard objective lenses and λ is the wavelength) and it can be broken via super-resolution techniques.[23] With bacterial samples, resolutions on the order of 10–20 nm are possible with super-resolution fluorescence microscopy, which sensitively depends on the type of fluorophore used in labelling (e.g. the photon yield before bleaching), the excitation lasers and the imaging modality. The most common varieties of *super-resolution fluorescence microscopy* are STED (*stimulated emission depletion*, a non-linear optical effect is used to decrease the illumination volume of a confocal type scanning microscope, Figure 13.10b), STORM (*stochastic optical*

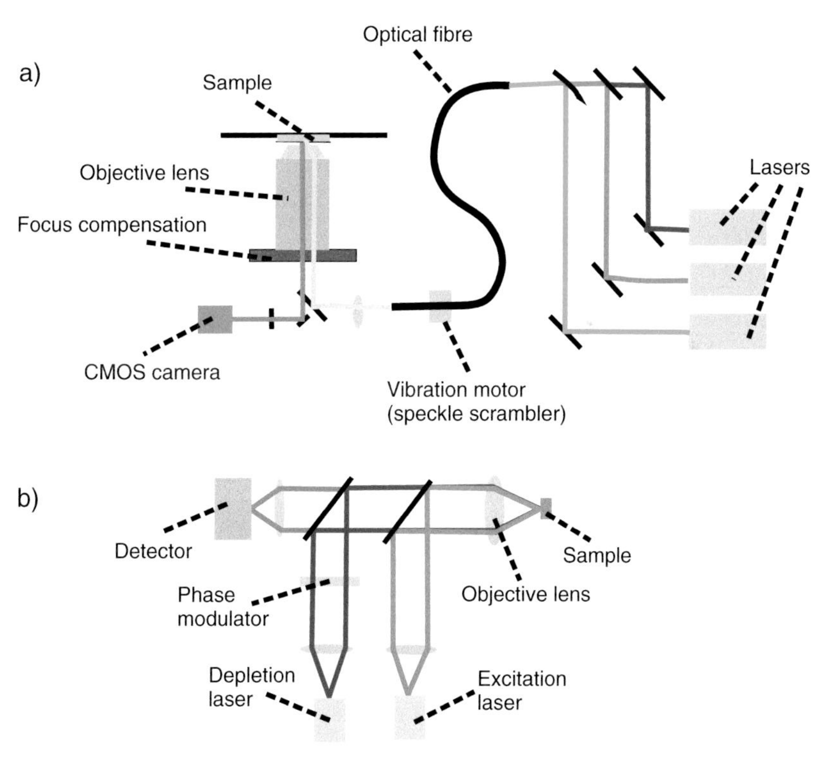

Figure 13.10 Schematic diagram of two prototypical varieties of super-resolution fluorescence microscopy: (a) *stochastic optical reconstruction microscopy* (STORM),[32] and (b) *stimulated emission depletion* (STED) *microscopy*.[23]

reconstruction microscopy, blinking fluorophores allow dot to dot images to be reconstructed, Figure 13.10a), *lattice sheet* and *structured illumination*. STED and STORM tend to have the highest spatial resolution, but relatively low time resolution. In general, there is a very large range of super-resolution techniques and many associated acronyms.[23] STORM has been used to explore the adhesive proteins of the EPS in *Vibrio cholerae* biofilms.[33] Other STORM studies have been used to image the capsular structure of *E. coli*[19] labelled with antibodies attached to synthetic fluorophores (Figure 13.9).

13.2.5 Optical Coherence Tomography

OCT uses the coherence length of a light source (super-luminescent diodes or pulsed lasers are normally used) to optically section a sample with an interferometer (Figure 13.11).[23] Fringes are only observed on the detector when the optical path length of the reference and sample arms match. Thus, a coherence gate is created that suppresses the multiple scattering that reduces image contrast in thick specimens. The main advantages of OCT are its ability to image through opaque objects (e.g. through up to 1–2 mm of human tissue) and its non-invasivity (infrared light is often used). OCT has thus been used to image biofilms.[34,35] Its resolution at ~1 μm is well below that of confocal microscopy, so individual bacterial cells will only just be resolved. However, OCT can be used in clinical settings *in vivo* (low-intensity infrared light is harmless), so it could be useful for diagnostic purposes, assuming the detection fibre can be placed sufficiently close to the position of the bacterial infections. Doppler

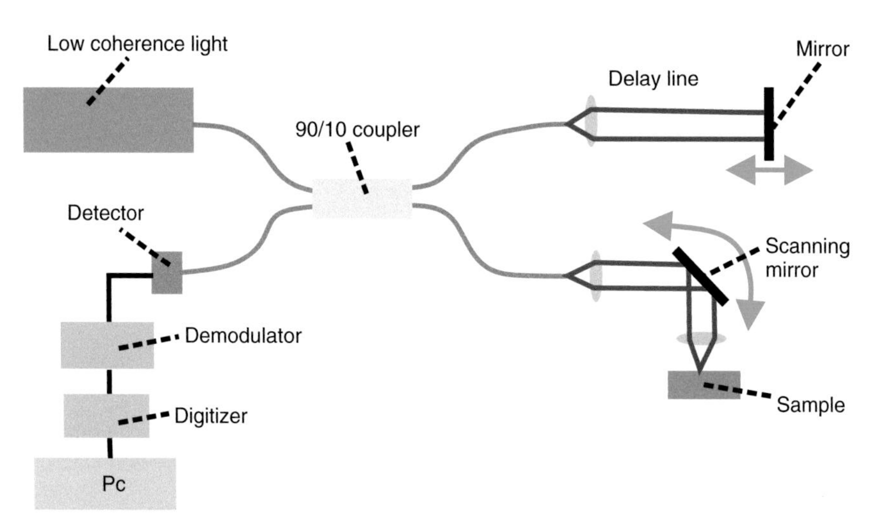

Figure 13.11 Schematic diagram of *optical coherence tomography* (OCT) apparatus that can be used to image structures deep within a sample e.g. 1–2 mm of opaque tissue, using infrared light. It uses a Michelson interferometer as a coherence gate to reduce the effects of multiple scattering. The low coherence light source is typically a super-luminescent infrared diode or a pulsed Ti:sapphire laser.

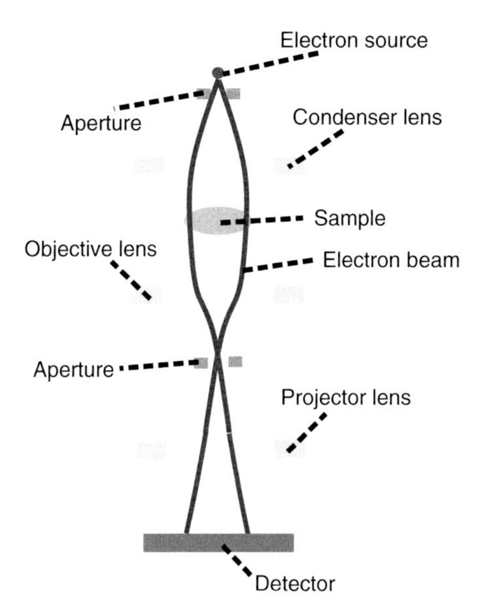

Figure 13.12 Schematic diagram of a *transmission electron microscope* used in cryoelectron microscopy, which is shown in cross-section. The path of the electron beam is shown with a continuous line.

shift OCT can be used to measure fluid flows in 3D at the micron length scale[36] and polarising OCT can quantify birefringent samples[17] e.g. liquid crystalline phases can be detected (Chapter 8).

13.2.6 Electron Microscopy

Electron microscopy (*scanning electron microscopy* [SEM] and *transmission electron microscopy* [TEM], including cryoTEM tomography) techniques can provide exquisite detail on images of cells, molecules and molecular aggregates in three dimensions with Angstrom resolution, due to the small de Broglie wavelength of electrons[37] (Figure 13.12). The major weaknesses of TEM are that samples need to be frozen and prepared in thin slices, which prevents *in vivo* studies. Specific labelling is more limited with TEM than fluorescence microscopy to determine the behaviour of the exact varieties of molecules imaged, but colloidal gold can be used with antibodies. SEM techniques have lower resolution than TEM and are restricted to sample surfaces (conveniently requiring less sample preparation than TEM), but still require high vacuums similar to TEM.

13.2.7 Other Microscopy Techniques

A wide range of other optical microscopy techniques can be effectively used with bacteria.[23] *Differential dynamic microscopy* (DDM, Section 1.5) offers an alternative to particle tracking to calculate the distributions of cell speeds, the fraction of motile cells and the diffusivities of bacteria.[38] DDM uses digital correlation of microscopy images to calculate the intermediate scattering function. The technique allows rapid ensemble averaging over many bacteria but is slightly limited in its dynamic range (limited by the acquisition rate of the camera to shortest time scales of $\sim 10^{-3}$ s compared with $\sim 10^{-9}$ s for conventional dynamic light scattering) and it depends on a parameterised model to describe the swimming of bacteria.

Quantitative phase microscopy can measure the growth rate of individual bacteria.[39] Bacterial masses can be calculated via their refractive indices.

13.3 Other Photonics Techniques

13.3.1 Remote Sensing

Remote sensing of bacteria has a range of applications in biodefense, food sciences, population monitoring and ecology. Remote measurements need to be non-invasive, safe and applicable over long distances (>1 m) e.g. satellite measurements of the milky sea phenomenon have been performed using fluorescent imaging.[40]

Laser-induced fluorescence (LIF) has been one successful variant of remote sensing (Figure 13.13),[41] in which the strong intrinsic fluorescence signal when bacteria are excited with a powerful laser can allow identification of bacteria over large distances e.g. km. A challenge however is to determine the species of bacteria at such long distances, since the fluorescence spectra only contain a few broad features. Remote sensing is important for detecting bacteria in aerosols e.g. bacterial contagions created by coughs or sneezes.

13.3.2 Quantum Sensing

Quantum sensing techniques can approach the limits of sensitivity that are physically permitted by quantum mechanics.[42] Nitrogen defects in nanodiamonds contain unpaired electronic spins and can be used as an ultrasensitive probe of their physical environment using the *Zeeman effect* i.e. the modulation of the splitting of the

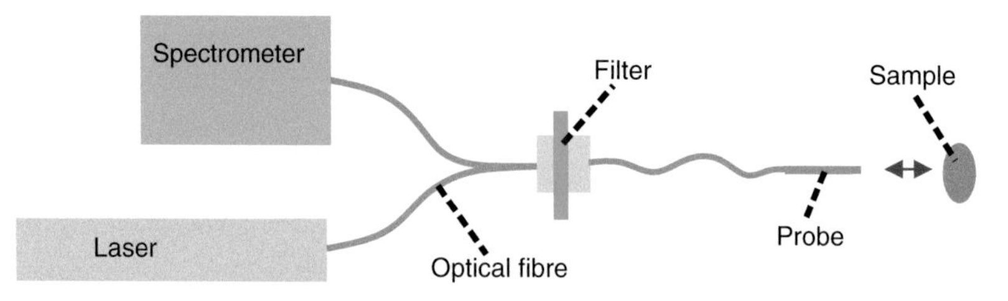

Figure 13.13 Schematic diagram of a spectrometer to measure *laser-induced fluorescence* emitted from a sample (a backscattering geometry is used).

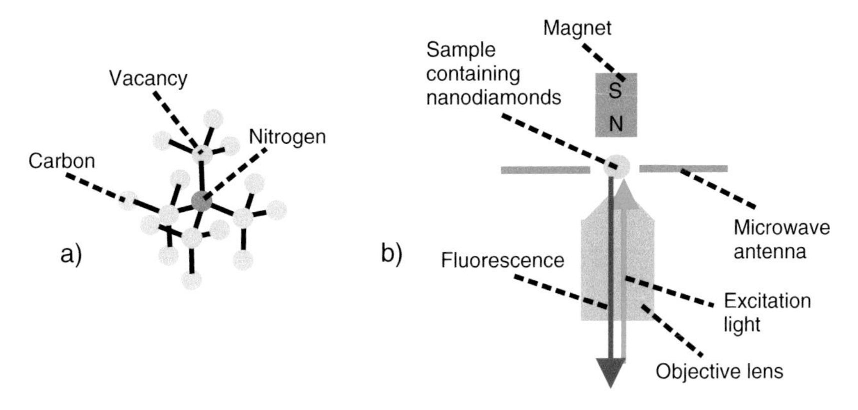

Figure 13.14 (a) Nitrogen defects in the tetrahedral lattice of a *nanodiamond* (carbon). Nitrogen defects in nanodiamonds cause them to strongly fluoresce under laser illumination with little photobleaching. (b) *Quantum sensing* with the nitrogen defects within the nanodiamonds is possible using the Zeeman effect. Nanodiamonds are placed in a sample and subjected to a magnetic field. A microwave antenna can alter the occupancy of the Zeeman split energy levels and thus modulate the resultant fluorescence. The fluorescence is measured with an inverted fluorescence microscope.

crystal's energy levels by an applied magnetic field (Figure 13.14). The Zeeman effect is sensitive to a range of other physical parameters via the effects on the splitting of energy levels e.g. temperature, pH and electrostatics can be measured. Nanodiamonds are compatible with live bacteria and they allow quantum sensing at room temperature. Nanodiamonds have also been used to measure Raman signals with live *E. coli*.[43] Furthermore, nanodiamond sensors have been used to image magnetic structures inside magnetotactic bacteria.[44]

In general, *quantum metrology* is being used to boost the performance of imaging and spectroscopy techniques.[45] Promising quantum techniques have used squeezed light to improve tracking experiments,[46] quantum correlations to improve localisation accuracy for imaging[47] and ghost imaging to provide non-invasive measurements.

13.3.3 Spectroscopy

Optical spectroscopy can provide specific molecular information on the contents of bacteria (e.g. the biomolecules they contain) and can thus provide diagnostic information. Commonly used spectroscopic techniques include *Raman* (Figure 13.15), *infrared, circular dichroism, terahertz* and *fluorescence*.[4,48] More detailed discussions are left to the specialist literature.[48] Raman scattering is an inelastic process with a small scattering cross-section (the process is ~10^6 times weaker than elastic Rayleigh scattering), whereas fluorescence is significantly larger (~10^2 times weaker than Rayleigh scattering).

Non-linear extensions of Raman spectroscopy (*coherent anti-Stokes Raman scattering, stimulated Raman scattering*, etc.) lead to increases in sensitivities, although the resultant spectra can be more complex to interpret and the high-power lasers needed are more expensive and can damage samples. Raman spectroscopy can be combined

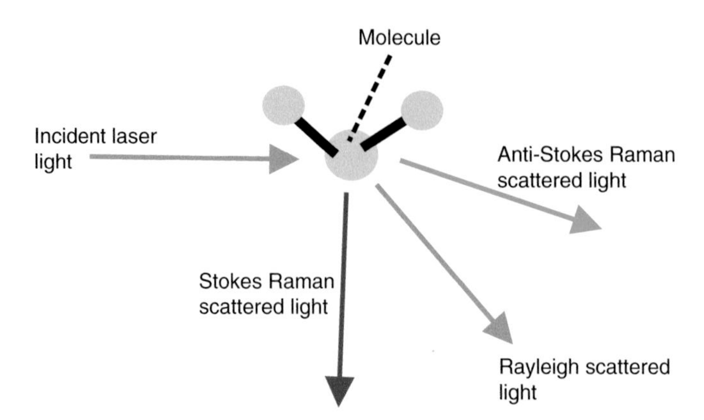

Figure 13.15 Schematic diagram of *Raman scattering* from a biomolecule that can be measured in a Raman spectroscopy experiment. Changes in the energies of the scattered photons are caused by changes in quantised vibrations of the biomolecule.

with imaging.[23] A combined automated microfluidic, optical tweezers and Raman microscopy instrument was able to sort bacterial cells of four different strains with 98.3 ± 1.7% accuracy.[49]

Surface plasmon resonance can be used to amplify signals from single bacteria and samples with lower molecular concentrations e.g. Raman spectroscopy using SERS (surface-enhanced Raman scattering), where noble metals on surfaces amplify signals using plasmon resonance. SERS has been used with biofilms.[50] It has also been used to detect and image quorum sensing molecules in *P. aeruginosa*[51] and it was extended to *in vivo* measurements with mice.

13.3.4 Zeta Sizer

Zeta sizers use photon correlation spectroscopy to detect the Doppler shifts of light scattering by particles oscillating in an electric field. The most commonly used commercial zeta sizers use phase analysis light scattering (PALS) to increase their sensitivity, where the optical signal is modulated in an interferometer and then electronically demodulated (Figure 13.16), to reduce the effects of low frequency detector noise.[52] Zeta sizers can thus be used to measure the zeta potential of colloids in the sample cell i.e. the electric potential at the point of slip defined by the hydrodynamic radius. The zeta potential varies with the ionic environment and the charge on bacterial cells and it can be used to differentiate between different strains e.g. capsular versus non-capsular strains.[53] Zeta sizers can also be used in a dynamic light scattering mode to provide correlation functions, which can be used to size the colloids.

13.3.5 Other Photonics Techniques

Optical tweezers and *magnetic tweezers* can allow the pN forces applied by molecular machines to be measured, including those from bacteria (Figure 13.17). Optical trapping of viruses and bacteria was first demonstrated by Ashkin and played a role in the

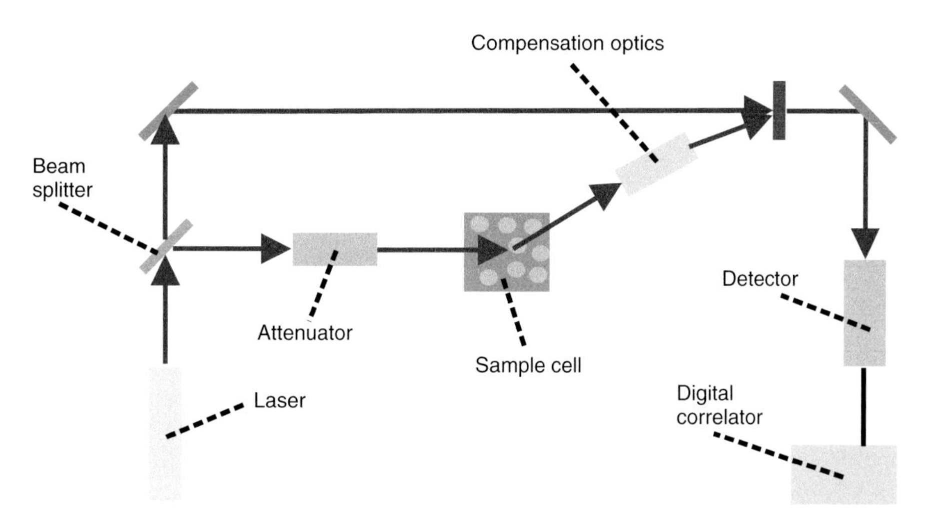

Figure 13.16 Schematic diagram of a *zeta sizer* that uses photon correlation spectroscopy (dynamic light scattering) to measure the zeta potential of colloids e.g. bacteria.[52] Interference of the signal passing through the sample and the reference beam makes the technique sensitive to the phase of the scattered light. Thus, smaller changes in the zeta potentials can be measured.

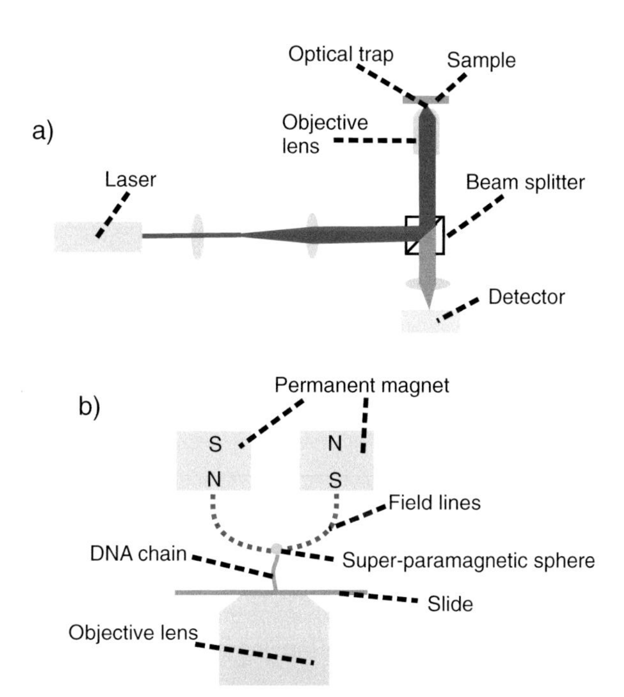

Figure 13.17 (a) *Optical tweezers* and (b) *magnetic tweezers* that can be used to manipulate single molecules and cells.

award of the Nobel Prize for Physics.[54] In whole-cell experiments, a challenge is to not fry the bacteria or detrimentally perturb their physiology, so the bacteria do not experience extreme oxidative stress, radiation damage or heat sterilisation. Cellular damage is not necessarily obvious by visual inspection and cell viability via division or live/dead assays (via intracellular potential dyes and lysis penetrating dyes) are necessary diagnostic methods to check for damage.

13.4 Miscellaneous Physical Techniques

13.4.1 Rheology and Microrheology

As discussed in Chapter 10, all biological materials are *viscoelastic* to some degree. Even prototypical Newtonian fluids, such as water, have elasticity at very short picosecond time scales and their viscoelasticities have been measured in pulsed laser experiments.[55] *Rheology* is the study of the flow behaviour of materials and can be considered a generalised form of fluid mechanics. Biofilms and bacterial suspensions can demonstrate a wide range of rheological phenomena that are only starting to be understood with quantitative detail (Chapter 21).

It is challenging to make *non-invasive* rheological measurements on biofilms via traditional bulk rheology methods (Figure 10.1). Biofilm structure can be disturbed during sample preparation and the forces applied to measure their rheology disrupt their behaviour, because the materials are soft and fragile.

In contrast to bulk rheology, *microrheology* shows promise for non-invasive measurements of bacterial suspensions and biofilms. The first microrheology experiments on biofilms used particle tracking microrheology and the bacterial cells were used as probes (Figure 1.3).[8] Thermal energies caused the Brownian motion of the cells that was modified by the viscoelasticity of the biofilm. Since the energy spectrum of thermal displacements is known (it follows a white noise form with the magnitude given by the equipartition theory), the displacements of particles in response to these forces can be used to calculate the viscoelasticity of the cells' microenvironments.

Magnetic microrheology has also been used to study the viscoelasticity of *E. coli* biofilms in three dimensions.[56] Forces are applied to superparamagnetic spheres in an oscillating magnetic field and the viscoelasticity is calculated by measuring the spheres' responses in a microscope. Creep compliance experiments were performed i.e. the displacements of magnetic particles were measured in response to a constant stress. The adhesion of the biofilm to the surface was found to play a big role.

It is possible to use microfluidic geometries with bacteria to explore their non-linear viscoelasticity at micron length scales, such as the slot geometry and extensional flow.[57]

Atomic force microscopy measurements with microbead cantilevers were used to measure the viscoelasticity of *P. aeruginosa* biofilms using creep measurements[21] i.e. the strain applied by a cantilever to a biofilm was measured as a function of time at a fixed stress. Other biofilm microrheology experiments have been previously reviewed.[58] Bulk rheology experiments can also be performed on biofilms (Chapter 10[59]). Bulk measurements are much more invasive than microrheology measurements, since they disturb the biofilm structure, and they lack spatial sensitivity. However, bulk measurements help to understand the behaviour of biofilms at large length scales and with large imposed stresses, which is often practically important[60] e.g. in food processing.

Traction force microscopy (TFM) allows the forces applied by bacteria on surfaces to be explored.[61] Fiducial markers (e.g. colloidal spheres) are embedded in a gel and their displacements can be measured in response to the adhesive forces of the bacteria growing on the surface of the gel. Microrheology versions of TFM are possible.[58]

13.4.2 NMR/MRI

Nuclear magnetic resonance (NMR) techniques are standard tools in analytical chemistry for the structural elucidation of unknown molecules (Figure 13.18). High-resolution solution-state NMR can be used to molecularly fingerprint a very wide range of molecules in solution, since molecular bonding patterns shift nuclear resonant frequencies in characteristic ways. The resonance can be probed with radiowave absorption. In solid-state samples, dipolar broadening of nuclear resonances increases due to the interaction of neighbouring nuclei, which substantially reduces the resolution of the NMR spectra. Methods to help circumvent these issues are available e.g. cross-polarisation magic angle spinning or dynamic nuclear polarisation.[62] NMR experiments on intact bacteria, biofilms and model bacterial membranes thus require solid-state techniques to explore their behaviour due to the dipolar broadening.

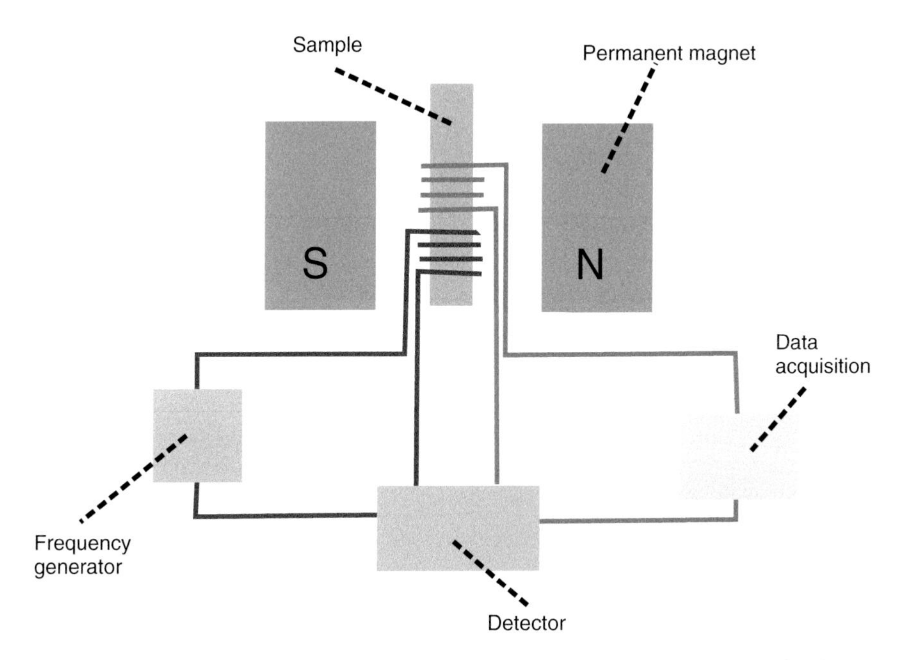

Figure 13.18 Schematic diagram of a *nuclear magnetic resonance* (NMR) experiment used to probe the behaviour of a solution-state sample.

Pulsed gradient spin echo (PGSE) NMR has been used to measure the transport of *P. aeruginosa* in a synthetic porous material made from 241 μm monodisperse colloidal beads.[63] Biofilm growth reduced the flow dynamics and anomalous transport models were needed to describe the data.

It is possible to observe the development and growth *in vivo* of large intraorganism biofilms using *magnetic resonance imaging* (MRI).[64] However, due to the low resolution of MRI (~mm), very large biofilms are needed, which would have a huge impact on the organism's health (less useful for diagnosis of early stages of infection) and the methodology is too insensitive and expensive for routine diagnosis.

13.4.3 SIMs and Mass Spectrometry

Mass spectrometry is another standard tool in analytical chemistry to deduce the structure of unknown molecules (Figure 13.19). Ions are created from the molecules and the ions are accelerated by an electric field, which has curved trajectories in a magnetic field due to the Lorentz force. Modern developments in mass spectrometry have increased the size of the ions that can be studied in molecular biology e.g. MALDI-TOF and electrospray techniques that allow small chemical modifications on large biomolecules to be measured, such as single phosphorylation events on large globular proteins.

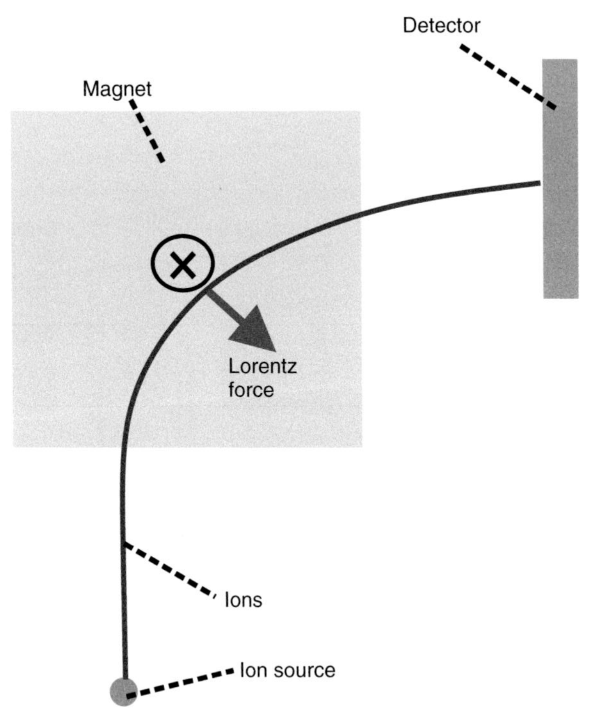

Figure 13.19 Schematic diagram of a *mass spectrometer* that can be used to measure the charge/mass ratios of ions. Ions are created from the molecules and ions are accelerated by an electric field (not shown), which have curved trajectories in a magnetic field due to the Lorentz force.

Mass spectrometry imaging (e.g. secondary ion mass spectrometry, SIMS) has recently been demonstrated with complex microbial communities and provides detailed molecular information as a function of position across a biofilm.[65] SIMS can create image resolutions down to 50 nm combined with mass spectrometry structural elucidation, but sample/probe interactions are very strong and destructive i.e. only *ex vivo* measurements are possible and samples need to be frozen and in a vacuum.

13.4.4 Mass Cantilever

Resonant cantilever methods can be used to weigh single bacteria,[66] which, perhaps surprisingly, is a very recent advance (Figure 13.20). The mass of individual bacteria can also be indirectly inferred from *quantitative phase microscopy* experiments (based on the refractive indices). Mass measurements are very useful to quantify the physiology of bacteria during growth.

13.4.5 Scattering Techniques

Scattering techniques (Figure 13.21) have historically been extensively used to characterise biofilm EPS and the individual components of bacteria (proteins, DNA, membranes, etc.) e.g. via small-angle X-ray/neutron scattering (SAXS/SANS) or X-ray/neutron crystallography, since they are standard tools for soft matter characterisation of gels and crystallography can provide Angstrom-resolution structures of crystalline biomolecules, which provide useful reference systems for *in vivo* structures.

Modern *X-ray techniques*, such as scanning microdiffraction, can characterise bacterial morphology inside biofilms, although the methodology is very invasive i.e. the bacteria are rapidly killed by this form of ionising radiation.[67] Extensive work has been performed on neutron and X-ray reflectivity to understand model bacterial

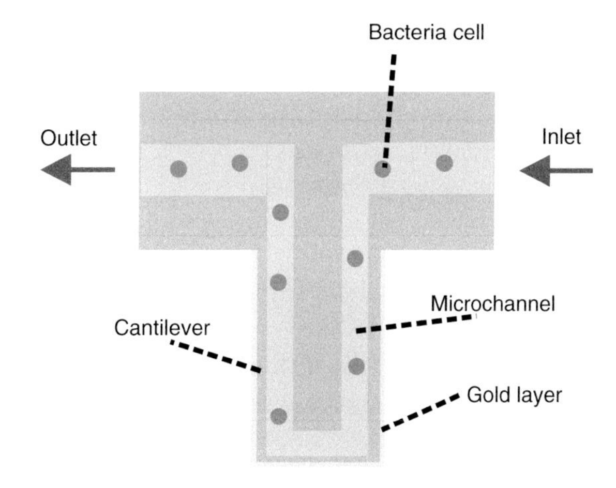

Figure 13.20 Schematic diagram of a *mass cantilever* that can be used to measure the mass of single bacteria via the cantilever's resonant frequency.[66]

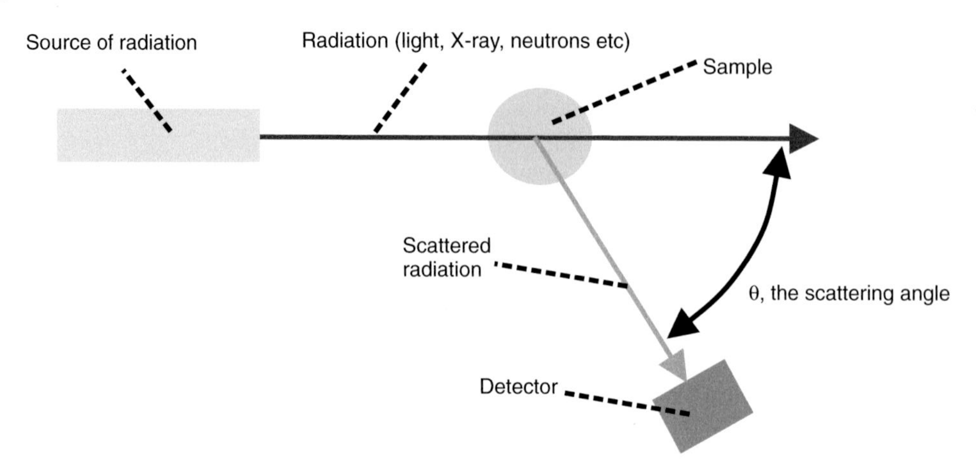

Figure 13.21 Schematic diagram of *scattering techniques* used to measure the structure of a sample. The radiation could be visible light, X-rays or neutrons. θ is the scattering angle.

membranes.[68] Such studies can quantify the interactions of surface-active molecules with the membrane. SAXS has demonstrated that the mesoscopic structure of the EPS from marine bacteria depends on the pH.[69] The gels swelled as the pH varied with increasing mesh sizes and aggregate sizes.

13.4.6 Magnetic Measurements

Super-conducting quantum interference devices (SQUIDS) have been used (by combining them with optical microscopes) to measure the minute magnetic fields from individual magnetotactic bacteria.[70] These SQUID techniques are awkward and bulky due to the requirements of ultra-low temperatures for the superconductors. More modern techniques have used the Zeeman effect in *nanodiamond defects* (Section 13.3.3) and *spin valves* to measure magnetic structures inside bacteria. These measurements are not restricted to such low temperatures, facilitating *in vivo* studies.[71]

Optical birefringence measurements can be used to measure the average magnetic moments of bacteria when they align in magnetic fields.[72] Furthermore, the growth of intracellular magnetosomes can be studied using *ferromagnetic resonance spectroscopy* and a docking mechanism was postulated to explain the stabilisation of newly nucleated superparamagnetic particles inside the bacteria.[73]

13.5 Molecular Biology Techniques

High throughput techniques, such as optical imaging or molecular biology methodologies (e.g. cytometry or genetic techniques), can produce such huge data sets that they lend themselves to machine learning techniques (Chapter 14). This is expected to be a large growth area in the field of bacterial biophysics and can allow many subtle correlations to be established e.g. those between molecular biology and phenotype, the language of bacterial metabolites and the structure of globular proteins from primary sequences. Computational modelling in genomics is already a well-developed field[74] using classical machine learning techniques, but the introduction of deep learning neural networks will accelerate the development of this field.

13.5.1 Binding Assays

Standardised microfluidic platforms have been created for the determination of MICs (*minimum inhibitory concentrations* i.e. a parameter to quantify the potency of antimicrobials, Chapter 24) for 98 biofilms in parallel.[75] Carbohydrate-binding arrays have been developed to detect bacteria.[76]

Enzyme-linked immunosorbent assays (ELISA) can be used to quantify the biomass of a bacterial species in a biofilm and the proteins they produce (Figure 13.22). Antigens from the bacteria are attached to a surface, which is then exposed to antibodies and combined with a fluorescent label to allow readout. A similar technology is used in pregnancy tests, although ELISA in contrast is often used with an array of different antibodies to detect multiple antigens.

Patterns of gene expression in biofilms can be followed using *DNA microarrays*.[77] However, it is challenging to do these measurements non-invasively.

Magnetoimmunology is a standard technique to separate different strains of bacteria i.e. magnetic colloids covered with antibodies can bind to specific strains of bacteria and be separated using a permanent magnet. *Bacteriophage assays* can also be used to detect different strains of bacteria[78] e.g. capsular strains of *E. coli*.

13.5.2 Electrophoresis and Dielectrophoresis

Electrophoresis (the mobility of a polyelectrolyte in a constant electric field, often confined within a gel) has a long history for the separation of biopolymers e.g. for the sizing of proteins and nucleic acids to allow sequencing[79] (Figure 13.23a). *Dielectrophoresis* has been used to differentiate between strains of *E. coli*[80] (Figure 13.23b). An alternating current electric field was applied across a microfluidic

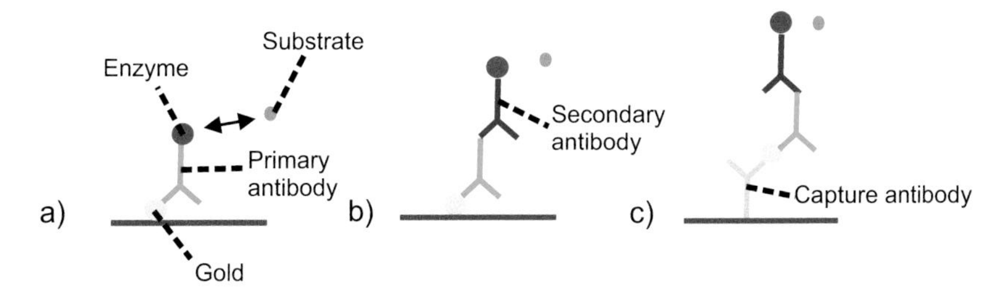

Figure 13.22 *Binding assays* can be used to identify a variety of metabolites in molecular biology. Three standard *ELISA* (enzyme-linked immunosorbent assay) techniques are shown: (a) direct, (b) indirect and (c) sandwich.

a)

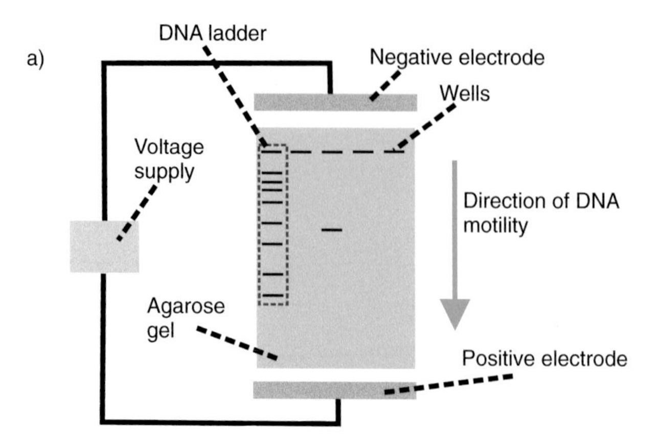

b)

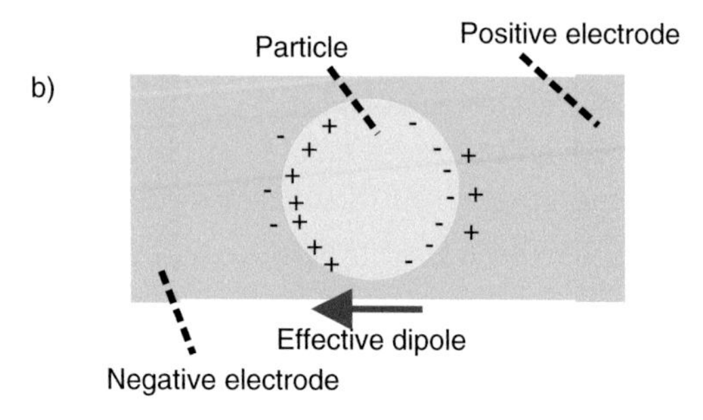

Figure 13.23 (a) Schematic diagram of an *electrophoresis* and (b) a *dielectrophoresis* experiment that can be used to explore bacterial biophysics.

geometry over frequencies of 1–10^9 Hz and different dielectric responses allowed identification of the different strains. Dielectrophoretic motility is found to vary with the charge content of the bacterial surfaces and is thus a sensitive measure of the serotype of capsules, although the effect is blurred slightly by the size polydispersity of the cells. Whole-cell microelectrophoresis experiments also show different mobilities for different strains of bacteria.[81]

13.5.3 Other Electrical Phenomena

Microelectrodes have been used to spin bacteria in an electric field to understand the action of flagellar motors.[82] Microelectrodes have also been used to measure the ingress of chemicals into biofilms.[83]

Microelectrode arrays and *patch clamps* have been used to measure the electrical potentials of bacterial cells (Chapters 3 and 29). A range of techniques can be employed to probe the conductivities of bacteria and biofilms e.g. *electrical impedance spectroscopy*.

Electroporation can be used to facilitate gene transfer into bacteria by creating holes in the bacterial membranes.[84] It is also an effective mechanism to kill bacteria.

Suggested Reading

Gunning, P. A.; Kirby, A. R.; Morris, V. J., *Atomic Force Microscopy for Biologists*. Imperial College Press: 1999. Lots of useful tips for sample preparation during AFM experiments.

Mersk, J., *An Introduction to Optical Microscopy*, 2nd ed. Cambridge University Press: 2019. Optical microscopes are primary tools to study the behaviour of bacteria and confocal microscopy is the standard method used with biofilms. The optical physics of such microscopes is covered in detail.

Waigh, T. A., Advances in the microrheology of complex fluids. *Reports on Progress in Physics* **2016**, *79* (7), 074601. Reviews some applications of microrheology in complex fluids which include biofilms.

Waigh, T. A., *The Physics of Living Processes*. Wiley: 2014. Contains a review of some standard experimental techniques important in biological physics.

Zaccai, N. R.; Serdyuk, I. N.; Zaccai, J., *Methods in Molecular Biophysics: Structure, Dynamics, Function for Biology and Medicine*, 2nd ed. Cambridge University Press: 2017. Broad review of physical techniques used in molecular biology experiments.

References

1. Jensen, H. J., *Complexity Science: The Study of Emergence*. Cambridge University Press: 2023.
2. Waigh, T. A., *The Physics of Living Processes*. Wiley: 2014.
3. Leake, M. C., *Single-Molecular Cellular Biophysics*. Cambridge University Press: 2013.
4. Zaccai, N. R.; Serdyuk, I. N.; Zaccai, J., *Methods in Molecular Biophysics*. Cambridge University Press: 2017.
5. Locke, A.; Fitzgerald, S.; Mahadevan-Jansen, A., Advances in optical detection of human associated pathogenic bacteria. *Molecules* **2020**, *25* (22), 5256.
6. Xu, Y.; Dhaouadi, Y.; Stoodley, P.; Ren, D., Sensing the unreachable: Challenges and opportunities in biofilm detection. *Current Opinion in Biotechnology* **2020**, *64*, 79–84.
7. Hart, J. W.; Waigh, T. A.; Lu, J. R.; Roberts, I. S., Microrheology and spatial heterogeneity of *Staphylococcus aureus* biofilms modulated by hydrodynamic shear and biofilm-degrading enzymes. *Langmuir* **2019**, *35* (9), 3553–3561.
8. Rogers, S. S.; van der Walle, C.; Waigh, T. A., Microrheology of bacterial biofilms in vitro: *Staphylococcus aureus* and *Pseudomonas aeruginosa*. *Langmuir* **2008**, *24* (23), 13549–13555.
9. Novick, A.; Szilard, L., Experiments with the chemostat on spontaneous mutations of bacteria. *Proceedings of the National Academy of Sciences of the United States of America* **1950**, *36* (12), 708–719.

10. McBain, A. J., *In vitro* biofilm models. In *Advances in Applied Microbiology*, Laskin, A. I., Ed. Elsevier: 2009; Vol. 69, pp. 100–126.
11. Capuccino, J. G.; Welsh, C., *Microbiology: A Laboratory Manual*. Pearson: 2018.
12. Torok, E.; Moran, E.; Cooke, F., *Oxford Handbook of Infectious Diseases and Microbiology*. Oxford University Press: 2016.
13. Woyke, T.; Doud, D. F. R.; Schulz, F., The trajectory of microbial single-cell sequencing. *Nature Methods* **2017**, *14* (11), 1045–1054.
14. Balagadde, F. K.; You, L.; Hansen, C. L.; Arnold, F. H.; Quake, S. R., Long-term monitoring of bacteria undergoing programmed population control in a microchemostat. *Science* **2005**, *309* (5731), 137–140.
15. Connell, J. L.; Ritschdorff, E. T.; Whiteley, M.; Shear, J. B., 3D printing of microscopic bacterial communities. *Proceedings of the National Academy of Sciences of the United States of America* **2013**, *110* (46), 18380–18385.
16. Waigh, T. A.; Rau, C., X-ray and neutron imaging with colloids. *Current Opinion in Colloid and Interface Science* **2012**, *17* (1), 13–22.
17. deBoer, J. F.; Milner, T. E.; vanGemert, M. J. C.; Nelson, J. S., Two-dimensional birefringence imaging in biological tissue by polarization-sensitive optical coherence tomography. *Optics Letters* **1997**, *22* (12), 934–936.
18. Pasquina-Lemonche, L.; et al., The architecture of the Gram positive bacterial cell wall. *Nature* **2020**, *582* (7811), 294–297.
19. Phanphak, S.; Georgiades, P.; Li, R.; King, J.; Roberts, I. S.; Waigh, T. A., Super-resolution fluorescence microscopy study of the production of K1 capsules by *Escherichia coli*: Evidence for the differential distribution of the capsule at the poles and the equator of the cell. *Langmuir* **2019**, *35* (16), 5635–5646.
20. Morris, V. J.; Kirby, A. R.; Gunning, A. P., *Atomic Force Microscopy for Biologists*, 2nd ed. Imperial College Press: 2009.
21. Lau, P. C. Y.; Dutcher, J. R.; Beveridge, T. J.; Lam, J. S., Absolute quantitation of bacterial biofilm adhesion and viscoelasticity by microbead force spectroscopy. *Biophysical Journal* **2009**, *96* (7), 2935–2948.
22. Esteban-Ferrer, D.; Edwards, M. A.; Fumagalli, L.; Juarez, A.; Gomila, G., Electric polarization properties of single bacteria measured with electrostatic force microscopy. *ACS Nano* **2014**, *8* (10), 9843–9849.
23. Mertz, J., *Introduction to Optical Microscopy*. Cambridge University Press: 2019.
24. Drescher, K.; Dunkel, J.; Nadell, C. D.; Bassler, B. L., Architectural transitions in *Vibrio cholerae* biofilms at single-cell resolution. *Proceedings of the National Academy of Sciences of the United States of America* **2016**, *113* (14), E2066–E2072.
25. Hauth, J.; Chodorski, J.; Wirsen, A.; Ulber, R., Improved FRAP measurements on biofilms. *Biophysical Journal* **2020**, *118* (10), 2354–2365.
26. Golding, I.; Paulsson, J.; Zawilski, S. M.; Cox, E. C., Real-time kinetics of gene activity in individual bacterial cells. *Cell* **2005**, *123* (6), 1025–1036.
27. Zhao, X.; Hilliard, L. R.; Mechery, S. J.; Wang, Y.; Bagwe, R. P.; Jin, S.; Tan, W., A rapid bioassay for single bacterial cell quantitation using bioconjugated nanoparticles. *Proceedings of the National Academy of Sciences of the United States of America* **2004**, *101* (42), 15027–15032.

28. Aglott, J. W.; Hardie, K. R., Fluorescent nanosensors reveal dynamics in pH gradients during biofilm formation. *npj Biofilms and Microbiomes* **2021**, *7*, 50.
29. Li, R.; Georgiades, P.; Cox, H.; Phanphak, S.; Roberts, I. S.; Waigh, T. A.; Lu, J. R., Quenched stochastic optical reconstruction microscopy (qSTORM) with graphene oxide. *Scientific Reports* **2018**, *8* (1), 16928.
30. Shi, H.; Shi, Q.; Grodner, B.; Lenz, J. S.; Zipfel, W. R.; Brito, I. L.; de Vlaminck, I., Highly multiplexed spatial mapping of microbial communities. *Nature* **2020**, *588* (7839), 676–681.
31. Celli, J. P.; et al., *Helicobacter pylori* moves through mucus by reducing mucin viscoelasticity. *Proceedings of the National Academy of Sciences of the United States of America* **2009**, *106* (34), 14321–14326.
32. Cox, H.; Georgiades, P.; Xu, H.; Waigh, T. A.; Lu, J. R., Self-assembly of mesoscopic peptide surfactant fibrils investigated by STORM super-resolution fluorescence microscopy. *Biomacromolecules* **2017**, *18* (11), 3481–3491.
33. Berk, V.; Fong, J. C. N.; Dempsey, G. T.; Develioglu, O. N.; Zhuang, X.; Liphardt, J.; Yildiz, F. H.; Chu, S., Molecular architecture and assembly principles of *Vibrio cholerae* biofilms. *Science* **2012**, *337* (6091), 236–239.
34. Wagner, M.; Horn, H., OCT in biofilm research: A comprehensive review. *Biotechnology and Bioengineering* **2017**, *114* (7), 1386.
35. Gierl, L.; Stoy, K.; Faina, A.; Hora, H.; Wagner, M., An open-source robotic platform that enables automated monitoring of replicate biofilm cultivation using OCT. *npj Biofilms and Microbiomes* **2020**, *6* (1), 18.
36. White, B. R.; et al., *In vivo* dynamic human retinal blood flow imaging using ultra-high-speed spectral domain optical Doppler tomography. *Optics Express* **2003**, *11* (25), 3490–3497.
37. Murphy, D. B.; Davidson, M. W., *Fundamentals of Light Microscopy and Electronic Imaging*, 2nd ed. Wiley: 2012.
38. Martinez, V. A.; et al., Differential dynamic microscopy: A high-throughput method for characterizing the motility of microorganisms. *Biophysical Journal* **2012**, *103* (8), 1637–1647.
39. Popescu, G., *Quantitative Phase Imaging of Cells and Tissues*. McGraw-Hill: 2011.
40. Miller, S. D.; Haddock, S. H. D.; Elvidge, C. D.; Lee, T. F., Detection of a bioluminescent milky sea from space. *Proceedings of the National Academy of Sciences of the United States of America* **2005**, *102* (40), 14181–14184.
41. Hill, S. C.; et al., Real-time measurement of fluorescence spectra from single airborne biological particles. *Field Analytical Chemistry and Technology* **1999**, *3* (4–5), 221–239.
42. Giovannetti, V.; Lloyd, S.; Maccone, L., Advances in quantum metrology. *Nature Photonics* **2011**, *5* (4), 222–229.
43. Chao, J. L.; Perevedentseva, E.; CHung, P. H.; Liu, K. K.; Cheng, C. Y.; Chang, C. C.; Cheng, C. L., Nanometer-sized diamond particle as a probe for biolabelling. *Biophysical Journal* **2007**, *93* (6), 2199–2208.
44. Le Sage, D.; et al., Optical magnetic imaging of living cells. *Nature* **2013**, *496* (7446), 486.

45. Taylor, M. A.; Janousek, J.; Daria, V.; Knittel, J.; Hage, B.; Bachor, H. A.; Bowen, W. P., Biological measurement beyond the quantum limit. *Nature Photonics* **2013**, *7* (3), 229–233.
46. Taylor, M. A.; Bowen, W. P., Quantum metrology and its application in biology. *Physics Reports* **2016**, *615* (5700), 1–59.
47. Lubin, G.; Oron, D.; Rossman, U.; Tenne, R.; Yallapragada, V. J., Photon correlations in spectroscopy and microscopy. *ACS Photonics* **2022**, *9*, 2891–2904.
48. Parson, W. W., *Modern Optical Spectroscopy*, 2nd ed. Springer: 2015.
49. Lee, K. S.; et al., An automated Raman-based platform for the sorting of live cells by functional properties. *Nature Microbiology* **2019**, *4* (6), 1035–1048.
50. Hill, E. H.; Marzan, L. M., Toward plasmonic monitoring of surface effects on bacterial quorum sensing. *Current Opinion in Colloid and Interface Science* **2017**, *32*, 1–10.
51. Bodelon, G.; et al., Detection and imaging of quorum sensing in *P. aeruginosa* biofilm communities by surface-enhanced resonance Raman scattering. *Nature Materials* **2016**, *15* (11), 1203–1211.
52. McNeil-Watson, F.; Tscharnuter, W.; Miller, J., A new instrument for the measurement of very small electrophoretic mobilities using phase analysis light scattering (PALS). *Colloids and Surfaces A* **1998**, *140* (1–3), 53–57.
53. Waz, N. T.; et al., Influence of the polysaccharide capsule on the bactericidal activity of indolicidin on *Streptococcus pneumoniae*. *Frontiers in Microbiology* **2022**, *13*, 898815.
54. Ashkin, A.; Dziedzic, J. M., Optical trapping and manipulation of viruses and bacteria. *Science* **1987**, *235* (4795), 1517–1520.
55. Uthe, B.; Sader, J. E.; Pelton, M., Optical measurement of the picosecond fluid mechanics in simple liquids generated by vibrating nanoparticles: A review. *Reports on Progress in Physics* **2022**, *85* (10), 103001.
56. Galy, O.; Latour-Lambert, P.; Zrelli, K.; Ghigo, J. M.; Beloin, C.; Henry, N., Mapping of bacterial biofilm local mechanics by magnetic microparticle actuation. *Biophysical Journal* **2012**, *103* (6), 1400–1408.
57. Haward, S. J., Microfluidic extensional rheometry using stagnation point flow. *Biomicrofluidics* **2016**, *10* (4), 043401.
58. Waigh, T. A., Advances in the microrheology of complex fluids. *Reports on Progress in Physics* **2016**, *79* (7), 074601.
59. Geisel, S.; Secchi, E.; Vermant, J., Experimental challenges in determining the rheological properties of bacterial biofilms. *Interface Focus* **2022**, *12* (6), 20220032.
60. Pavlovsky, L.; Younger, J. G.; Solomon, M. J., *In situ* rheology of *Staphylococcus epidermidis* bacterial biofilms. *Soft Matter* **2013**, *9* (1), 122–131.
61. Sabass, B.; Koch, M. D.; Liu, G.; Shaevitz, J. W., Force generation by groups of migrating bacteria. *Proceedings of the National Academy of Sciences of the United States of America* **2017**, *114* (28), 7266–7271.
62. Separovic, F.; Hofferek, V.; Duff, A. P.; McConville, M. J.; Sani, M. A., In-cell DNP NMR reveals multiple targeting effect of antimicrobial peptide. *Journal of Structural Biology: X* **2022**, *6* (12), 100074.

63. Seymour, J. D.; Gage, J. P.; Codd, S. L.; Gerlach, R., Anomalous fluid transport in porous media induced by biofilm growth. *Physical Review Letters* **2004**, *93* (19), 198103.
64. Van de Vyver, H.; et al., A novel mouse model of *Staphylococcus aureus* vascular graft infection: Noninvasive imaging of biofilm development in vivo. *The American Journal of Pathology* **2017**, *187* (2), 268.
65. Dunham, S. J. B.; Ellis, J. F.; Sweedler, J. V., Mass spectrometry imaging of complex microbial communities. *Accounts of Chemical Research* **2017**, *50* (1), 96–104.
66. Etayash, H.; Khan, M. F.; Kaur, K.; Thundat, T., Microfluidic cantilever detects bacteria and measures their susceptibility to antibiotics in small confined volumes. *Nature Communications* **2016**, *7*, 12947.
67. Azulay, D. N.; et al., Multiscale X-ray study of *Bacillus subtilis* biofilms reveals interlinked structural hierarchy and elemental heterogeneity. *Proceedings of the National Academy of Sciences of the United States of America* **2022**, *119* (4), e2118107119.
68. Gong, H.; et al., Aggregated amphiphilic antimicrobial peptides embedded in bacterial membranes. *ACS Applied Materials and Interfaces* **2020**, *12* (40), 44420–44432.
69. Dogsa, I.; Kriechbaum, M.; Stopar, D.; Laggner, P., Structure of bacterial extracellular polymeric substances at different pH values as determined by SAXS. *Biophysical Journal* **2005**, *89* (4), 2711–2720.
70. Chemla, Y., A new study of bacterial motion: Superconducting quantum interference device microscopy of magnetotactic bacteria. *Biophysical Journal* **1999**, *76* (6), 3323–3330.
71. Hsiao, W. W. W.; Hui, Y. Y.; Tsai, P. C.; Chang, H. C., Fluorescent nanodiamond: A versatile tool for long-term cell tracking, super-resolution imaging, and nanoscale temperature sensing. *Accounts of Chemical Research* **2016**, *49* (3), 400–407.
72. Rosenblatt, C.; Torres de Araujo, F. F.; Frankel, R. B., Birefringence determination of magnetic moments of magnetotactic bacteria. *Biophysical Journal* **1982**, *40* (1), 83–85.
73. Faivre, D.; Fischer, A.; Garcia-Rubio, I.; Mastrogiacomo, G.; Gehring, A. U., Development of cellular magnetic dipoles in magnetotactic bacteria. *Biophysical Journal* **2010**, *99* (4), 1268–1273.
74. Zvelebil, M. J.; Baum, J. O., *Understanding Bioinformatics*. Garland Science: 2007.
75. Ceri, H.; Olson, M. E.; Stremick, C.; Read, R. R.; Morck, D.; Buret, A., The calgary biofilm device: New technology for rapid determination of antibiotic susceptibilities of bacterial biofilms. *Journal of Clinical Microbiology* **1999**, *37* (6), 1771–1776.
76. Evanko, D., Measuring a bacteria's sweet tooth. *Nature Methods* **2005**, *2* (2), 88–89.
77. Whiteley, M.; Bungera, G.; Bumgarner, R. E.; Passek, M. R.; Teltzel, G. M.; Lory, S.; Greenberg, E. P., Gene expression in *P. aeruginosa* biofilms. *Nature* **2001**, *413* (6858), 860.

78. Edgar, R.; et al., High-sensitivity bacterial detection using biotin-tagged phage and quantum-dot nanocomplexes. *Proceedings of the National Academy of Sciences of the United States of America* **2006**, *103* (13), 4841–4845.
79. Viovy, J. L., Electrophoresis of DNA and other polyelectrolytes: Physical mechanisms. *Reviews of Modern Physics* **2000**, *72* (3), 813–872.
80. Castellarnau, M.; Errachid, A.; Madrid, C.; Juarez, A.; Samitier, J., Dielectrophoresis as a tool to characterize and differentiate isogenic mutants of *Escherichia coli*. *Biophysical Journal* **2006**, *91* (10), 3937–3945.
81. Dague, E.; Duval, J.; Jorand, F.; Thomas, F.; Gaboriaud, F., Probing surface structures of *Shewanella spp.* by microelectrophoresis. *Biophysical Journal* **2006**, *90* (7), 2612–2621.
82. Berg, H. C.; Turner, L., Torque generated by the flagellar motor of *Escherichia coli*. *Biophysical Journal* **1993**, *65* (5), 2201–2216.
83. Lewandowki, Z.; Beyenal, H., *Fundamentals of Biofilm Research*. CRC Press: 2013.
84. Chen, C.; Smye, S. W.; Robinson, M. P.; Evans, J. A., Membrane electroporation theories: A review. *Medical and Biological Engineering and Computing* **2006**, *44* (1–2), 5–14.

14 Machine Learning

A vast number of bacterial species, ~10^{12}, are thought to exist on planet Earth and advanced computational techniques will be needed to handle the vast amount of data required to make their description a tractable problem. Data sets could be genomic (their DNA sequences), connectomic (the interactions of their gene circuits) or other varieties of omics (proteomic, glycomic, lipidomic, etc.). Data could also describe phenotypic identification based on high throughput microscopy or spectroscopy data that can be used to classify bacteria based on cell morphology, cellular aggregation or motility.

Machine learning (ML) provides a range of new tools that can be applied to bacterial biophysics to automate data analysis, such as the segmentation of images or the quantification of time series data. The field can be separated into classical and modern stages of development i.e. traditional ML techniques[1] versus deep learning neural networks (NNs).[2] Both subfields provide valuable computational tools that can be successfully applied to understand the behaviour of bacteria. Physically constrained models based on ML are particularly powerful i.e. physical laws provide constraints via Bayesian priors to help optimise the amount of information that can be extracted from data sets.[3]

Thomas Bayes' theorem is generally valid for the conditional probabilities measured during an experiment,

$$P(A/B) = \frac{P(A)P(B/A)}{P(B)}, \tag{14.1}$$

where $P(A/B)$ is the probability of the event A given B has happened, $P(B/A)$ is the probability of B given A has happened, $P(A)$ is the probability of A and $P(B)$ is the probability of B. This simple equation gives rise to the Bayesian interpretation of experimental science (a new paradigm[4]) and provides a quantitative framework to incorporate prior information into probabilistic predictions of experimentally determined variables.[1,3] Bayesian techniques are used widely in both ML and physics e.g. to determine the states of prestress in peptide gels.[5,6]

Deep learning NNs are a relatively recent development in ML (a 2018 Turing award) and can provide superhuman performance in a number of applications that previously were considered unautomatable e.g. image segmentation or speech recognition. Furthermore, some long standing intractable problems in biophysics suddenly have some new tools that have met with some startling successes e.g. to tackle the glass transition,[7] protein folding[8] or non-Markovian particle tracking.[9]

With NNs, the architecture chosen is crucial for their performance. Deep-learning NNs have multiple layers of neurons between the inputs and outputs, which are found

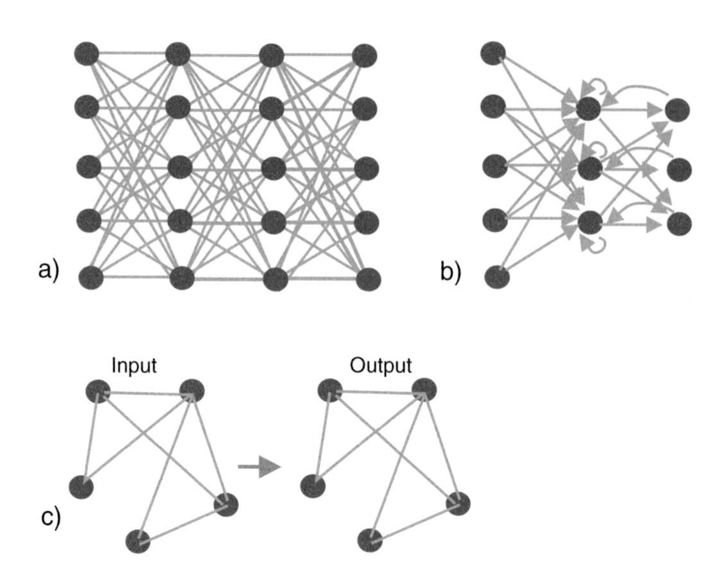

Figure 14.1 Comparison of three different neural network architectures. (a) *Feedforward neural network* (FNN), (b) *recurrent neural network* (RNN) and (c) *graph neural network* (GNN).[10,11]

to greatly increase the range of behaviours they can describe. The artificial neurons are connected together using non-linear activation functions and the non-linearities are crucial for the ability of the networks to flexibly perform a wide range of calculations. Linear activation would negate the use of extra layers of neurons, since the deep layers would become redundant. The *universal approximation theorem* states that a very wide range of non-linear functions can in turn be described by deep learning networks i.e. deep learning NNs can act as generalised non-linear function generators. Training of NNs is a key stage that determines their abilities and is typically performed using the back propagation algorithm.[10] Standard deep learning architectures include *convolutional neural networks* (CNNs), *feedforward neural networks* (FNNs), *generative adversarial networks* (GANs), *recurrent neural networks* (RNNs), *graph NNs*, *transformers* and *reinforcement learning NNs*[10] (Figure 14.1). The field is rapidly developing and each architecture has its own strengths and weaknesses.

Although deep learning NNs can be very effective, many older ML tools are still competitive for some tasks e.g. *Gaussian mixture models* are widely used in spectroscopy. Some of the classical ML tools have the advantage that their algorithms are more intuitive, which can facilitate rigorous statistical mechanics applications. Specifically, Bayesian ML techniques highlight the stochastic nature of the phenomena modelled and provide a powerful methodology to include physical constraints on models.[3] In general, NNs predominantly act as non-linear black boxes that can quickly perform classification tasks, although it is not necessarily clear how they perform the tasks.

ML can be used to discover new phases of matter based on subtle combinations of non-standard order parameters, which can be useful for medical diagnostics as well as fundamental studies.[12] ML is also very good in classification problems on

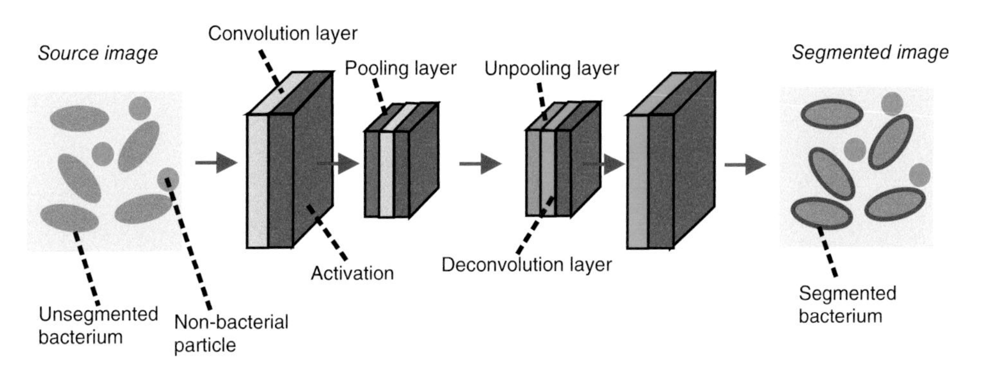

Figure 14.2 Image segmentation of bacteria (orange) using a *convolutional neural network* (CNN).

multidimensional data sets e.g. mixture models with multivariate data. For example, they can be used to define phases of swarming bacteria.[13]

Convolutional neural networks were originally inspired by the architectures of neural processing circuits in animal retinas.[14] CNNs can accurately describe the spatial relationships between pixels in two-dimensional (2D) or three-dimensional (3D) images and can automatically learn reduced abstract representations of key features in images to allow robust image classification. CNNs are a very effective type of deep learning used for segmenting objects during image analysis (Figure 14.2), but can be more generally used e.g. for pattern detection in time series data. Once the objects are segmented in images using CNNs, their coordinates can be linked together to form tracks (Chapter 2). Segmentation from CNNs is thus good for tracking non-Gaussian shaped objects, such as bacteria,[15] but a slight reduction in performance is expected compared with software that assumes Gaussian-shaped objects under some circumstances (an informative prior has been lost, in the terminology of Bayesian analysis, due to the generalisation of the particles' shapes). CNNs are particularly useful for the segmentation of aspherical particles, which is more challenging for Gaussian trackers.[16] CNNs can be naturally generalised to 3D image segmentation and thus allow 3D tracking. Another advantage is that CNNs require less parameter optimisation than Gaussian trackers, which can be very time-consuming. CNNs have been used to segment bacteria embedded in biofilms,[17] which is a demanding application due to the overlapping images of the bacteria and the reduction in image contrast due to multiple scattering.

As well as the segmentation of images, the tracks themselves created from segmented movies can also be segmented using deep learning NNs. Thus, tracks of cells can be segmented in terms of different types of non-Markovian motion,[9,18] which solves some important questions in the classification of cell motility (Figure 14.3). FNNs were used for segmentation of tracks based on fractional Brownian motion models, which subsequently led to the development of multi-fractal models for bacterial motion due to spatial and temporal heterogeneity (Section 2.2). Other models for anomalous transport could also be used to train the NNs (Chapter 2), e.g. Lévy walks, dependent on the system analysed.

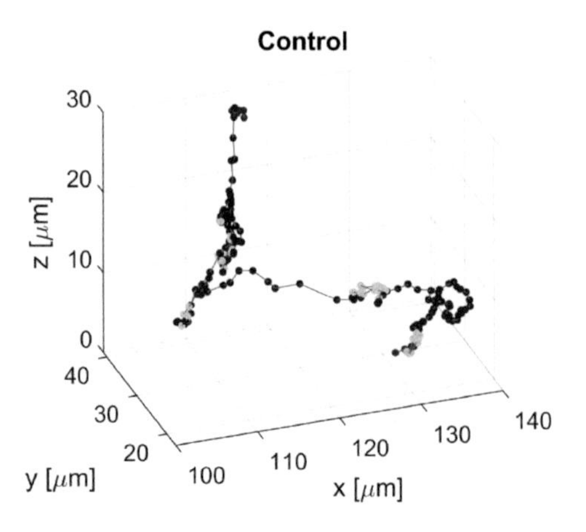

Figure 14.3 *Dynamic segmentation* of the non-Markovian motility of cells is possible using feedforward deep learning neural networks. Tracks were obtained from hemocytes in developing drosophila embryos. The dark violet regions indicate persistent motion ($H > 0.5$, where H is the Hurst exponent), whereas the yellow regions are anti-persistent motion ($H < 0.5$).[19] A similar analysis has been subsequently performed with bacteria (*E. coli*).

Neural networks were used to analyse images from quantitative phase microscopy of growing bacteria on agar plates.[20] The differences between images were analysed using a CNN to detect the growth of the bacteria and care was taken to suppress non-bacterial objects in the images. A high accuracy was found in identifying species based on their growth kinetics, which was in part determined by the length of the lag phase of the bacteria i.e. this was a key feature identified by the CNN (Section 16.1). Understanding the lag phase is important in a wide range of applications including food science.[21] NNs have also been used to identify bacterial and fungal pathogens from blood cultures.[22]

Convolutional neural networks were used to count bacterial cells in optical microscopy images and were found to be less prone to error and much faster than alternative techniques.[23] CNNs were also used to differentiate between aggregates of myxobacterial species (fruiting bodies) based on optical microscopy images.[24] Automated identification has been performed of optical microscopy images of the fruiting bodies of 30 different genera of *Myxococcales* using a CNN.[25] Furthermore, CNNs have been used to detect biofilms in nasal samples.[26]

There are a large number of applications of ML in molecular biology, such as sequence alignment (finding similar sequences in DNA sequences from different sources), gene detection, analysing structure/function relationships, gene expression, building phylogenetic tress, evolutionary biology and predicting secondary and tertiary structures of proteins.[27] Unsupervised ML has been used to search for independent modules in the transcriptome of *Bacillus subtilis*.[28] CNN models have been used to predict the properties of small peptide antimicrobials and have shown success in *de novo* peptide design.[29] Deep learning has also been used to repurpose known drugs to see if they will function well against bacteria.[30] NNs have also been used to predict signalling peptides in both Gram-positive and Gram-negative bacteria.[31]

Numerous applications have been found for deep learning NNs in *photonics techniques* for both image analysis and spectroscopy (Raman, infrared and hyperspectral).[32] CNNs have been trained on the Raman spectra of infections to directly assign the appropriate antibiotics (Section 13.3.3). Spores have been separated from vegetative cells based on Raman spectra. The identification of bacteria has also been possible using Fourier transform infrared (FTIR) spectroscopy with ML. Early work with NNs was used to identify urinary tract infections (UTIs) based on both Raman and FTIR data.[33] Deep learning with surface-enhanced Raman scattering (SERS) discriminated antibiotic responses at an order of magnitude smaller concentrations of antibiotics than conventional minimum inhibitory concentrations (MICs) and thus can more easily detect antibiotic-resistant bacteria.[34]

ML has also shown promise with *mass spectrometry* data (Section 13.4.3, e.g. MALDI-TOF) for microbial identification and antimicrobial susceptibility testing.[35] Imaging flow cytometry data of eukaryotic cell particle interactions was also analysed by a CNN.[36] It is expected that this methodology will also be effective to study bacterial/eukaryotic cell interactions.

Motility mechanisms can be used to classify bacteria (Chapter 2). A simple portable phase contrast microscope was used to classify tracks and was found to perform well in the simple task of identifying motile bacteria versus inanimate objects (based on the amplitude of motion).[37] Very simple segmentation was performed for the tracks and more advanced deep learning NNs would be needed to identify bacterial species with higher confidence based on their motility.[9]

Other varieties of NNs have also been used with bacteria e.g. GANs were used to classify pathogenic marine bacteria with Raman spectroscopy.[38] GANs can help reduce the amount of training data required in supervised learning.[39] GANs could be used to simulate the motility patterns of bacteria to improve statistics for modelling. They could also predict biofilm structures in different geometries if well-defined textures observed in training data are generalised.

Attention networks have produced industry-leading results for the folding patterns of proteins[8] (alphafold), comparable to the resolutions available from crystallography experiments. Folded structures of the entire range of genomically encoded proteins are now available for a wide range of organisms including species of bacteria.[40]

Applications of ML in active matter have been reviewed.[41] Highlights include the classification of dynamic phases in rafting bacteria (Figure 14.4). *Reinforcement learning* can be used to understand goal-oriented motility e.g. the search strategies of microorganisms. Reinforcement learning has also been used to optimise multi-agent-based models.[42] Collective dynamics in interacting populations have also been explored. An interesting analogy exists for swarms of bacteria to distributed control systems and it could be used for optimising strategies for nanobots used to clean bacterial infections (Chapter 30).

Automatic differentiation (AD) is a very effective tool for solving partial differential equations that is used to train NNs in the backpropagation algorithm.[43] AD is finding many applications in fluid mechanics and may thus be applied to bacteria.

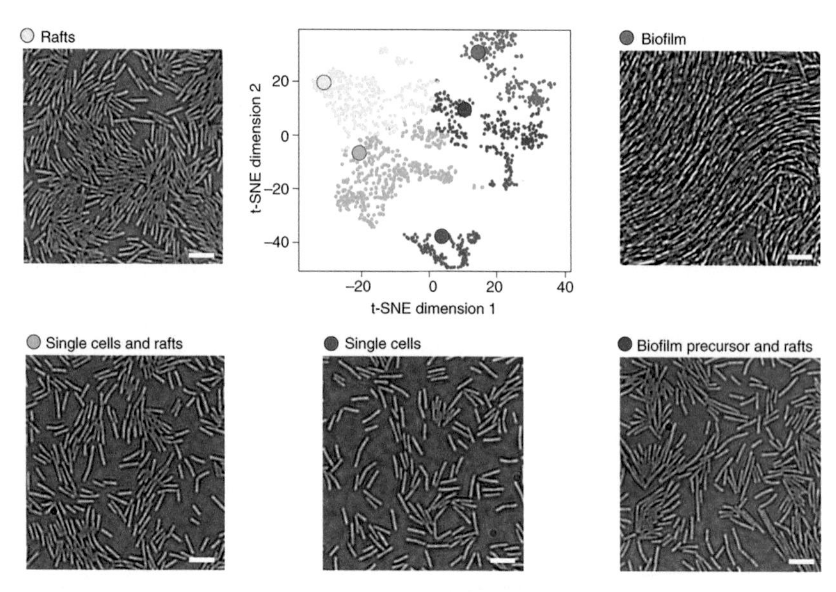

Figure 14.4 *Automatic classification* of motile phases of bacteria was possible using ML techniques combined with CNN segmentation.[41] Separate phases were discovered containing: rafts, single cells and rafts, single cells, biofilm precursors and rafts and biofilm.

Suggest Reading

Bishop, C. M., *Pattern Recognition and Machine Learning*. Springer: 2008. Accessible introduction to classical machine learning.

Chollet, F., *Deep Learning with Python*. Manning: 2018. Another practical introduction to neural network coding, slightly simpler than the Geron approach.

Geron, A., *Hands on Machine Learning with Scikit-learn, Kera and TensorFlow*. O'Reilly: 2019. The go to manual for practical aspects for coding machine learning algorithms and neural networks.

Goodfellow, I.; Bengio, Y.; Courville, A., *Deep Learning: Adaptive Computation and Machine Learning*. MIT Press: 2017. Good theoretical overview of deep learning including generative adversarial networks.

Murphy, K. P., *Probabilistic Machine Learning: An Introduction*. MIT Press: 2023. Excellent introduction to ML with an emphasis on Bayesian techniques.

Nielsen, A., *Practical Time Series Analysis: Prediction with Statistics and Machine Learning*. O'Reilly: 2019. Considers modern techniques to interpret time series data sets.

Zvelebil, M.; Baum, J. O., *Understanding Bioinformatics*. Garland Science: 2007. Excellent introduction to classical computational techniques in molecular biology.

References

1. Bishop, C. M., *Pattern Recognition and Machine Learning*. Springer: 2006.
2. Goodfellow, I.; Bengio, Y.; Courville, A., *Deep Learning*. MIT Press: 2016.
3. Murphy, K. P., *Probabilistic Machine Learning: An Introduction*. MIT Press: 2022.
4. Jaynes, E. T.; Bretthorst, G. L., *Probability Theory: The Logic of Science*. Cambridge University Press: 2003.
5. Cox, H.; Xu, H.; Waigh, T. A.; Lu, J. R., Single-molecule study of peptide gel dynamics reveals states of prestress. *Langmuir* **2018**, *34* (48), 14678–14689.
6. Cox, H.; Cao, M.; Xu, H.; Waigh, T. A.; Lu, J. R., Active modulation of states of prestress in self-assembled short peptide gels. *Biomacromolecules* **2019**, *20* (4), 1719–1730.
7. Bapst, V.; et al., Unveiling the predictive power of static structure in glassy systems. *Nature Physics* **2020**, *16* (4), 448–454.
8. Jumper, J.; et al., Highly accurate protein structure prediction with Alphafold. *Nature* **2021**, *596* (7873), 583–589.
9. Han, D.; Korabel, N.; Chen, R.; Johnston, M.; Gavrilova, A.; Allan, V. J.; Fedotov, S.; Waigh, T. A., Deciphering anomalous heterogeneous intracellular transport with neural networks. *eLife* **2020**, *9*, e52224.
10. Drori, I., *The Science of Deep Learning*. Cambridge University Press: 2022.
11. Geron, A., *Hands-on Machine Learning with Scikit-learn, Keras and TensorFlow*. O'Reilly: 2019.
12. Hartmann, R.; Singh, P. K.; Pearce, P.; Mok, R.; Song, B.; Diaz-Pascual, F.; Dunkel, J.; Drescher, K., Emergence of three-dimensional order and structure in growing biofilms. *Nature Physics* **2019**, *15* (3), 251–256.
13. Jeckel, H.; et al., Learning the space-time phase diagram of bacterial swarm expansion. *Proceedings of the National Academy of Sciences of the United States of America* **2019**, *116* (5), 1489–1494.
14. Dowling, J. E., *The Retina: An Approachable Part of the Brain*. Belknap Harvard: 2012.
15. Newby, J. M.; Schaefer, A. M.; Lee, P. T.; Forest, M. G.; Lai, S. K., Convolutional neural networks automate detection for tracking of submicron-scale particles in 2D and 3D. *Proceedings of the National Academy of Sciences of the United States of America* **2018**, *115* (36), 9026–9031.
16. Helgadottir, S.; Argua, A.; Volpe, G., Digital video microscopy enhanced by deep learning. *Optica* **2019**, *6* (4), 506.
17. Zhang, M.; Zhang, J.; Wang, Y.; Wang, J.; Achimovich, A. M.; Acton, S. T.; Gahlmann, A., Non-invasive single-cell morphometry in living bacterial biofilms. *Nature Communications* **2020**, *11* (1), 6151.
18. Korabel, N.; Waigh, T. A.; Fedotov, S.; Allan, V. J., Non-Markovian intracellular transport with sub-diffusion and run-length dependent detachment rate. *PLOS One* **2018**, *13* (11), e0207436.

19. Korabel, N.; Clemente, G. D.; Han, D.; Feldman, F.; Millard, T. H.; Waigh, T. A., Hemocytes in *Drosophila melanogaster* embryos move via heterogeneous anomalous diffusion. *Communications Physics* **2022**, *5* (1), 269.
20. Wang, H.; et al., Early detection and classification of live bacteria using time lapse coherent imaging and deep learning. *Light: Science and Applications* **2020**, *9* (1), 118.
21. Swinnen, I. A. M.; Bernaerts, K.; Dens, E. J. J.; Geeraerd, A. H.; van Impe, J. F., Predictive modelling of the microbial lag phase: A review. *International Journal of Food Microbiology* **2004**, *94* (2), 137–159.
22. Maquelin, K.; et al., Prospective study of the performance of vibrational spectroscopies for rapid identification of bacterial and fungal pathogens recovered from blood cultures. *Journal of Clinical Microbiology* **2003**, *41* (1), 324–329.
23. Tamiev, D.; Furman, P. E.; Reuel, N. F., Automated classification of bacterial cell sub-populations with CNNs. *PLOS One* **2020**, *15* (10), e0241200.
24. Sajedi, H.; Mohammadipanah, F.; Pashaei, A., Image-processing based taxonomy analysis of bacterial macromorphology using machine-learning model. *Multimedia Tools and Applications* **2020**, *79* (43–44), 32711–32730.
25. Sajedi, H.; Mohammadipanah, F.; Pashaei, A., Automated identification of myxobacterial genera using CNN. *PLOS One* **2019**, *9* (1), 18238.
26. Dimauro, G.; Deperte, F.; Maglietta, R.; Bove, M.; La Gioia, F.; Reno, V.; Simone, L.; Gelardi, M., A novel approach for biofilm detection based on CNN. *Electronics* **2020**, *9* (6), 881.
27. Zvelebil, M. J.; Baum, J. O., *Understanding Bioinformatics*. Garland Science: 2007.
28. Rychel, K.; Sastry, A. V.; Palsson, B. O., Machine learning uncovers independently regulated modules in the *Bacillus subtilis* transcriptome. *Nature Communications* **2020**, *11* (1), 6338.
29. Yan, J.; Bhadra, P.; Li, A.; Sethiya, P.; Qin, L.; Tai, H. K.; Wong, K. H.; Siu, S. W. I., Deep-Am PEP30: Improve short antimicrobial peptide predictions with deep learning. *Molecular Therapy: Nucleic Acids* **2020**, *20*, 882–894.
30. Stokes, J. M.; et al., A deep learning approach to antibiotic discovery. *Cell* **2020**, *180* (4), 688–702.
31. Bendtsen, J. D.; Nielsen, H.; von Heijne, G.; Brunak, S., Improved prediction of signal peptides: Signal P 3.0. *Journal of Molecular Biology* **2004**, *340* (4), 783–795.
32. Fang, J.; Swain, A.; Unni, R.; Zhang, Y., Decoding optical data with machine learning. *Lasers and Photonics Reviews* **2021**, *15* (2), 2000422.
33. Goodacre, R.; Timmins, E. M.; Burton, R.; Kaderbhai, N.; Woodward, A. M.; Kell, D. B.; Rooney, P. J., Rapid identification of urinary tract infection bacteria using hyperspectral whole-organism fingerprinting and artificial neural networks. *Microbiology* **1998**, *144* (Pt 5), 1157–1170.
34. Thrift, W. J.; et al., Deep learning analysis of vibrational spectra of bacterial lysate for rapid antimicrobial susceptibility testing. *ACS Nano* **2020**, *14* (11), 15336–15348.
35. Weis, C. V.; Jutzeler, C. R.; Borgwardt, K., Machine learning for microbial identification and antimicrobial susceptibility testing on MALDI-TOF mass spectra: A systematic review. *Clinical Microbiology and Infections* **2020**, *26* (10), 1310–1317.

36. Mochalova, E. N.; Kotov, I. A.; Rozenberg, J. M.; Nikitin, M. P., Precise quantitative analysis of cell targeting by particle-based agents using imaging flow cytometry and convolutional neural network. *Cytometry* **2020**, *97* (3), 279–287.
37. Riekeles, M.; Schirmak, J.; Schulze-Makuch, D., Machine learning algorithms applied to identify microbial species by their motility. *Life* **2021**, *11* (1), 44.
38. Yu, S.; Li, H.; Li, X.; Fu, Y. V.; Liu, F., Classification of pathogens by Raman spectroscopy combined with generative adversarial networks. *Science of the Total Environment* **2020**, *726*, 138477.
39. Foster, D., *Generative Deep Learning: Teaching Machines to Paint, Write, Compose and Play*. O'Reilly: 2019.
40. alphafold, G. Alpha fold protein structure database. alphafold.ebi.ac.uk.
41. Cichos, F.; Gustavsson, K.; Mehlig, B.; Volpe, G., Machine learning for active matter. *Nature Machine Intelligence* **2020**, *2*, 94–103.
42. Hou, H.; Gan, T.; Yang, Y.; Zhu, X.; Liu, S.; Guo, W.; Hao, J., Using deep reinforcement learning to speed up collective cell migration. *BMC Bioinformatics* **2019**, *20* (Suppl 18), 571.
43. Baydin, A. G.; Pearlmutter, B. A.; Radul, A. A.; Siskind, J. M., Automatic differentiation in machine learning: A survey. *Journal of Machine Learning Research* **2018**, *18* (153), 1–43.

PART II

SINGLE BACTERIA

Following the reductionist (bottom-up) approach, this section attempts to describe the properties of bacteria based on the behaviour of single cells. The first chapter (Chapter 15) is a crash course in bacterial microbiology for physical scientists. The following chapters investigate some key phenomena with single cells where physics can contribute to their understanding, focusing on the properties of individual cells: growth, motility, chemotaxis, molecular machines and capsules.

15 An Introduction to Bacteria

It is important for biophysicists to understand as much biology as possible, so they can elaborate on the simple colloidal models for bacteria introduced in Chapter 7 and better understand the impact of the complex internal structures of bacteria on measurements (the hidden variables). The challenge is for researchers to be reasonably well informed about the underlying molecular biology, without being totally overwhelmed by the complexity, obstructing progress with the physics.

15.1 Bacterial Cellular Biology

There are three branches to the tree of life on planet Earth and they are labelled *archaea*, *bacteria* and *eukaryote*[2] (Figure 15.1). Practically, this classification scheme is based on the DNA sequences of protein factories (ribosomal subunits) in the organisms' genomes and, as taxonomy goes, it is a very robust scheme. It moves beyond more qualitative phenotypic classification schemes (e.g. the bacteria are purple, so let's call them purple bacteria) and provides a coherent explanation for the interconnected nature of life due to evolution on planet Earth. All known organisms on planet Earth fall into one of these three classifications and they can then be placed into one of the three regions on an interconnected evolutionary tree. Two of the branches are types of prokaryotes (archaea and bacteria) and both can produce biofilms. Single-celled eukaryotes can also form biofilms, such as fungi (think of yeast for bread[3] or beer production), although the majority of eukaryotes do not.

Bacteria (Figure 15.2a) are living organisms and their DNA is contained within the cytoplasm of their cells, which is a single compartment delimited by their external cellular membrane (with no specialised subcompartments), whereas eukaryotes have compartmentalised DNA as well as an external membrane (and often other specialised compartments that are collectively called organelles) i.e. each eukaryotic cell has at least one internal membrane-covered nucleus that contains its DNA (Figure 15.2c). In addition to the much smaller size of typical bacteria (diameters of ~μm, an order of magnitude smaller than eukaryotic cells), this is the main characteristic used to differentiate bacteria from eukaryotes i.e. look at one of their cells under a microscope and if a nucleus is seen in it, the cell is almost certainly eukaryotic. Archaea superficially resemble bacteria in their size and morphology (Figure 15.2b).

Bacteria can be compared with other cells from microbial and multicellular organisms. Table 15.1 shows a standard bacterium (*Escherichia coli*) compared with yeast,

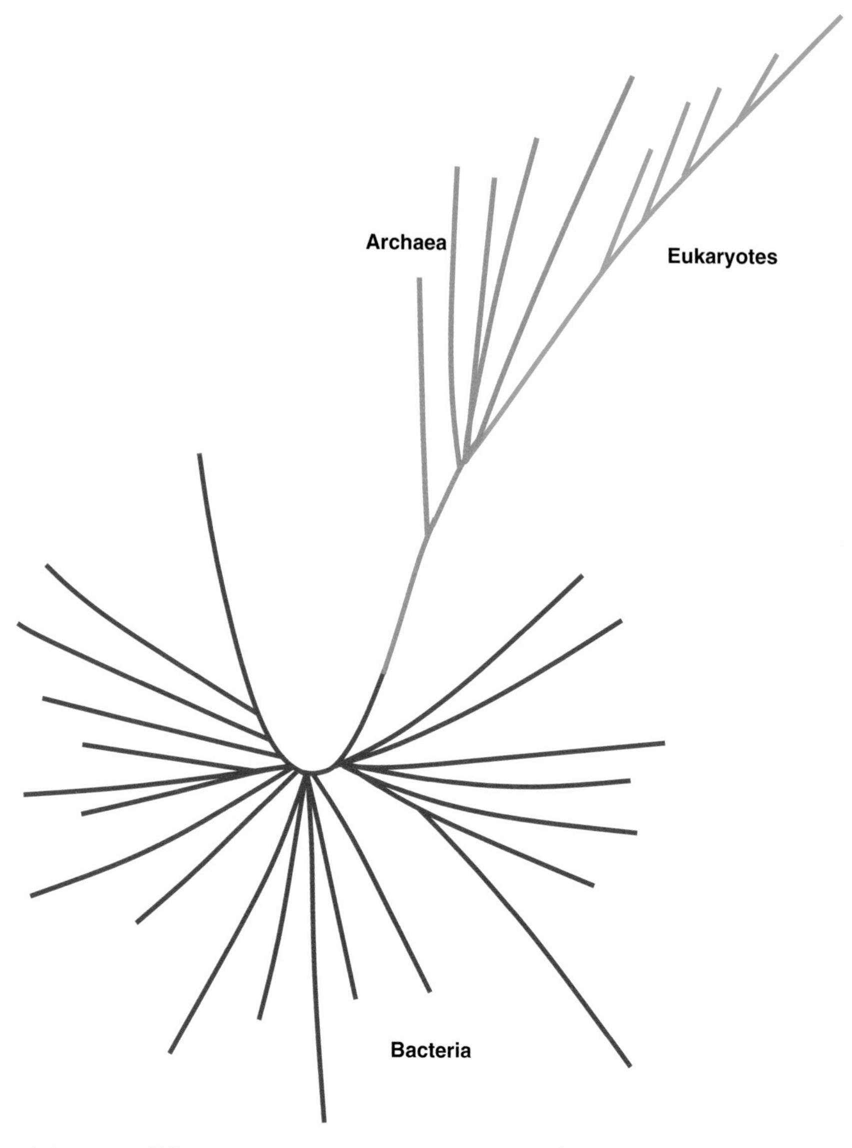

Figure 15.1 The evolutionary tree of life on Earth highlighting species of microorganism[1] (humans, animal and plants are found in the small blue Eukaryotes section on the top right).

human, malaria and worm cells. The most striking feature in this comparison is the relatively small size of the bacterial cells. Exceptions to this rule do exist, such as recently discovered marine bacterial cells that are up to 2 cm in length, but these are rare.[4]

Bacteria are the most successful life form on planet Earth in terms of biomass (only second to plants), variety and the extent of habitats they have colonised.[5] There are estimated to be 4–6 × 10^{30} prokaryotic cells (bacteria and archaea) on planet Earth based on biomass calculations.[6] Such huge numbers are hard to comprehend; they are

Table 15.1 Comparison of the sizes of different types of microorganisms with human cells, including bacteria (*E. coli*), fungi (yeast), human cells, malaria and worms.

Type of organisms	Size of cells	Size of organisms	Notes
Bacterium (e.g. *E. coli*)	1–2 μm	Single cells	Biofilms are not considered separate organisms.
Human	~20 μm, although neurons can be much larger (up to 1 m)	0.2–2 m	Multi-cellular organisms, ~200 types of cell differentiate during morphogenesis.
Saccharomyces cerevisiae	5–10 μm	Single cells	Yeast for bread and beer making.
Plasmodium falciparum	1–80 μm dependent on stage of life cycle	Single cells	Malaria causing eukaryotic parasite.
Trypanosoma brucei	8–50 μm	Single cells	Eukaryotic parasite that causes sleeping sickness, which depends on symbiotic bacteria.

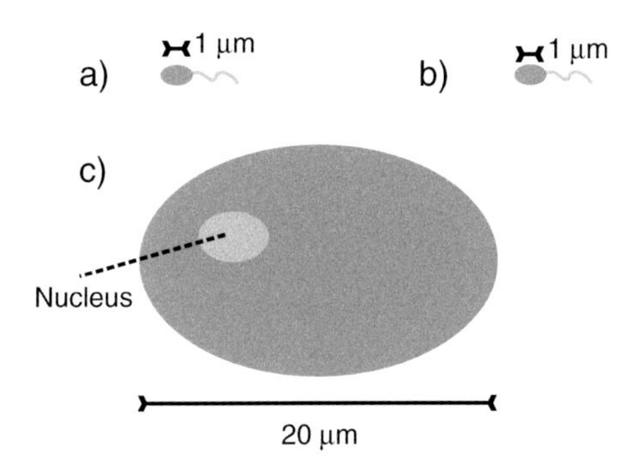

Figure 15.2 Rough schematic diagram showing the relative size of a standard (a) bacterial cell (e.g. *Escherichia coli*), (b) archaeal cell and (c) eukaryotic cell (e.g. human epithelial). Many exceptions to this sizing occur e.g. axon cells in blue whales can be 40 m in length.

on the order of the diameter of the known observable universe measured in millimetres. Archaea are situated between bacteria and eukaryotes in terms of their taxonomy (Figure 15.2b). Superficially, they resemble bacteria in that they are relatively small single-celled microorganisms, although they have some biochemical features in common with eukaryotes (e.g. the enzymes they use for transcription and translation of their DNA) and some unique features (e.g. ether lipids in their membranes).

Normally, bacteria divide and form identical clones of themselves. They can have sex with one another in the form of horizontal gene transfer (HGT), although there are no male/female phenotypes and HGT tends to be much less frequent than clonal division. A memorable quote is that 'bacteria may not have sex often, but when they do, it can be really good'.[7] However, mutation and DNA uptake (from a diverse range of sources including HGT) can also lead to bacterial evolution, which due to the high

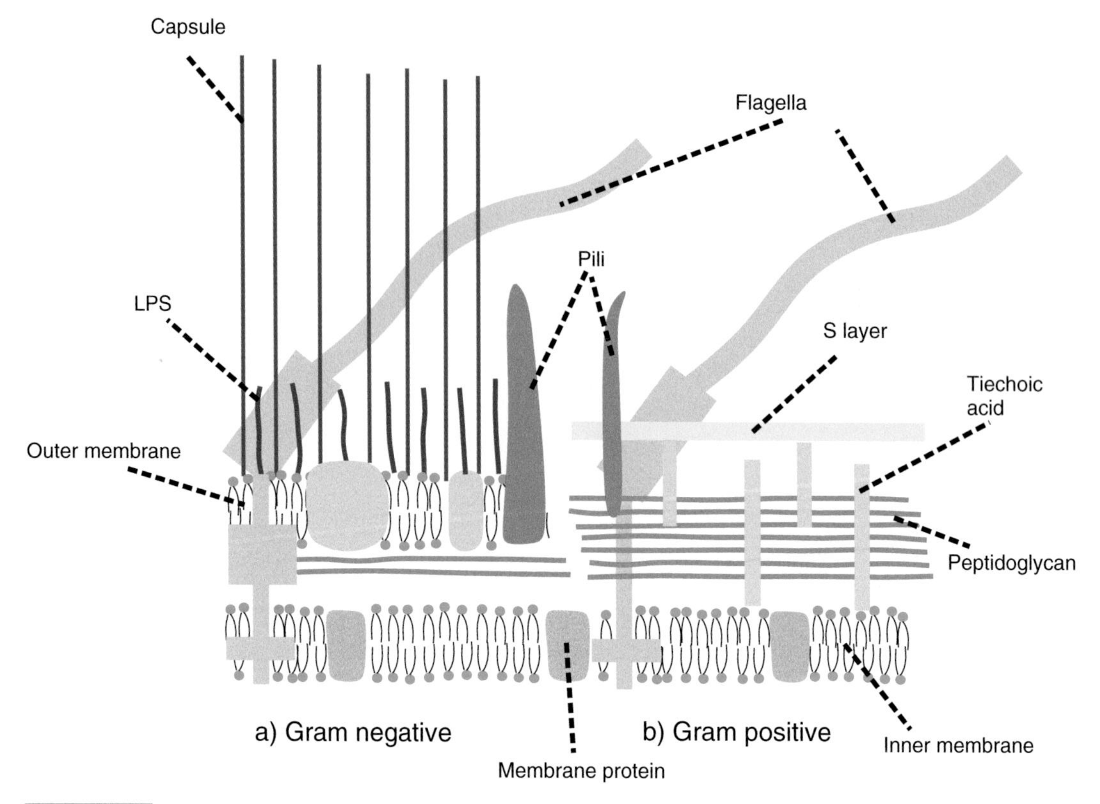

Figure 15.3 Schematic diagram of the structures of the membranes of typical (a) *Gram-negative* bacteria and (b) *Gram-positive* bacteria.[10]

division rate of bacteria can lead to significant changes occurring to their genetics in real time. Darwin's theory of evolution can thus be used to create quantitative theories that are tested against bacterial experiments performed in real time.[8,9]

Traditionally, bacteria were classified based on their phenotypes (the observed characteristics) and two of the most obvious descriptions of them under a microscope are whether they are *Gram positive* or *Gram negative* (Figure 15.3). This classification scheme is dependent on whether the bacteria have one or two external cell membranes (which are called Gram positive or Gram negative, respectively) and thus can hold on to a Gram stain. Gram-positive cells are further subdivided into whether they have *acid-fast* cell walls due to mycolic acids, which are an important virulence factor in some medically important bacteria, such as *Mycobacterium spp.*, *Nocardia spp.* and *Corynebacterium spp.*[11]

Other membrane-associated phenotypes include *S layers* (Figure 15.4a), which are hard, semi-crystalline armoured membranes that occur with some prokaryote species (both Gram positive and negative) and most archaea. The opposite extreme is provided by capsular bacteria in which polymeric brushes attached to their external membranes provide very soft repulsive mechanical responses (Figure 15.4b, Chapter 4).

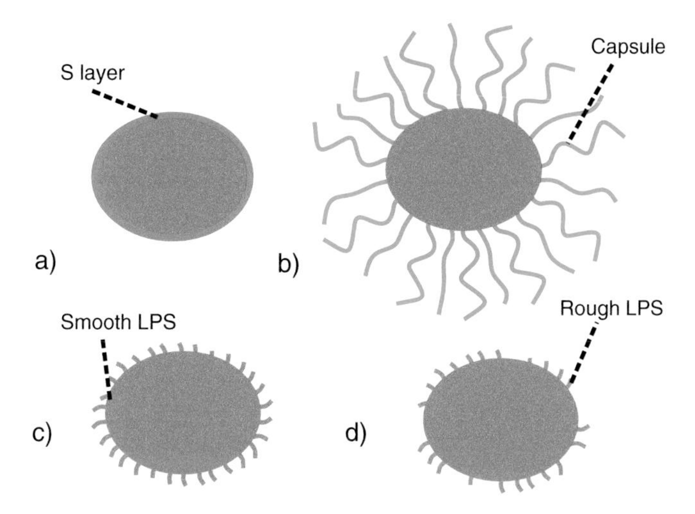

Figure 15.4 Comparison of some *surface coatings* of bacterial cells: (a) S-layer, (b) capsule, (c) smooth lipopolysaccharide (LPS) and (d) rough lipopolysaccharide (also Figure 20.2).

Capsules commonly occur in both Gram negative and positive bacteria, but are rarer with archaea. The crystalline structure of S layers for *Caulobacter crescentus* was reconstructed to 3.7 Å resolution using cryogenic transmission electron microscopy (cryo-TEM).[12] In this case, the S layer is bound to the lipopolysaccharide embedded in the external membrane. Recent high-resolution atomic force microscopy experiments indicate Gram-positive bacterial cell walls (*Staphylococcus aureus* and *Bacillus subtilis*) have porous gel-like structures on their surfaces.[13] Lipopolysaccharides (both rough and smooth, Figures 15.4c and d, Chapter 20) are common on the surfaces of bacteria and are differentiated from capsules due to their sub-optical wavelength size (lipopolysaccharides are invisible in traditional optical microscopy experiments) and their specific biochemistry.

Some of the most common morphologies of single bacterial cells are shown in Figure 15.5, which also include budding and stalked bacteria, sheathed bacteria and magnetic bacteria. Thus, bacteria often adopt morphologies that have simple symmetric shapes, such as spherical, ellipsoidal, rod-like and helical. Their cells can be rigid or flexible.

Bacteria commonly stick to one another and form a wide range of autoaggregation states, such as filamentous morphologies.[15] Some examples of aggregate morphologies are shown in Figure 15.6 and often the information is included in the Latin names of the bacterial species (Table 15.2). Such states of aggregation should be considered separately from the biofilm phenomenon. For example, *S. aureus* has both distinct aggregated and surface-attached (biofilm) phases,[16] which are driven by separate genetic programmes. The word *Staphylococcus* means clustered. The aggregated state of *S. aureus* is an important virulence factor in blood infections. Other standard names for bacteria that imply aggregation states are *diplococcus* (diplo = pair), *streptococcus* (strepto = chain), *tetrad* (packets of four) and *sarcina*

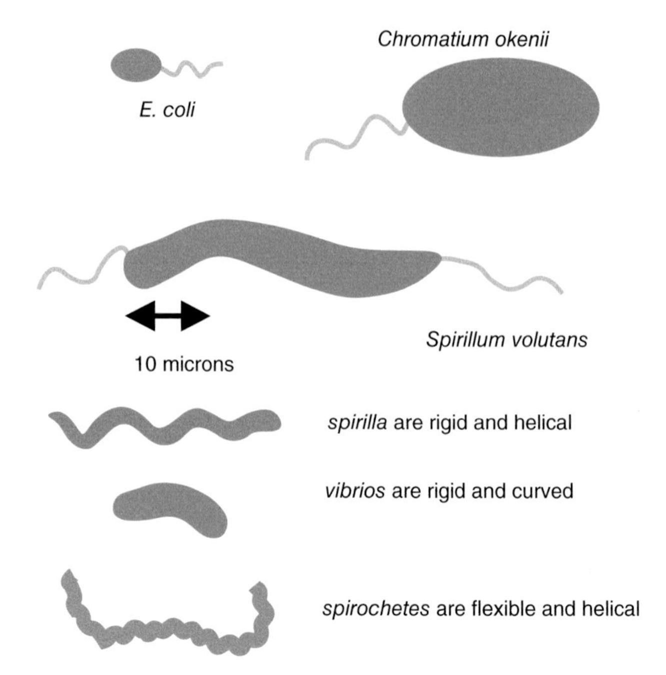

Figure 15.5 Schematic diagram showing the *shapes* of some standard bacterial cells (not to scale). Spiral-shaped bacteria can be flexible or rigid.

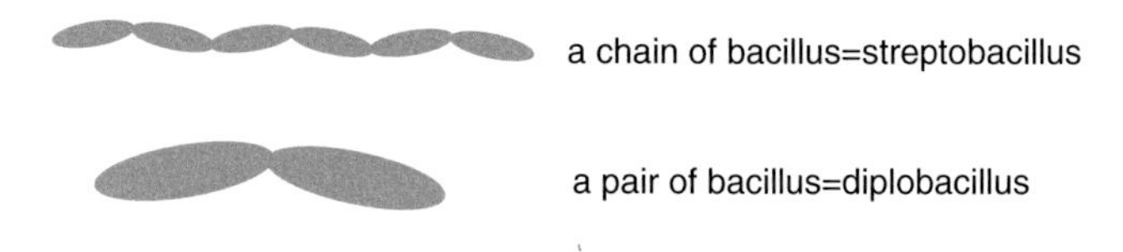

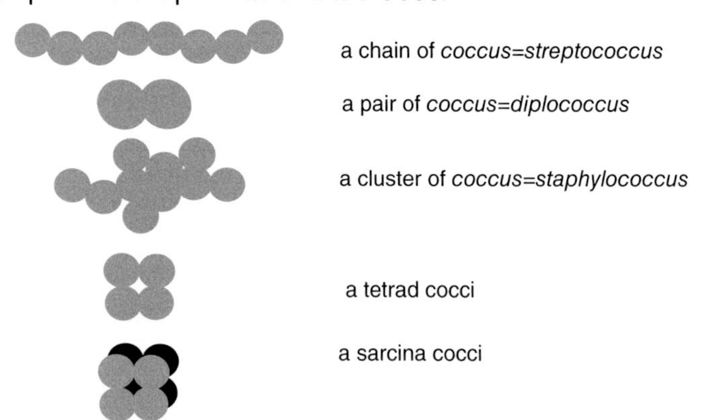

Figure 15.6 Shapes of *bacterial aggregates* include: (a) *bacilli* which are rod-shaped, and (b) *cocci* which are spherical.[14]

(packets of eight). These names are also found in aggregates of rod-shaped bacilli e.g. *diplobacillus* and *streptobacillus*. Complicated branched super-structures can also form from the end-on-end aggregation of bacteria into fibres e.g. in *streptomycetes*. Cord-like structures are typical of *Mycobacterium tuberculosis* and are formed from fibrous aggregates of bacteria that associate side by side into a hierarchical structure. Clearly, the shapes of bacterial aggregates are often characteristic of a particular bacterial species, although the physical environment can also play an important role in aggregate morphology.

Clumping in *S. aureus* provides a physical barrier to phagocytosis (the mechanism by which white blood cells wrap around foreign entities) and can also modulate the chemistry of blood clots. The clumping mechanism also makes the bacteria more antibiotic resistant.[16] It is hard to study monomeric *S. aureus* in experiments as a result of its tendency to clump e.g. for quantitative experiments to study their sedimentation (Section 17.2). Curiously, defensins from the innate immune system of humans also cause bacteria to aggregate and with some bacterial infections, this can help the infected organisms to clear the infection. Similarly, flocculation is used during water treatment in sewage plants to remove the bacterial load. The coagulants used for flocculation are normally aluminium salts, iron salts or positively charged polyelectrolytes e.g. polyamines.

Bacterial colonies can form intricate *fractal-like structures* when grown on agar gels, reminiscent of patterns created by diffusion-limited aggregation (DLA)[17] with colloids (Chapter 7) i.e. patterns that are formed by colloidal particles that can diffuse and stick to one another irreversibly upon contact to form fractal aggregates e.g. soot particles can resemble DLA aggregates. The patterns in which the aggregate morphologies of bacteria grow depend on the growth rate, which in turn is a function of the culture conditions, such as peptone and agar concentrations. The interplay between colony growth and hydrodynamics is considered in more detail in Chapter 21.

Bacteria can also be differentiated based on whether they can metabolise oxygen i.e. *aerobic* versus *anaerobic* bacteria. More energy is typically available to aerobic bacteria when metabolising sugars with the help of oxygen, but they must cope with a higher level of oxidative damage to the molecules in their cells and thus higher levels of stress. This presents an important challenge when culturing bacteria, since oxygen must be fastidiously occluded from the environment to grow anaerobic bacteria.

Haploid organisms only have a single set of chromosomes. Bacteria are mostly haploid (humans in contrast are diploid with two sets of chromosomes), but there are occasional diploid examples e.g. *Neisseria* species.[18]

The *phylogenic diversity* of the bacteria in an environment is different from the number of species present. For example, soil has low phylogenic diversity (i.e. the bacteria are closely related and fulfil similar functions), but contains a large number of different species. Salinity is one of the most important factors determining phylogenic diversity over a broad range of environmental niches, and it is more important than temperature or pH, although these factors also play a role.[19]

Many more bacterial shapes are known beyond those in Table 15.1, including Y-shaped structures[20] and asymmetric structures. However, most bacteria resemble

Table 15.2 Typical sizes, shapes and aggregation states of standard strains of bacteria. Different strains within a species can show substantial variability. The density of *E. coli* is ~1.105 g/mL.

Bacterium	Shape	Average size	Aggregation **states**	Description
E. coli	Cylinder with hemispherical end caps.	2.5 μm length, 0.8 μm diameter.	Typically, non-clustered but clustered states are possible.	Gram –ve. Pathogen/ Commensal.
S. aureus	Slightly oblate sphere.	1 μm diameter.	Clustered.	Gram +ve. Pathogen.
Pseudomonas aeruginosa	Cylinder with hemispherical end caps.	1.5–3 μm length, 0.8 μm diameter.	Clustered.	Gram –ve. Pathogen.
Shewanella oneidensis	Rod-shaped.	2.5 μm length, 0.55 μm diameter.	Typically, non-clustered but clustered states are possible.	Gram –ve. Conducting marine bacterium.
B. subtilis	Rod-shaped.	3 μm length, 0.75 μm diameter.	Fibrous end on end clustering.	Gram +ve. Soil bacterium.
Deinococcus radiodurans	Spherical.	2.5 μm diameter.	Clustered in tetrads.	Gram +ve. World's toughest bacterium?
Mycobacterium tuberculosis	Rod-shaped.	3 μm length, 0.35 μm diameter.	Cording.	Neither Gram +ve nor –ve.
Thiomargarita namibiensis	Ellipsoidal cocci.	750 μm length.	Giant fibrous aggregates.	Very large bacterium.
Streptomyces	Asymmetric division – filaments.	1.25 μm diameter.	Filamentous.	Gram +ve. Source of antibiotics. Infrequent pathogen.

reasonably symmetrical colloidal particles on length scales above 100 nm and have sizes on the order of a few microns.

Deinococcus radiodurans R1 is an extremely radiation-resistant bacterium. The genetics of the species have been probed in detail and it has a wide battery of systems to protect its DNA from radiation damage, including DNA repair, DNA damage export, desiccation and starvation recovery. For example, *D. radiodurans* specimens were found to have survived on the outside of a space vessel for over three years[21] in a vacuous environment with substantial radiation exposure and the bacterium was originally discovered when it caused persistent infections during the gamma ray sterilisation of food (tins of spam). Such hardy bacteria can be useful for bioremediation studies as a result[22] e.g. in mines with heavy metal pollution.

Bacteria have often coevolved with the *immune system* of host organisms. This is highlighted by the development of stealth strategies of pathogenic bacteria to circumvent the immune responses of the host, so that the infection proceeds unnoticed and septic shock does not occur (a large uncontrolled immune response that can lead to

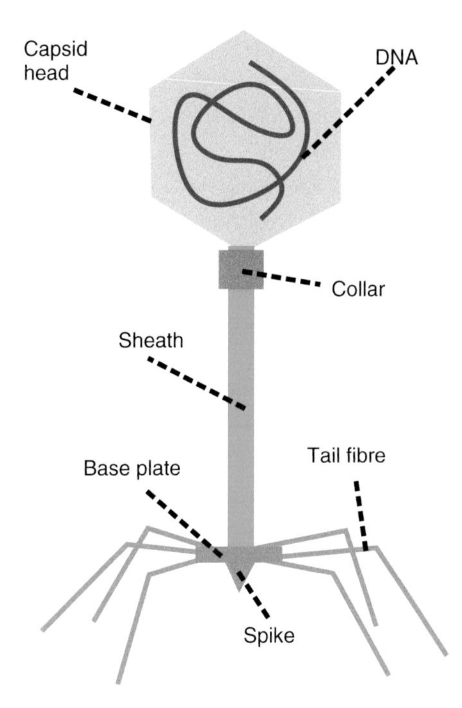

Figure 15.7 Schematic diagram of *T4 Bacteriophage*, which is a virus that commonly infects *E. coli*.

the death of the host and is also often unfavourable for the bacteria). Complex communication strategies also occur with symbiotic bacteria to avoid the host's immune responses e.g. nitrogen-fixing bacteria in the roots of leguminous plants[23] or the migration of luminescent *Vibrio fischeri* bacteria into the bobtail squid.

Viruses are the largest repository of genetic diversity on planet Earth. They are the most abundant biological entities[23] (~10^{31} individuals, outnumbering bacteria 10:1) and provide a very effective mechanism for moving genetic material between organisms and can thus drive evolution. Most viruses are bacteriophages that infect bacteria, followed by the viruses of archaea and then eukaryotes. Bacteriophages tend to be much smaller (~100 nm) and simpler than bacteria (Figure 15.7). Viruses in general are considered to be non-living (mostly obligate parasites), since they cannot reproduce alone.

Huge levels of carnage are involved with viral infections of microorganisms e.g. it is estimated that 5–50% of bacteria are killed every day in sea water due to viral infections. Thus, protection against viral infection is a major driver of bacterial evolution, which has shaped the activity of capsules (bacteriophages are exquisitely sensitive to the serotype of the capsules they target, i.e. their specific biochemistry, and bacteriophages have been developed as standardised tests for capsular strains) and biofilms (biofilm extracellular polymeric substance can block viral infections by adsorption of bacteriophage[24]). Bacteriophage can also transfer genetic material between bacteria and this mechanism also helps to drive bacterial evolution. To establish rigorous connections between viruses, bacterial cells and human cells in diseases, the challenge is predominantly a practical one due to the complexity and slower rate of division for eukaryotic cells, which makes robust experimentation more time-consuming. The interaction of viruses with biofilms in human disease is a vast developing field.

In addition to the huge range of species presented by bacteria, an additional practical problem is that the nutrients supplied to bacteria can change which of the genetic programmes they run.[14] This can have a very noticeable impact on the phenotypes the bacteria exhibit, such as the molecules they synthesise in their cells and extrude into their biofilms e.g. *S. aureus* exudes amyloids in nutrient-deficient culture media in to their biofilms, whereas they can produce polysaccharide intercellular adhesion (PIA) in more nutritious media.[25]

Bacteria are packed full of nanomachines that have been sensitively optimised by evolution. Much can be learnt by would-be nanotechnologists from the designs of nanomachines in bacteria (Chapter 19) e.g. how to create an efficient motor or nanosyringe. Pili are multi-purpose nanomachines that act as adhesive appendages attached to the surfaces of some bacteria and they can lead to motility (they can be retracted). The name F pili is given to pili used by Gram-positive bacteria for DNA transfer during conjugation, but they also have other roles. Bacteria can have multiple types of pili e.g. *Thermus thermophilus* produces two types of type IV pili.[26]

Bacteria are often motile (Chapter 17), which is commonly driven by the activity of flagella, pili or by shape changes of the bodies of the bacteria. Suspensions of bacteria can therefore exist in active fluid-like states in which flows are internally driven. Otherwise, sedimentation will necessarily lead to the formation of granular or colloidal precipitates from bacterial suspensions, since most bacteria are denser than water. Occasional exceptions include some bacteria that have the bacterial equivalent of a swim bladder full of gas i.e. they carefully control gas bubbles in vesicles to modulate their buoyancy.[27]

15.1.1 Sporation

Bacterial spores are bacterial cells in which the metabolism has almost completely been switched off. The spore cells have distinct morphologies compared with normal cells e.g. *B. subtilis* spores are much smaller spherical particles compared with the normal extended rod-like cells (Figure 15.8). Spores are extremely resistant to antimicrobials because of their low rates of metabolism. Sporation is thus an important virulence factor and commonly occurs in types of *Bacillus* and *Clostridium* bacteria e.g. *B. subtilis*, *Bacillus anthrax* and *Clostridium difficile*.

With *B. subtilis*, the sporation process is triggered by nutrient limitation and in the short term, it also leads to cannibalism i.e. ~50% of the bacteria switch to sporation and release toxins that kill and then eat their non-sporating siblings.[28] The process of cannibalism continues until there are no siblings left to be eaten and then spore formation becomes irreversible. Fratricide also occurs with *Streptococcus pneumoniae* (not associated with sporation), but here it is used to release virulence factors from dying cells that improve the prospects for the remaining bacteria.

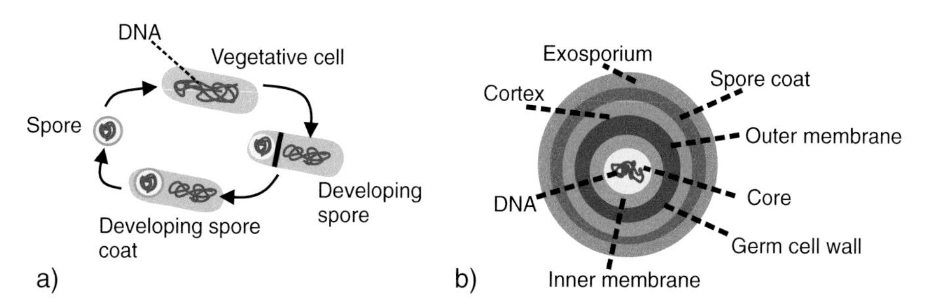

Figure 15.8 Schematic diagrams of *bacterial spores*. (a) Growth of spores in a *B. subtilis* cell. (b) Morphology of a spherical spore cell.

Sporation can occur in biofilms. Waves of gene expression are known to occur across time and space in *B. subtilis* biofilms as they coordinate the transition to the spore state.[29] The wavefronts have been measured using fluorescence microscopy.

15.2 Bacteria in Disease

Historically, a major driver of bacterial research has been medicine. Many diseases have a direct bacterial origin e.g. gonorrhoea, acne and stomach ulcers. If the causative bacteria can be destroyed, often the disease can be cured. Other subtler medical problems can also occur, so this is not necessarily a complete solution e.g. when bacteria create poisonous molecules and killing the bacteria does not neutralise the poison, such as *botulism*, or when killing the bacteria can cause a huge release of bacterial molecules (particularly lipid A from their membranes) causing *sepsis*. In sepsis, it is the uncontrolled immune response of the patients, not the direct activity of the bacteria, that can cause death. More exotic issues are seen with scarlet fever, where it is a virus that infects the bacteria, which in turn are modified to cause the disease; the bacteria on their own are benign and do not cause the illness without the virus. This phenomenon is also found with diphtheria and cholera in which the main toxin is produced by a bacteriophage i.e. a virus that infects bacteria.

There is a question mark hanging over amyloid diseases (e.g. Parkinson's, Alzheimer's and Creutzfeldt–Jakob's disease) and their connection with bacterial infections. Many bacteria do produce amyloids in their biofilms[30] (although they are chemically distinct from the amyloids that are directly implicated in the disease) and there have been high-profile experiments showing a positive correlation of bacterial infections with Alzheimer's[31] and Parkinson's[32] disease in model animal systems. Whether the bacteria are the primary cause of these diseases is still open for debate; they may be one of a range of factors that predispose an organism to the diseases (immunological factors can positively correlate with a diverse range of diseases). One recent hypothesis is that intracellular eukaryotic amyloids in human cells have a protective function against infections from both viruses and bacteria[33,34] (bacteria are also thought to use amyloids to protect against bacteriophage[24]). Amyloid diseases in humans could thus be initiated as a response to bacterial infections and the progression of the disease is then due to the dysregulation of the metabolic control of amyloid production.

Antibiotic resistance is also a major driver for medical research into bacteria and their biofilms. Many advances in medicine would become impossible without good antimicrobials e.g. general surgery, organ transplants, the treatment of immunocompromised individuals (e.g. HIV) and cancer treatment.[35] Bacteria naturally possess a range of mechanisms for rapid evolution to negate antibiotics, since antibiotics have been present in their environments for their entire history. For example, neighbouring microbes are in a constant arms race with one another as they compete for limited resources and many produce antimicrobials e.g. penicillin is produced by fungi to kill bacteria that compete for resources in their natural environments. Other examples of microbial competition include *B. subtilis* that kills and cannibalises its neighbours under low food conditions[36] and a broad range of bacteria that inject poisonous molecules into their neighbours using nanosyringes[37] (Chapter 19). Typically, bacteria in biofilms require two or three orders of magnitude higher concentrations of antibiotics to kill them and cure the infections, when compared with non-biofilm-forming strains. Bacterial biofilms are thus much more antibiotic-resistant than individual bacteria alone (Chapter 24).

15.3 Other Applications of Bacteria

In general, moving beyond just bacteria that directly colonise humans, there are a vast range of applications in which bacteria, and thus biofilms, play an important role. These include biofouling (e.g. boats, pipes and factories), waste treatment (e.g. sewage and clean water), climate control, souring of oil wells, electricity generation, food creation, food spoilage and medicine (e.g. genetically engineered *E. coli* can be used to create medicines). This covers a huge range of disciplines, such as marine biology, naval architecture, chemical engineering, biotechnology, food science, regenerative medicine and ecology. Bacteria are thus seen to play both positive roles (e.g. recycling and water treatment) as well as negative ones (e.g. biofouling and degradation of materials).

A large number of bacteria that cause human disease can exist in the soil e.g. *Pseudomonas aeruginosa* and *B. anthrax*. Soil ecosystems provide a large reservoir of bacterial populations with a very high species diversity and can be useful for screening the novel natural products they make e.g. antimicrobials. Finding model porous materials that emulate soils for reproducible bacterial culture is still under active research.[38] The subject of soil structure and content is a complex, important, but specialist area.[23] For example, critical factors include pore size, pore horizons (layers of soil with distinct physical, chemical and biological characteristics), cation exchange, pH, particle size and aggregation states. Sand-packed columns are a standard system used by the oil industry to understand the growth of marine bacteria that can lead to the souring of oil wells e.g. bacteria that make sulphuric acid that corrodes steel pipes (Chapter 31).

Additional drivers of bacterial biophysics research include fuel cells and agrochemical applications (Chapter 31). Nitrogen-fixing bacteria that form biofilms in association with plants are a classic example of symbiosis.[23,39] Most plant-eating animals depend on bacteria for digestion (ruminants have an additional stomach to house the bacteria), since they are required to digest cellulose. Bacteria thus are seen to play a key role in food and agriculture for the usage and recycling of a wide range of vital nutrients.

15.4 Bacterial Biodiversity

It is estimated (based on mass calculations) that there are around 3.8×10^{13}, i.e. 38 trillion, microbial symbionts on every human on planet Earth and the majority are found as intestinal flora (~1–2 kg in the large intestines).[40] There is thus greater than one hundred times the diversity in terms of DNA in microbes per person than in the human cells that form the individuals who form the microbes' home[41] and a factor of 1.3–3.0 more bacterial cells than human cells in every individual (the bacteria are much smaller than human cells and their genomes are more varied, although much smaller). A case can thus be made to think of humans as super-organisms containing bacterial cells, eukaryotic cells (mostly human and fungal – with the occasional parasite, such as worms and malaria) and archaeal cells. Completely abiotic humans would not be very healthy e.g. abiotic mice suffer from a series of health issues including developmental, immunological and vitamin deficiencies.[42] Microbial communities have been analysed from a wide range of individuals and a wide variety of body locations.[43] There is a huge diversity of these microorganisms.

Modern estimates for bacterial biodiversity on planet Earth are still the subject of controversy, but there are thought to be a total of between 3 billion and 1 trillion species[44] (3×10^9–1×10^{12}) of living organisms and the vast majority of them are bacteria[45] (viruses are not classified as living). Only around 30 000 species of bacteria have been formally characterised in detail[46–48] and there are still important questions left with even the best understood organisms e.g. capsule production (Chapter 20) and electrical communication in *E. coli* (Chapter 29). Species diversity was found to be orders of magnitude greater in soils and sediments than in aqueous environments, so bacterial species are distributed in a heterogeneous manner around planet Earth, although bacteria exist in virtually all environments.[49] Naturally occurring abiotic environments are extremely rare (i.e. bacteria-free niches) and probably only occur in living organisms with active management (immune) systems or in environments with extreme physical behaviours e.g. inside the magma of volcanoes. As high throughput molecular biology techniques improve in tandem with machine learning methods, it may be possible to analyse the huge quantity of data required to describe the full range of bacterial species on planet Earth and it is then expected that significant inroads can be made into the study of these vast numbers of unexplored microorganisms. Major efforts are already underway to understand some of the enormous amounts of bacterial *dark matter* on planet Earth.[1,50,51]

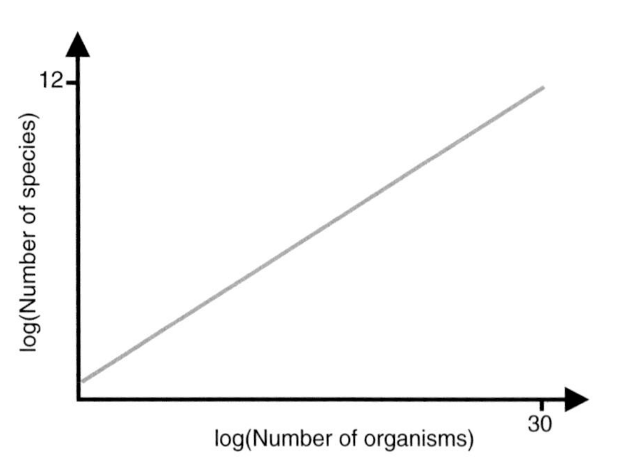

Figure 15.9 Scaling of the number of species of living organisms on planet Earth as a function of the number of organisms.[44] Most species are prokaryotes (bacteria and then archaea in terms of relative importance).

Returning to the question of how many different types of bacteria are there on planet Earth, it is found that this is a surprisingly difficult question to answer. There might be expected to many of them due to the estimated 4–6 × 10^{30} total number of bacteria on planet Earth, which is based on measurements of their mass. The most recent estimate for the number of bacterial species is around 10^{12} (Figure 15.9), but this number could easily be revised by orders of magnitude based on more extensive surveys and it sensitively depends on the statistical model used to interpret the data[44,52,53] i.e. the probability distribution is assumed to be lognormal, the scaling relationships follow power laws and the experimentally sampled diversity is free of any major biases (Figure 15.9).

Another complication in answering the diversity question is the difficulty of defining separate species with bacteria. Ten to twenty percent variations commonly occur between the sequences of the genomes of bacteria that are defined as the same species e.g. with *E. coli*.[54] This is a much broader variation than that observed in eukaryotes, due to the possibilities for HGT, plasmid exchange and rapid mutation with bacteria. For example, humans are only 4% different from chimpanzees in terms of their genomes and there is a maximum of 0.6% variation (~20 million base pairs) between two human beings[55] and on average, humans have much smaller variations.

High throughput genetic sequencing technology has enabled the completion of the first census for the *human gut microbiome project*. Most bacterial strains in human guts are found to be resident for decades forming very stable communities.[56] Sudden changes in these populations are often associated with aging, disease or abrupt lifestyle changes e.g. a sudden switch from an omnivorous to a vegetarian diet. Quantitative studies of the gut microbiome align well with the fields of probiotics[57] (e.g. live bacteria in yoghurts), nutraceuticals (molecules specifically ingested to encourage certain strains of bacteria) and faecal transplants. Faecal transplants are becoming standard treatments for *C. difficile* infections in the human intestines and are being tested to treat a range of other gastrointestinal and neurological conditions e.g. *colitis*. A challenge is to define the microbiota of an ideal healthy individual as a donor for the faeces and the merits of the treatments are still hotly debated. A key question with probiotics is whether the bacteria survive the harsh conditions inside the stomach (e.g. the acidic conditions) to repopulate the intestines (e.g. where they are then disrupted by bile salts) with more favourable species of bacteria.

15.5 Summary

Bacteria are extremely complex colloids. They are predominantly haploid cellular organisms (they have single sets of chromosomes), although rarer diploid species are known and many bacteria possess non-chromosomal DNA in the form of plasmids. Bacteria are the most common form of life on planet Earth and demonstrate a huge amount of genetic diversity e.g. this implies a wider range of molecules (e.g. proteins) are made by bacteria than any other living organism. The majority of bacteria form biofilms and the majority of bacteria spend the majority of their time in these biofilms.[49] Bacterial capsules, slimes and biofilms are important virulence factors for bacteria in human diseases and are considered in detail in this book.

Suggested Reading

Microbiology

Barton, L. L., *Structural and Functional Relationships in Prokaryotes*. Springer: 2005.

Kim, B. H.; Gadd, G. M., *Prokaryotic Metabolism and Physiology*, 2nd ed. Cambridge University Press: 2019.

Madigan, M.; Martinko, J.; Bender, K.; Buckley, D.; Stahl, D., *Brock Biology of Microorganisms*, 14th ed. Pearson: 2015.

White, D.; Drummond, J.; Fuqua, C., *The Physiology and Biochemistry of Prokaryotes*, 4th ed. Oxford University Press: 2012.

Wilson, B. A.; Salyers, A. A.; Whitt, D. D.; Winkler, M. E., *Bacterial Pathogenesis: A Molecular Approach*. ASM Press: 2011.

Wilson, M.; McNab, R.; Henderson, B., *Bacterial Disease Mechanisms: An Introduction to Cellular Microbiology*. Cambridge University Press: 2002.

Biophysics

Boal, D., *Mechanics of the Cell*, 2nd ed. Cambridge University Press: 2012. Interesting consideration of bacterial cell walls. Hoop stresses could be the predominant failure mechanism for bacterial cell walls, similar to how sausages burst longitudinally when you cook them in a frying pan.

Mazzo, M. G., The physics of biofilms – an introduction. *Journal of Physics D: Applied Physics* 2016, 49 (20), 203001.

Waigh, T. A., *The Physics of Living Processes*. Wiley: 2014. A modern physical toolbox approach to biological physics.

Genetics

Dale, J. W.; Park, S. F., *Molecular Genetics of Bacteria*, 5th ed. Wiley-Blackwell: 2010. Approachable introduction to bacterial genetics.

References

1. Hug, L. A.; et al., A new view of the tree of life. *Nature Microbiology* **2016**, *1*, 16048.
2. Woese, C. R.; Kandler, O.; Wheelis, M. L., Towards a natural system of organisms: Proposal for the domains archaea, bacteria, and eucarya. *Proceedings of the National Academy of Sciences of the United States of America* **1990**, *87* (12), 4576–4579.
3. Reynolds, T. B.; Fink, G. R., Baker's yeast, a model for fungal biofilm formation. *Science* **2001**, *291* (5505), 878–881.
4. Volland, J. M.; et al., A centimeter-long bacterium with DNA contained in metabolically active, membrane-bound organelles. *Science* **2022**, *376* (6600), 1453–1458.
5. Costerton, J. W.; Lewandowki, Z.; Caldwell, D. E.; Korber, D. R.; Lappin-Scott, H. M., Microbial biofilms. *Annual Review of Microbiology* **1995**, *49*, 711–745.
6. Whitman, W. B.; Coleman, D. C.; Wiebe, W. J., Prokaryotes: The unseen majority. *Proceedings of the National Academy of Sciences of the United States of America* **1998**, *95* (12), 6578–6583.
7. Johnsen, P. J.; Dubnau, D.; Levin, B. R., Episodic selection and the maintenance of competence and natural transformation in *Bacillus subtilis*. *Genetics* **2009**, *181* (4), 1521–1533.
8. Barrick, J. E.; Yu, D. S.; Yoon, S. H.; Jeong, H.; Oh, T. K.; Schneider, D.; Lenski, R. E.; Kim, J. F., Genome evolution and adaptation in a long-term experiment with *Escherichia coli*. *Nature* **2009**, *461* (7268), 1243–1247.
9. Elena, S. F.; Lenski, R. E., Evolution experiments with microorganisms: The dynamics and genetic bases of adaptation. *Nature Reviews Genetics* **2003**, *4* (6), 457–469.
10. King, J. E.; Roberts, I. S., Bacterial surfaces: Front lines in host-pathogen cell-surface interactions. In *Biophysics of Infection*, Leake, M. C., Ed.; Springer: 2016; pp. 129–156.
11. Torok, E.; Moran, E.; Cooke, F., *Oxford Handbook of Infectious Diseases and Microbiology*. Oxford University Press: 2016.

12. von Kugelgen, A.; Tang, H.; Hardy, G. H.; Kureisaite-Ciziene, D.; Brun, Y. V.; Stansfeld, P. J.; Robinson, C. V.; Bharat, T. A. M., *In situ* structure of an intact lipopolysaccharide-bound bacterial surface layer. *Cell* **2020**, *180* (2), 348–358.
13. Pasquina-Lemonche, L.; et al., The architecture of the Gram-positive bacterial cell wall. *Nature* **2020**, *582* (7811), 294–297.
14. Capuccino, J. G.; Welsh, C., *Microbiology: A Laboratory Manual*. Pearson: 2018.
15. Trunk, T.; Khalil, H. S.; Leo, J. C., Bacterial autoaggregation. *AIMS Microbiology* **2018**, *4* (1), 140–164.
16. Haaber, J.; Cohn, M. T.; Frees, D.; Andersen, T. J.; Ingmer, H., Planktonic aggregates of *Staphylococcus aureus* protect against common antibiotics. *PLOS One* **2012**, *7* (7), e41075.
17. Ben-Jacob, E.; Cohen, I.; Levine, H., Cooperative self-organization of microorganisms. *Advances in Physics* **2010**, *49* (4), 395–554.
18. Tobiason, D. M.; Seifert, H. S., Genomic content of *Neisseria* species. *Journal of Bacteriology* **2010**, *192* (8), 2160–2168.
19. Lozupane, C. A.; Knight, R., Global patterns in bacterial diversity. *Proceedings of the National Academy of Sciences of the United States of America* **2007**, *104* (27), 11436–11440.
20. Young, K. D., The selective pressure of bacterial shape. *Microbiology and Molecular Biology Reviews* **2006**, *70* (3), 660–703.
21. Kawaguchi, Y.; et al., DNA damage and survival time course of deinococcal cell pellets during 3 years of exposure to outer space. *Frontiers in Microbiology* **2020**, *11*, 2020.
22. White, O.; et al., Genome sequence of the radioresistant bacterium *Deinococcus radiodurans* R1. *Science* **1999**, *284* (5444), 1571–1577.
23. Pepper, I.; Gerba, C. P.; Gentry, T. J., *Environmental Microbiology*, 3rd ed. Academic Press: 2015.
24. Vidakovic, L.; Singh, P. K.; Hartmann, R.; Nadell, C. D.; Drescher, K., Dynamic biofilm architecture confers individual and collective mechanisms of viral protection. *Nature Microbiology* **2018**, *3* (1), 26–31.
25. Peschel, A.; Otto, M., Phenol-soluble modulins and staphylococcal infections. *Nature Reviews Microbiology* **2013**, *11* (10), 667–673.
26. Neuhaus, A.; et al., Cryo-electron microscopy reveals two distinct type IV pili assembled by the same bacterium. *Nature Communications* **2020**, *11* (1), 2231.
27. Walsby, A. E., Gas vesicles. *Microbiological Reviews* **1994**, *58* (1), 94–144.
28. Bassler, B. L.; Losick, R., Bacterially speaking. *Cell* **2006**, *125* (2), 237–246.
29. Srinivasan, S.; Vladescu, I. D.; Koehler, S. A.; Wang, X.; Mani, M.; Rubinstein, S. M., Matrix production and sporulation in *Bacillus subtilis* biofilms localize to propagating wave fronts. *Biophysical Journal* **2018**, *114* (6), 1490–1498.
30. Schwartz, K.; Syed, A. K.; Stephenson, R. E.; Rickard, A. H.; Boles, B. R., Functional amyloids composed of phenol soluble modulins stabilize *Staphylococcus auerus* biofilms. *PLOS Pathogens* **2012**, *8* (6), e1002744.

31. Dominy, S. S.; et al., *Porphyromonas gingivalis* in Alzheimer's disease brains: Evidence for disease causation and treatment with small-molecular inhibitors. *Science Advances* **2019**, *5* (1), eaau3333.
32. Sampson, T. R.; et al., Gut microbiota regulate motor deficits and neuroinflammation in a model of Parkinson's disease. *Cell* **2016**, *167* (6), 1469–1480.
33. Abbott, A., Are infections seeding some cases of Alzheimer's disease. *Nature* **2020**, *587* (7832), 22–25.
34. Kumar, D. K. V.; et al., Amyloid-beta peptide protects against microbial infection in mouse and worm models of Alzheimer's disease. *Science Translational Medicine* **2016**, *8* (340), 340ra72.
35. Davies, S.; Grant, J.; Catchpole, M., *The Drugs Don't Work: A Global Threat*. Penguin: 2013.
36. Gonzalez-Pastor, J. E., Cannibalism: A social behavior in sporulating *Bacillus subtilis*. *FEMS Microbiology Reviews* **2011**, *35* (3), 415–424.
37. Galan, J. E.; Collmer, A., Type III secretion machines: Bacterial devices for protein delivery into host cells. *Science* **1999**, *284* (5418), 1322–1328.
38. Ginn, T. R.; Wood, B. D.; Nelson, K. E.; Scheibe, T. D.; Murphy, E. M.; Clement, T. P., Processes in microbial transport in natural subsurface. *Advances in Water Resources* **2002**, *25* (8), 1017–1042.
39. Rinaudi, L. V.; Giordano, W., An integrated view of biofilm formation in rhizobia. *FEMS Microbiology Letters* **2010**, *304* (1), 1–11.
40. Sender, R.; Fuchs, S.; Milo, R., Revised estimates for the number of human and bacteria cells in the body. *PLOS Biology* **2016**, *14* (8), e1002333.
41. Gill, S. R.; et al., Metagenomic analysis of the human distal gut microbiome. *Science* **2006**, *312* (5778), 1355–1359.
42. Dominguez-Bello, M. G.; Godoy-Vitorino, F.; Knight, R.; Blaser, M. J., Role of the microbiome in human development. *Gut* **2019**, *68* (6), 1108–1114.
43. The Human Microbiome Project Consortium, Structure, function and diversity of healthy human microbiome. *Nature* **2012**, *486*, 207–214.
44. Locey, K. C.; Lennon, J. T., Scaling laws predict global microbial diversity. *Proceedings of the National Academy of Sciences of the United States of America* **2016**, *113* (21), 5970–5975.
45. Larsen, B. B.; Miller, E. C.; Rhodes, M. K.; Wiens, J. J., Inordinate fondness multiplied and redistributed: The number of species on Earth and the new pie of life. *Quarterly Review of Biology* **2017**, *92* (3), 229–265.
46. Torsvik, V.; Ovreas, L.; Thingstad, T. F., Prokaryotic diversity – magnitude, dynamics and controlling factors. *Science* **2002**, *296* (5570), 1064–1066.
47. Dykhuizen, D., Species number in bacteria. *Proceedings of California Academy of Sciences* **2005**, *56* (6), 62–71.
48. Boone, D. R., Garrity, G. M., Castenholz, R. W., *Bergey's Manual of Systematic Bacteriology*. Springer: 2005.
49. Flemming, H. C.; Wuertz, S., Bacteria and archaea on Earth and their abundance in biofilms. *Nature Reviews Microbiology* **2019**, *17* (4), 247–260.

50. Castelle, C. J.; Banfield, J. F., Major new microbial groups expand diversity and alter our understanding of the tree of life. *Cell* **2018**, *172* (6), 1181–1197.
51. Parks, D. H.; Rinke, C.; Chuvochina, M.; Chaumeil, P. A.; Woodcroft, B. J.; Evans, P. N.; Hugenholtz, P.; Tyson, G. W., Recovery of nearly 8,000 metagenome-assembled genomes substantially expands the tree of life. *Nature Microbiology* **2017**, *2* (11), 1533–1542.
52. Lennon, J. T.; Locey, K. C., More support for Earth's massive microbiome. *Biology Direct* **2020**, *15* (1), 5.
53. Louca, S.; Mazel, F.; Doebeli, M.; Partrey, L. W., A census based estimate of Earth's bacterial and archeal diversity. *PLOS Biology* **2019**, *17* (2), e30000106.
54. Souza, V.; Rocha, M.; Valera, A.; Eguiarte, L. E., Genetic structure of natural populations of *Escherichia coli* in wild hosts on different continents. *Applied Environmental Microbiology* **1999**, *65* (8), 3373–3385.
55. Autan, A.; et al., A global reference for human genetic variation. *Nature* **2015**, *526* (7571), 68–74.
56. Faith, J. J.; et al., The long-term stability of the human gut microbiota. *Science* **2013**, *341* (6141), 1237439.
57. Seifert, A.; Kashi, Y.; Livney, Y. D., Delivery to the gut microbiota: A rapidly proliferating research field. *Advances in Colloid and Interface Science* **2019**, *274*, 102038.

16 Growth of Bacteria

The growth of bacteria is highlighted in the field of mathematical biology[1] and the subject is considered in more detail in Chapter 22 from the perspective of the dynamics of interacting populations. In this chapter, an introduction is given to growth phenomena that emphasise the behaviour of individual cells and their cellular biology from a physical perspective.

Growth statistics are one of the first practical issues encountered when handling bacteria in a laboratory[2] e.g. provided a particular strain of bacteria and a set of growth conditions, can sufficient numbers of bacteria be grown to perform some robust experiments? Furthermore, is it possible to cause the bacteria's metabolism to function reproducibly during growth? It is much more straightforward to work with standard laboratory strains for this reason, since there are standard protocols for their culture and a huge number of bacteria remain practically unculturable outside of their natural environments due to their fastidiousness.[3,4]

Historically, *pure culture* has predominantly been used to grow microorganisms in microbiology laboratories for quantitative studies of molecular biology which is an approach developed by Robert Koch in the 1880s. Single-cloned strains of bacteria are grown (e.g. on an agar plate), which facilitates the understanding of molecule/phenotype relationships due to the relative simplicity of the experimental systems used. All the cloned bacteria are expected to act similarly, although stochastic effects can lead to phenotypic variability for the bacteria over their range of genetically permitted behaviours. *Mixed culture* in which multiple strains of microorganisms grow simultaneously is much more challenging to quantify, but it is also more realistic in real-world scenarios, since microorganisms rarely occur in monocultures in the real world. One method for mixed culture uses a Winogradsky column.[5]

Many modern approaches to culture microorganisms are based on microfluidic apparatus and they can be very robust platforms to understand bacterial growth due to the well-defined hydrodynamics, since the nutrient availability, signalling molecules, effluent fluxes and so on are then better defined. Bacterial aggregation and biofilm formation can make microfluidics more challenging, since the bacteria can block small capillaries used for pumping. However, with careful maintenance, microfluidic devices can provide excellent platforms to explore dilute cells, clustering and biofilm formation. Due to space constraints (<mm), microfluidic studies are normally constrained to early-stage biofilms. The biofilms formed in very small microfluidic geometries (microclusters) are found to be softer than those grown in standard slides in atomic force microscopy indentation experiments,[6] so there appears to be a length scale for biofilms below which the films stop being representative of those on the

macroscale. In general, there may be a critical radius at which bacterial aggregates are considered to be a biofilm based on their mechanics.

Three-dimensional printing provides a radical new method to study bacterial growth[7] (Section 13.1.2). For example, living bacterial composites can be created in hydrogels,[8] such as calcium cross-linked alginates with *Escherichia coli*. Another modern printing approach used a simple optical mask to project a template onto a genetically modified array of bacteria. C-di-GMP levels were modified and patterned biofilms were created (10 μm features could be templated).[9] Three-dimensional (3D) printing is still reasonably expensive, but it may be the future of sample preparation in bacterial research, when combined with microfluidic geometries, as costs decrease. 3D printing also provides a convenient method to create microfluidic geometries for cell culture, although currently photolithography tends to be preferred for microfluidics due to its higher resolution.

16.1 Mathematical Models of Growth

A good modern account of the history of biophysics experiments in bacterial growth already exists,[10] which focuses on the regulation of the sizes of bacterial cells. Typically, the growth of bacteria as they are cultured in a laboratory is classified via four phases (Figure 16.1): a *lag phase* (I), an *exponential growth phase* (II), a *plateau phase* (III) and a *death phase* (IV). A simple successful model of bacterial growth is a sigmoidal relationship due to Monod.[11] According to this model, the growth rate of a bacterial culture is given by

$$\frac{dN}{dt} = \lambda_{\max} \frac{s}{s + K_m}, \tag{16.1}$$

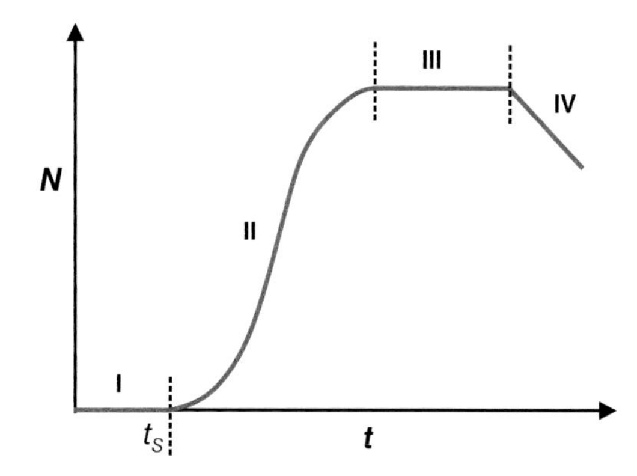

Figure 16.1 Typical phases of the growth of a population of bacteria in pure culture. The number of bacteria (N) is shown as a function of time (t). The regimes of growth are I the *lag phase*, II the *exponential phase*, III the *plateau phase* and IV the *death phase*.

where $\frac{dN}{dt}$ is the rate of change of bacterial number (N), s is the concentration of the growth-limiting substrate, λ_{max} is the maximum growth rate and K_m is the saturation constant. The Monod model can approximate regimes II and III in Figure 16.1, since it describes a sigmoidal relationship similar to that observed.

Equation (16.1) follows the same form as the standard Michaelis–Menten model for enzyme kinetics, so it could be motivated by invoking an enzymatic limiting state in the growth process. However, in practice, the Monod equation is just a simple fit function, since the majority of single enzymes needs more sophisticated models to be properly quantified (e.g. the Hill model) and bacteria are better described using the kinetics of complex gene networks containing a large number of interacting enzymes (the systems biology approach, Chapter 11).

Often, there are surprises for the kinetics in growth experiments in even the simplest scenarios due to a host of incompletely understood stochastic molecular events.[12] For example, careful image analysis with combined fluorescence and phase contrast microscopy images of individual *E. coli* cells indicates they can have either bilinear or trilinear growth rates for their length (Figure 16.2).[13]

Variability in the growth rates of genetically identical (cloned) bacteria has important implications for the consideration of persister cells that are found to be resistant to antibiotics (Chapter 24).[14] Low growth rates are closely associated with persistence.

It is found that the growth rate of cells is strongly correlated with their initial buoyant mass in microchannel resonator experiments with individual bacteria.[15] Similar results relating growth rate to dry mass were found using spatial light interference microscopy, which is based on a quantitative analysis of refractive indices.[16] Osmotic stress also regulates the division rate of bacteria and Gram-negative species

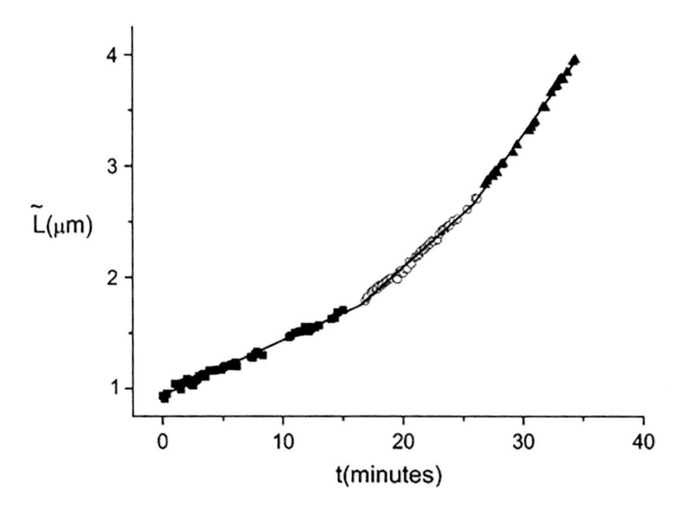

Figure 16.2 Plot of bacterial length ($\tilde{L}$) as a function of time (t) for *E. coli*[13] from the analysis of microscopy images (fluorescence and phase contrast). Three linear regimes can be observed. Reprinted from Reshes G., Vanounou S., Amari T., Wei X. Cell shape dynamics in *E. coli Biophysical Journal*, 2008, 94, (1), 251–264, with permission from Elsevier.

(e.g. *E. coli*) are less sensitive than Gram positive (*Bacillus subtilis* and *Staphylococcus aureus*),[17] presumably due to the differences in how their outer cell walls help regulate osmotic stress. *Sizer*, *adder* and *timer* models have been introduced to explain the elongation rate of single cells, but there is still no consensus on which is the most successful.[10,18] In sizer models, the cells monitor their own size and divide above a target value. With adder models, cells add a fixed size increment. With timer models, the cells grow for a fixed amount of time. With *B. subtilis*, evidence was found for both adder and timer mechanisms.[19]

For pure culture, the two standard mechanisms to grow bacteria in a laboratory are *batch culture* (e.g. growing them on agar plates) and *continuous culture* (e.g. growing them in a chemostat). Batch culture tends to be more characteristic of naturally occurring growth, but it is less reproducible, since the chemistry of such closed systems quickly diverges from steady-state conditions. The alternatives are considered in more detail in Section 13.1.

Fundamental molecular biology answers to questions on cell growth and division are becoming available, such as what molecular processes determine division during the separation of the dividing chromosomes and the membranes of cells. However, reference should be made to the microbiology literature for specific details,[20,21] since the unfolding stories are exceedingly complex and sensitively depend on the species of bacterium considered.

Bacteria contain homologues of both actin and tubulin (which have been explored in detail with eukaryotic cells as components of their cytoskeletons), which help determine the shape of their cells.[22,23] In *E. coli*, the position of cellular division is determined by the interactions of Min C, Min D and Min E proteins. They have been modelled using a self-organised reaction–diffusion framework.[24] Rapid pole-to-pole oscillations of the Min proteins determine the point of division of the cell membrane in *E. coli* by FtsZ (a tubulin analogue[25,26]). Confusingly, another model laboratory organism, *B. subtilis*, has a completely different mechanism to determine the point of division and contains different non-oscillating Min proteins.[27] In general, the division mechanisms of bacteria appear to be extremely complex and diverse.

Simple scaling relationships have been developed for the size of individual bacteria.[10] An idealised age probability distribution for exponentially growing bacteria[10] is given by

$$\varphi(a) = \frac{2\ln 2}{\tau} 2^{-a/\tau}, \quad 0 \le a < \tau, \tag{16.2}$$

where a is the age of the bacteria and τ is the duration of an experiment. The length probability distribution ($\rho(l)$) for the bacteria is

$$\rho(l) = \frac{l_d}{l^2}, \tag{16.3}$$

where l_d is the characteristic length scale. However, more sophisticated models are needed to accurately model the underlying molecular biology and stochastic nature of single-cell growth e.g. the multi-linear growth rates for *E. coli* shown in Figure 16.2.[13]

To model the four regimes that are classically defined for bacterial growth in more detail (Figure 16.1), they will be considered in turn. The *lag phase* is relatively straightforward to describe mathematically, the start time in the algebra just needs to be delayed (subtract off t_s from t), although a fundamental calculation of the size of the delay time would require some sophisticated systems biology modelling (Chapter 11) and it is practically important e.g. the time food can be stored in the fridge before substantial bacterial growth spoils the food depends sensitively on the lag phase.[28]

Exponential growth (regime II on Figure 16.1) can be motivated from several alternative starting assumptions and previously we considered the Monod approach (Equation (16.1)), which depends on the substrate concentration (exhaustion of finite resources). A model, which approximates the number of bacteria as a continuous variable, assumes the population of bacteria is not limited by the availability of resources and the growth rate $\left(\frac{dN}{dt}\right)$ is proportional to the number of individuals (N) present is

$$\frac{dN}{dt} = \mu N, \tag{16.4}$$

where μ is a constant, implying a constant rate of division of the bacteria. With the initial conditions $t = 0$ and $N = N_0$, where the lag phase in Figure 16.1 has been ignored for simplicity, the solution to Equation (16.4) is

$$N = N_0 e^{\mu t}. \tag{16.5}$$

The doubling time (τ) can be straightforwardly calculated (just substitute $N = 2N_0$ and $t = \tau$ in Equation (16.5)),

$$\tau = \frac{\ln 2}{\mu}. \tag{16.6}$$

E. coli strains can be very fast growing e.g. the doubling times can be as little as 20 minutes. In the exponential growth phase, this means a single bacterium will divide into a mass equivalent to an aircraft carrier within one day and that of planet Earth after 2 days. Therefore, there must inevitably be a limit on the exponential growth processes of all microorganisms, since finite resources quickly become exhausted.

Difference models provide an alternative mathematical formalism to describe growth curves. These models describe discrete numbers of bacteria as opposed to continuous differential equations, such as Equation (16.4).[29] They tend to be slightly more cumbersome to solve analytically, but they are more accurate for small numbers of organisms[1] e.g. difference models circumvent the awkward fiction of fractional numbers of organisms and can handle extinction events more accurately as a result.

A slightly more realistic approach based on continuous differential equations compared with Equation (16.4) is provided by the *logistic equation*, which can capture both the exponential and plateau regimes (phases II and III), since the model has a finite carrying capacity through the introduction of an additional variable (K, called the carrying capacity),

$$\frac{dN}{dt} = \mu\left(1 - \frac{N}{K}\right)N. \tag{16.7}$$

Solving this equation subject to the same boundary conditions as the exponential growth model gives the solution,

$$N = \frac{K}{1 + \left(\frac{K}{N_0} - 1\right)e^{-\mu t}}. \tag{16.8}$$

Thus, in the limit of long times ($t \to \infty$), the number of bacteria tends to K, the carrying capacity of the system ($N \to K$), and the model exhibits a plateau regime (phase III) i.e. the population growth is limited to a finite value (K) and the population saturates at this carrying capacity.

Other simple analytical fit functions can also describe regimes II and III in Figure 16.1, which include the *Gompertz growth law*,

$$N = N_0 e^{-\frac{\mu}{\alpha}\left(1 - e^{-\alpha t}\right)}, \tag{16.9}$$

where N_0 is the initial number of bacteria, μ and α are constants. Another alternative is an *allometric scaling law*,

$$N = N_0\left(1 - \left(1 - e^{-1/3}\right)e^{-\alpha t}\right)^3, \tag{16.10}$$

where N_0 is again the initial number of bacteria and α is a constant. The choice of these simple analytical models to fit experimental results is more a question of computational simplicity than any rigorous fundamental molecular motivation based on the underlying growth processes.[30]

A pedagogic description of the logistic growth of bacteria (Equation 16.7) in a chemostat is given by Edelstein–Keshet in her classic book, ‘Mathematical Models in Biology’.[1] Assuming Michaelis–Menten kinetics for the rate of bacterial growth on nutrient availability leads to a more sophisticated model based on two equations,

$$\frac{dN}{dt} = \left(\frac{K_{mac}C}{K_n + C}\right)N - \frac{FN}{V}, \tag{16.11}$$

$$\frac{dC}{dt} = -\alpha\left(\frac{K_{mac}C}{K_n + C}\right)N - \frac{FC}{V} + \frac{FC_0}{V}, \tag{16.12}$$

where N is the number of bacteria in the chamber, C is the mass of nutrients in the chamber, F is the volume flow rate, V is the volume of the chamber, α is a reciprocal yield constant, K_n is the Michaelis constant and $K_{\max}$ is the maximum velocity of the enzyme reaction. The original reference explores the range of solutions of these coupled non-linear differential equations and gives a stability analysis.[1] Oscillatory solutions occur for the number of bacteria for some parameter ranges and this phenomenon is observed in experiments. It is worth spelling out this effect: the number of bacteria during growth does not need to settle down to steady state behaviour in a chemostat, they can oscillate over long time periods and this behaviour is not restricted to chemostats. For example, oscillatory growth is observed for *B. subtilis* biofilms, although a more complex non-linear model based on electrical communication is invoked as an explanation.[31] Furthermore, planktonic *E. coli* demonstrated oscillations in growth in a micro chemostat.[32]

16.2 Mathematical Models of Death

To understand bacterial growth in the long-time limit necessitates an understanding of *bacterial death* (regime IV). Modelling the kinetics of bacterial death is much more challenging than growth.[30] Bacteria can age due to the accumulation of errors in protein folding and damage to DNA, which can eventually lead to death, but a wide range of other molecular processes can also contribute. Mechanisms to kill harmful infections via chemotherapy (antiseptics or antibiotics) or physical treatments (temperature, UV, light, pH, salinity, etc.) tend to be of great practical importance, so models to describe them are considered in detail in Chapter 24. The death of bacteria in response to antibiotics is important to gauge the efficacy of antibiotics for the treatment of disease. To illustrate the complexity of the underlying molecular biology, the general stress response pathways of bacteria can modulate the kill rate in response to a wide range of environmental stresses (including antibiotics), since starvation can decrease both the rates of growth and aging.[33]

Some progress has been made in understanding the death of bacteria due to ionising radiation, but gaps exist in connecting kill rates with the underlying molecular behaviour.[34] On a single cell level, some of the statistical phenomena with respect to bacterial death are similar to the death of eukaryotic cells due to ionising radiation and the modelling of this phenomenon is particularly important with radiation therapies for the treatment of cancerous tumours[30] e.g. it is a non-Markovian process with an increasing hazard rate (Chapter 1).

Experimentally, the *death rate* ($h(t)$, the hazard rate, Chapter 1) of *E. coli* due to aging was found to follow a Gamma Gompertz–Makeham model[30,33] (Figure 16.3).

$$h(t) = \lambda + \frac{bs}{1+(\beta-1)e^{-bt}}, \tag{16.13}$$

where b is the aging rate, t is time, s is a heterogeneity parameter, λ is an age-independent parameter and β controls transitions between the regimes. This is a convenient parameterisation of the death rate, but a fundamental derivation based on the underlying molecular biophysics is lacking.

Nutrient starvation and then death are often issues for bacteria in biofilms. One mechanism the bacteria use to alleviate issues is to create channels to allow nutrients to enter the films more easily via diffusion[35] (Chapter 23). Another mechanism is based on switching to cellular states with lower metabolisms, where the metabolism is almost completely switched off. Like primitive plant seeds, *spore states* are sometimes formed (Chapter 15). This is an important mechanism for antibiotic resistance, since

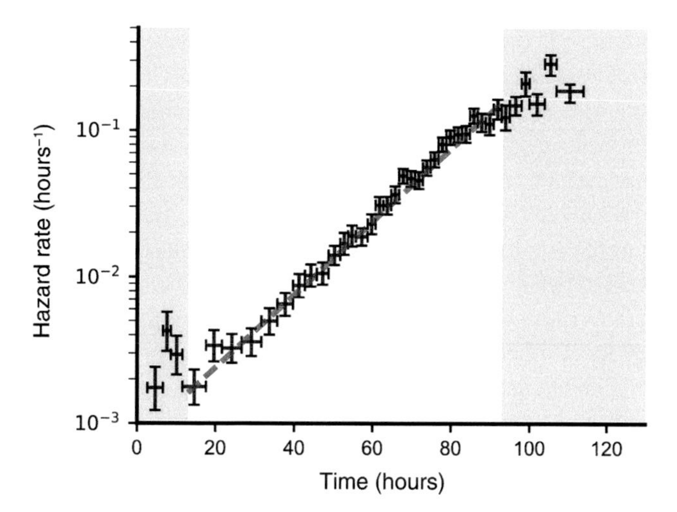

Figure 16.3 The *death rate* of *E. coli* (the hazard rate) follows a Gompertz law,[33] Equation (16.13) as a function of time.

the majority of antibiotics target metabolically active dividing cells. In addition to spores, other low metabolism bacterial cells are thought to occur called *persister cells*, although whether they are due to distinct genetic programmes is still the subject of debate (Chapter 24).

16.3 Measuring Growth Curves

The counting of bacteria to calculate accurate growth curves can be automated using high throughput microscopy and image analysis software, although it can be time-consuming to do it accurately. Traditionally, a Petroff Hauser counting chamber would have been used with a microscope in which bacteria are counted by hand.[4] Coulter counters are electronic devices that automatically quantify bacterial numbers via changes in impedance as bacteria pass through an aperture, but they are unable to differentiate between cells and other particulate matter e.g. dust. Quebec colony counters can be used as an alternative to count viable cells, which is another microscopy method and uses dark field contrast to facilitate the identification of individual bacteria. A standard alternative to directly counting bacteria is to measure the turbidity of a suspension by measuring growth curves using plate readers i.e. perform *optical density* (OD) measurements (Figure 13.2).

Optical density measurements are made following standard protocols for the growth of planktonic bacterial species and are based on the *Beer–Lambert* law i.e. the attenuation of light due to single scattering from a suspension of bacteria causes an exponential decay of the transmitted light intensity (T) with sample thickness,

$$T = I_0 e^{-\tau} = I_0 10^{-\varepsilon lc}, \tag{16.14}$$

where τ is the optical depth, I_0 is the incident light intensity, ε is the molar attenuation coefficient of the bacterial suspensions, c is the bacterial concentration and l is the optical length. The concentrations of bacteria need to be restricted to low values in these measurements, since multiple scattering negates the use of the Beer–Lambert law, which is based on a single scattering approximation. Thus, bacterial concentrations above a threshold value quickly make OD measurements poor qualitative approximations. Dilution of samples alleviates many of these issues if done rapidly and substantial aggregation does not occur. OD measurements can be made with plate counters (with multi-well plates, this allows parallel measurements to be made) or spectrometers combined with larger volume cuvettes.

The availability of suitable media often limits the range of bacteria that can be cultured.[4] Also, the choice of media can change the genetic expression of the bacteria i.e. the same bacteria grown in different media can have different patterns of genetic expression.[36]

Similar to eukaryotic cells (e.g. used in tissue engineering for replacement human tissue), there is an issue that bacterial culture techniques are fairly artificial. Bacterial cultures are often grown on top of agar gels and are thus restricted to growth in two dimensions (they can be quasi-two-dimensional (2D) if the agar concentration is very low, so the bacteria slightly penetrate the gel). 3D culture is more representative of many naturally occurring biofilms e.g. *B. subtilis* biofilms growing in damp soil. As a rarer example of 3D culture of non-biofilm bacteria, the growth of *E. coli* in silica gels has been considered.[37] Growth in optically transparent porous media can facilitate photonics measurements and it is found to increase bacterial viability. The culture of bacteria inside polymeric hydrogels is also possible, though it is less common than on the surfaces of such gels e.g. agar.

It is interesting to consider bacterial growth *in vivo*. Bacterial populations in the human gut are thought to be stable for many years e.g. 5 years.[38] Sudden changes in population are associated with disease, aging or abrupt changes in lifestyle choices e.g. vegetarianism.

Diauxic growth (cellular growth in two phases, Figure 16.4) can occur in bacteria when the available nutrients suddenly changes e.g. when the bacteria are grown in a mixture of sugars and one source of food is exhausted. The dip in the orange growth curve is due to the time needed for the bacteria to switch between the separate genetic circuits that are required to metabolise the different sugars. Most famously diauxic growth was observed in the PhD research of Jacques Monod who changed the growth media from glucose to lactose with a monoculture of *E. coli*.

Suggested Reading

Allen, R. J.; Waclaw. B., Bacterial growth: A statistical physicist's guide. *Reports on Progress in Physics* **2018**, *82* (1), 016601.

Cappuccino, J. G.; Welsh, C., *Microbiology: A Lab Manual*, 11th ed. Pearson: 2018. Considers the practical skills required to culture bacteria e.g. aseptic technique.

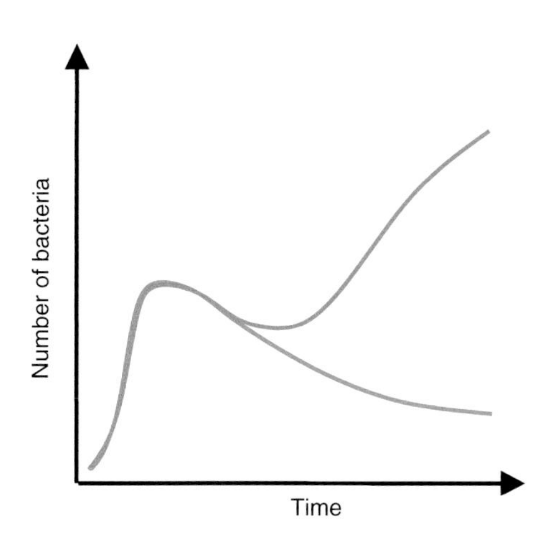

Figure 16.4 Growth curve of the number of bacteria as a function of time e.g. *Lactococcus lactis* or *E. coli*. *Diauxic growth* is observed when glucose in the growth media is exhausted and the bacteria switch to using lactose.[36] Orange, growth in media which contains *x*% glucose and *x*% lactose. Green, growth in media that only contains *x*% glucose.

Jun, S.; Si, F.; Pugatch, R.; Scott, M., Fundamental principles in bacterial physiology – history, recent progress, and the future with focus on cell size control. *Reports on Progress in Physics* **2018**, *81* (5), 56601.

Otto, S.; Day, T., *A Biologist's Guide to Mathematical Modelling in Ecology and Evolution*, Princeton: 2007.

Pham, V. H. T.; Kim, J., Cultivation of unculturable soil bacteria. *Trends in Biotechnology* **2012**, *30* (9), 475–484.

Scott, M.; Hwa, T., Bacterial growth laws and their applications. *Current Opinion in Biotechnology* **2011**, 22, 559–565.

References

1. Edelstein-Keshet, L., *Mathematical Models in Biology*. SIAM: 2005.
2. Allen, R. J.; Waclaw, B., Bacterial growth: A statistical physicist's guide. *Reports on Progress in Physics* **2018**, *82* (1), 016601.
3. Kaeberlein, T.; Lewis, K.; Epstein, S. S., Isolating 'uncultivable' microorganisms in pure culture in a simulated natural environment. *Science* **2002**, *296* (5570), 1127–1129.
4. Capuccino, J. G.; Welsh, C., *Microbiology: A Laboratory Manual*. Pearson: 2018.
5. Dyer, B. D., *A Field Guide to Bacteria*. Comstock Publishing Associates: 2003.
6. Kundakad, B.; Sevious, T.; Liang, Y.; Rice, S. A.; Kjelleberg, S.; Doyle, P. S., Mechanical properties of the superficial biofilm layer determine the architecture of biofilm. *Soft Matter* **2016**, *12* (26), 5718–5726.
7. Schaffner, M.; Ruhs, P. A.; Coulter, F.; Kilcher, S.; Studart, A. R., 3D printing of bacteria into functional complex materials. *Science Advances* **2017**, *3*, eaao6804.

8. Lehner, B. A. E.; Schmieden, D. T.; Meyer, A. S., A straightforward approach for 3D bacterial printing. *ACS Synthetic Biology* **2017**, *6* (7), 1124–1130.
9. Huang, Y.; Xia, A.; Yang, G.; Jun, F., Bioprinting living biofilms through optogenetic manipulation. *ACS Synthetic Biology* **2018**, *7* (5), 1195–1200.
10. Jun, S.; Si, F.; Pugatch, R.; Scott, M., Fundamental principles in bacterial physiology – history, recent progress, and the future with focus on cell size control: A review. *Reports on Progress in Physics* **2018**, *81* (5), 056601.
11. Scott, M.; Hwa, T., Bacterial growth laws and their application. *Current Opinion in Biotechnology* **2011**, *22* (4), 559–565.
12. Osella, M.; Tans, S. J.; Lagomarsino, M. C., Step by step, cell by cell: Quantification of the bacterial cell cycle. *Trends in Microbiology* **2017**, *25* (4), 250–256.
13. Reshes, G.; Vanounou, S.; Fishov, I.; Feingold, M., Cell shape dynamics in *Escherichia coli*. *Biophysical Journal* **2008**, *94* (1), 251–264.
14. Banerjee, S.; Lo, K.; Daddysman, M. K.; Selewa, A.; Kuntz, T.; Dinner, A. R.; Scherer, N. F., Biphasic growth dynamics control cell division in *Caulobacter crescentus*. *Nature Microbiology* **2017**, *2*, 17116.
15. Godin, M.; et al., Using the bouyant mass to measure the growth of single cells. *Nature Methods* **2010**, *7* (5), 387–390.
16. Mir, M.; Wang, Z.; Shen, Z.; Bednarz, M.; Bashir, R.; Golding, I.; Prasanth, S. G.; Popescu, G., Optical measurement of cell-dependent cell growth. *Proceedings of the National Academy of Sciences of the United States of America* **2011**, *108* (32), 13124–13129.
17. Rojas, E. R.; Huang, K. C., Regulation of microbial growth by turgor pressure. *Current Opinion in Microbiology* **2018**, *42*, 62–70.
18. Facchetti, G.; Chang, F.; Howard, M., Controlling cell size through sizer mechanisms. *Current Opinion in Systems Biology* **2017**, *5*, 86–92.
19. Nordholt, N.; van Heerden, J. H.; Bruggeman, F. J., Biphasic cell-size and growth-rate homeostasis by single *Bacillus subtilis* cells. *Current Biology* **2020**, *30* (12), 2238–2247.
20. White, D.; Drummond, J.; Fuqua, C., *The Physiology and Biochemistry of Prokaryotes*, 4th ed. Oxford University Press: 2012.
21. Kim, B. H.; Gadd, G. M., *Prokaryotic Metabolism and Physiology*, 2nd ed. Cambridge University Press: 2019.
22. van den Ent, F.; Amos, L. A.; Lowe, J., Prokaryotic origin of the actin cytoskeleton. *Nature* **2001**, *413* (6851), 39–44.
23. Ramos-Leon, F.; Ramamurthi, K. S., Cytoskeletal proteins: Lessons learned from bacteria. *Physical Biology* **2022**, *19* (2), 021005.
24. Howard, A.; Rutenberg, A. D.; de Vet, S., Dynamic compartmentalization of bacteria: Accurate diffusion in *E. coli*. *Physical Review Letters* **2001**, *87* (27 Pt 1), 278102.
25. Whitley, K. D.; et al., FtsZ treadmilling is essential for Z-ring condensation and septal constriction initiation in *Bacillus subtilis* cell division. *Nature Communications* **2021**, *12* (1), 2448.

26. Raskin, D. M.; de Boer, P. A. J., Rapid pole-to-pole oscillation of a protein required for directing division to the middle of *Escherichia coli. Proceedings of the National Academy of Sciences of the United States of America* **1999**, *96* (9), 4971–4976.
27. Snyder, L.; Peters, J. E.; Henkin, T. M.; Champness, W., *Molecular Genetics of Bacteria*, 4th ed. American Society for Microbiology: 2013.
28. Bertrand, R. L., Lag phase is a dynamic, organized, adaptive and evolvable period that prepares bacteria for cell division. *Journal of Bacteriology* **2019**, *201* (7), e00697–18.
29. Otto, S.; Day, T., *A Biologist's Guide to Mathematical Modeling in Ecology and Evolution*. Princeton University Press: 2007.
30. Wittrup, K. D.; Tidor, B.; Hackel, B. J.; Sarkar, C. A., *Quantitative Fundamentals of Molecular and Cellular Bioengineering*. MIT Press: 2020.
31. Liu, J. et al., Coupling between distant biofilms and emergence of nutrient time-sharing. *Science* **2017**, *356* (6338), 638–642.
32. Balagaddle, F. K.; You, L.; Hansen, C. L.; Arnold, F. H.; Quake, S. R., Long-term monitoring of bacteria undergoing programmed population control in a microchemostat. *Science* **2005**, *309* (5731), 137–140.
33. Yang, Y.; Santos, A. L.; Xu, L.; Lotton, C.; Taddei, F.; Lindner, A. B., Temporal scaling of aging as an adaptive strategy of *Escherichia coli. Science Advances* **2019**, *5* (5), eaaw2069.
34. Confalonieri, F.; Sommer, S., Bacterial and archael resistance to ionizing radiation. *Journal of Physics: Conference Series* **2011**, *261* (1), 012005.
35. Birjinuik, A.; Billings, A. N.; Nance, E.; Hanes, J.; Ribbeck, K.; Doyle, P. S., Single particle tracking reveals spatial and dynamic organization of the *E. coli* biofilm matrix. *New Journal of Physics* **2014**, *16* (8), 085014.
36. Solopova, A.; van Gestel, J.; Weissing, F. J.; Bachmann, H.; Teusink, B.; Kok, J.; Kuipers, O. P., Bet-hedging during bacterial diauxic shift. *Proceedings of the National Academy of Sciences of the United States of America* **2014**, *111* (20), 7427–7432.
37. Nassif, N.; Bouvet, O.; Rager, M. N.; Roux, C.; Coradin, T.; Livage, J., Living bacteria in silica gels. *Nature Materials* **2002**, *1* (1), 42–44.
38. Faith, J. J.; et al., The long-term stability of the human gut microbiota. *Science* **2013**, *341* (6141), 1237439.

17 Motility

Strains of *Escherichia coli* bacteria are the prototypical system used to study bacterial motility because of their well-defined genetics and the detailed ground-breaking work of Berg and others.[1] As a result, most of the molecules involved in the motors that drive the motility of *E. coli* are understood in great detail and so too are the genetic circuits for sensory transduction that determine whether the motors are switched on or off in response to external stimuli.[1] In-depth studies of bacterial motility from other species are patchier, but good detailed work is available with *Pseudomonas aeruginosa*, *Bacillus subtilis* and a handful of other model organisms. Thus, many more important discoveries are still expected in the field of bacterial motility that contains an estimated 10^{12} different species of bacteria.

Complex large-scale flows occur in some macroscopic bacterial systems e.g. marine environments, sewerage pipes, aerosols in clouds and blood infections. It has been hypothesised that a back and forth strategy for motility is advantageous for marine bacteria to stay in the vicinity of food sources in turbulent waters.[2] However, modelling turbulence in fluids is notoriously difficult and studies of microorganisms in many large-scale flows are challenging as a result.

17.1 Swimming of Dilute Bacteria

Some generic motility strategies for microorganisms are shown in Figure 17.1. They include *swimming*, *gliding*, *twitching*, *sliding*, *darting*,[3] *modulated buoyancy*, *tank treading* and *walking*. Some bacteria hijack the motile mechanisms of other host cells e.g. *Listeria* uses actin polymerisation of host eukaryotic cells to move around. Other mechanisms of motility, such as *swarming*, occur at high bacterial concentrations (Chapter 21).

17.1.1 Flagellar Bacteria

Flagellar bacteria are the most common form of motile bacteria.[4] They can be further subdivided into peritrichous (with multiple flagella that are randomly attached to the cellular surfaces e.g. *E. coli*), polar monotrichous (with a single flagellum attached at one end e.g. *P. aeruginosa*), lophotrichous cells (a tuft of flagella occurs at a single pole e.g. *Photobacterium fischeri*) and polar amphitrichous bacteria (the flagella occur at either end e.g. *Ectothiorhodospira halochloris*).[4] Inevitably due to the extreme diversity

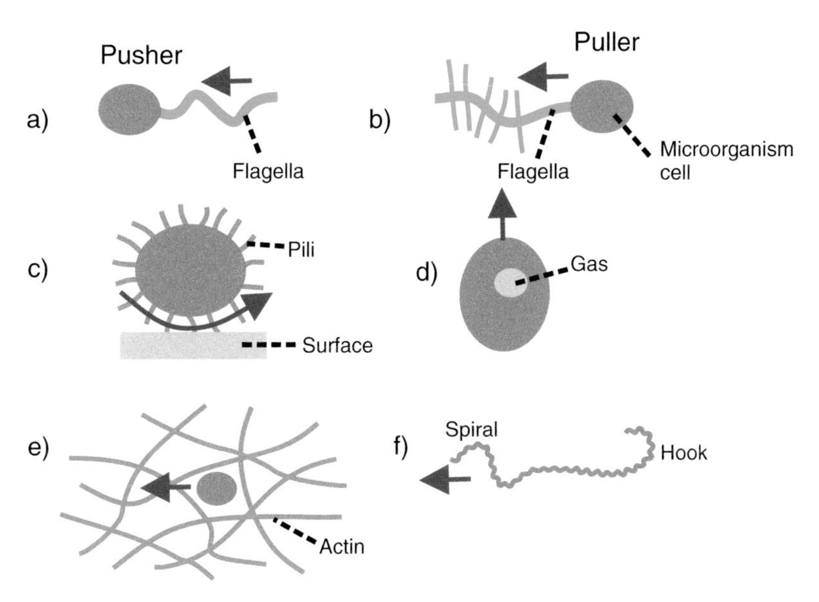

Figure 17.1 *Motility strategies* of microorganisms. (a) and (b) *swimming* via flagellar propulsion, pushers and pullers respectively, Equation (9.8), (c) *tank treading* with pili, (d) *active control of buoyancy*, (e) *actin self-assembly* and (f) *shape changes* of spirochetes. The microorganism cells are shown in orange.

of bacteria, exceptions to these categories exist, but these are the main varieties of flagellar bacteria.

Experimentally, it is found that bacteria with a single flagellum in most cases are restricted to running or reversing motions with a few able to perform run-reverse-flick manoeuvres e.g. *Vibrio alginolyticus*.[5] Multiflagellar bacteria demonstrate more sophisticated swimming behaviour, such as runs and tumbles (e.g. *E. coli*, Chapter 2) or the adoption of specific angles between runs.[6]

Escherichia coli are the most studied bacteria and thus peritrichous flagellar motility is relatively well understood (Figure 17.2). Counterclockwise motion of the flagella causes *E. coli* cells to move in approximately straight lines (they have high directional persistence), whereas clockwise motion causes the bacteria to tumble and randomise their direction (motion with lower persistence). A combination of counterclockwise and clockwise motion can allow the bacteria to navigate their environment in a stochastic manner (Chapter 2) and perform chemotaxis, since the bacteria are less likely to tumble when travelling up concentration gradients of chemoattractants and down concentration gradients of chemorepellants (Chapter 18).

Different strains of bacteria can have different numbers of flagella e.g. there are strains of *B. subtilis* with averages of 9, 26 or 41 flagella. A strong positive correlation was found between the number of flagella and the curvature of the swimming trajectories.[8]

The *swimming gaits* of individual bacteria can be highly variable, to the extent that it becomes possible to identify individual bacteria according to their gaits.[9] Whether

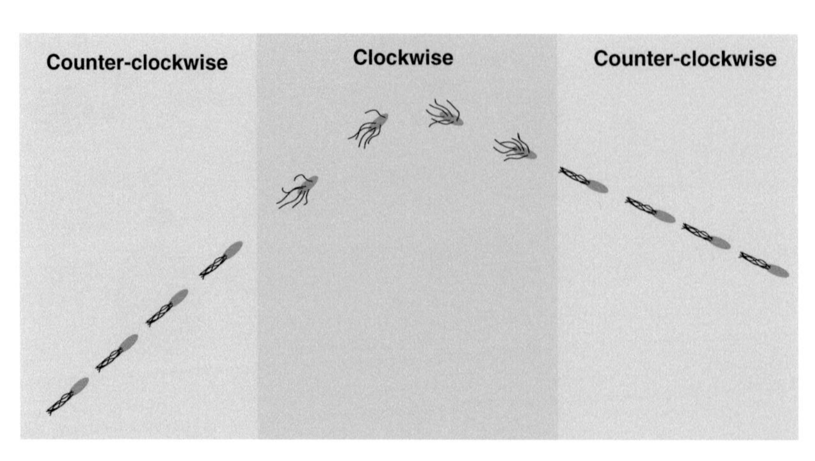

Figure 17.2 Motility of the prototypical flagellar bacterium, *E.coli*.[4] Counterclockwise rotation of the flagellar bundle causes bacteria to move in reasonably straight lines. Clockwise rotation of the flagellar motors (Section 19.3) causes the bacteria to tumble and change direction.[7]

the differences are wholly at the genetic level or due to changes in response to the environments (e.g. in sensory circuits) requires more research. A quantitative model for the sensory circuits in reasonable agreement with motility experiments is provided by the behavioural variability model (Chapter 2).[10]

Lots is known about the proteins in the sensory circuits of *E. coli*,[11] such as CheY (Chapter 18). A gene circuit involving CheY leads to the ultrasensitive activity of the bacterial motor that drives the flagella i.e. very small changes in chemoattractants can lead to very large changes in motility due to a very sharp sigmoidal response e.g. think of a Hill function for the sigmoidal kinetics of an enzyme $\left(f(c) = \frac{Bc^n}{c^n + X} \right)$ with a very large cooperativity parameter n, where c is the concentration of the chemoattractant and $f(c)$ determines the response of the bacterium.

Whether the flow field created by a single *E. coli* at low Reynolds number can be described by a force dipole or tripole[12] depends on the distance to a wall (Figure 17.3). Near to the wall, the bacterium acts as a tripole, whereas further away, it is better described with a dipole. The velocity (u) of fluid around a bacterium in the bulk is therefore given by the dipolar form,

$$\bar{u}(r) = \frac{A}{|r^2|}\left[3\left(\bar{r}.\bar{d}'\right)^2 - 1\right]\hat{r}, \quad A = \frac{lF}{8\pi\eta}, \quad \hat{r} = \frac{\bar{r}}{|r|}, \tag{17.1}$$

where d' is the unit vector in the swimming direction, r is the distance from the centre of the dipole, l is the dipole length, F is the dipole force, η is the viscosity of water and $\hat{r}$ is the unit vector. The flow field around a single *E. coli* is markedly different to larger eukaryotic microorganisms, such as *Volvox* and *Chlamydomonas*, in which cilial motions need to be considered. Such simplified dipole/tripole approximations for bacteria are often used to understand the collective hydrodynamics of interacting suspensions.[14] Whether they are accurate approximations for all varieties of flagellar

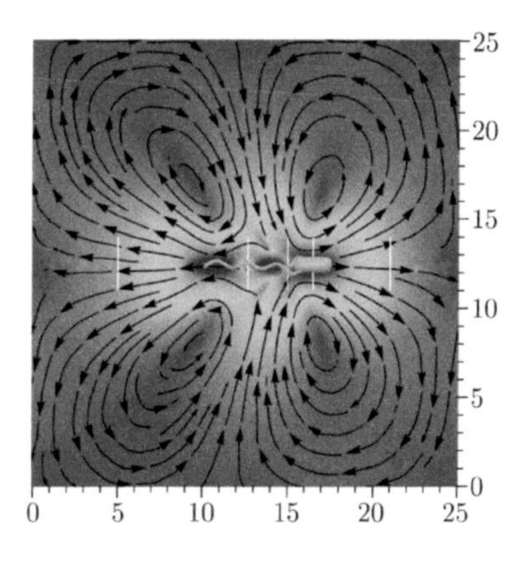

Figure 17.3 *Flow field* around an *E. coli* is well approximated by a dipole at large distances from a wall[12] with characteristic r^{-2} scaling of the hydrodynamic forces on the distance (r).[13]

bacteria is yet to be established. The sign of the dipole changes for pushers and pullers by definition, Chapter 9.

A question is how do non-motile bacteria (such as *Staphylococcus aureus*) or bacteria imbedded in a biofilm that are fixed in space obtain their nutrients. The answer is predominantly due to diffusion (Chapter 2), although bacteria growing near to flows (e.g. a *S. aureus* infection growing on a heart valve in contact with blood) can make use of convection to improve nutrient fluxes. Bacterial division can also occur preferentially in the direction of a source of food during growth, although this occurs relatively slowly.

Experiments have examined the torques of flagellar motors for forward versus backward motion i.e. pusher versus puller activity (Chapter 9) in terms of their molecular mechanisms.[15] The forward-backward torque anisotropy is found to be similar in *E. coli* and *Caulobacter crescentus*, but the puller motion requires more torque.

Comparative hydrodynamics calculations show that the flagellar behaviour observed in experiments with *E. coli* is the most energetically efficient possibility,[16] which indicates flagellar efficiency is an important parameter for bacterial fitness that is finely tuned by evolution. The efficiency of propulsion improves with flagellar diameter, which in part explains the occurrence of flagellar bundles e.g. eight flagella wind together and act as a single appendage in standard strains of *E. coli* during forward motion.

Positive rheotaxis is often seen for *E. coli* i.e. the bacteria swim upstream.[17] The rheotaxis of dilute suspensions of *E. coli* has been examined in a microfluidic apparatus.[18] The results were in good agreement with an active colloidal model that included the chirality of the flagella e.g. Jeffrey orbits (also observed for inanimate colloids under flow[19]) of the bacterial orientations were measured and predicted in the flow.

The flagellar motor of *E. coli* can be driven to rotate using the application of external electrical fields[20] (Chapters 3 and 19). It could also be a useful mechanism to actuate motors in synthetic bionanotechnology applications.

Pseudomonas putida in contrast to *E. coli* has two speeds of motility during its run phases, rather than one. *P. putida* alternates between two speeds of forward motion and then experiences sudden 180° switches in direction.[6] A more sophisticated statistical model is thus required to describe *P. putida* motility compared to *E. coli* (Chapter 2), although multi-fractal models may conveniently generalise for its description (three peaks would be expected in the distributions of generalised diffusion coefficients and anomalous exponents).

Hydrodynamics in microfluidic flows can be used to separate populations of both different-shaped bacteria and bacteria with different motility mechanisms.[21] Microfluidic geometries can enable the accurate measurement of bacterial transport parameters and improve the reproducibility of experiments.[22] Since the geometry is at such small length scales, the flow behaviour follows the predictions for low Reynolds number hydrodynamics (Equation (9.3)). Microfluidic flows lack conventional turbulence effects and are commonly laminar and stable, although exceptions are possible in strongly viscoelastic flows e.g. due to elastic turbulence.[23] Simplification of the background flow patterns in laminar flows facilitates measurements of the flow patterns of bacteria and can allow single bacterial effects to be isolated.

17.1.2 Non-flagellar Motility

The next most common form of motility after flagellar varieties are *spirochetes*, e.g. *Borrelia burgdorferi* that cause Lyme disease. In general, spirochetes are cork-screw-shaped bacteria that are motile due to the shape changes of their cells (Figure 17.4). Spirochetes can both crawl on surfaces and swim in the bulk. The transition between these two motile behaviours has been explored in detail with *Leptospira*.[24]

Spiroplasma is a helical bacterium that infects insects and plants. It swims by switching the sense of its helicity (i.e. its chirality) along its body. Hydrodynamic calculations point to an optimal occurrence of both the pitch angle of the helix (35.5°) and the inter-kink length i.e. the length between changes in helical handedness depends on the cell body length and an optimal value is chosen for swimming efficiency.[25]

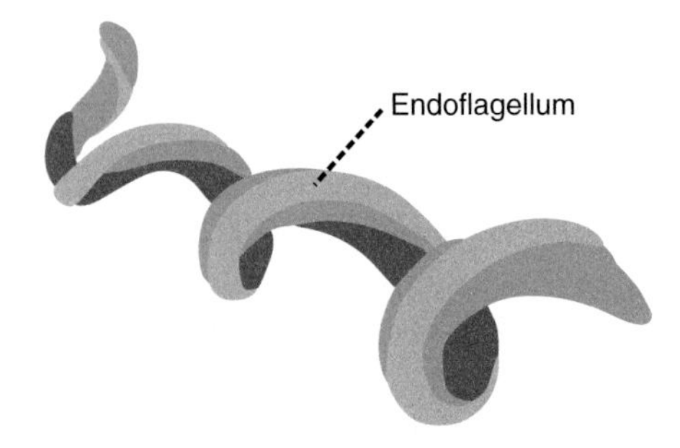

Figure 17.4 *Spirochetes* are cork-screw-shaped organisms that move using shape changes to their body. *Borrelia burgdorferi* is shown in the image. The endoflagellum drives the shape changes.

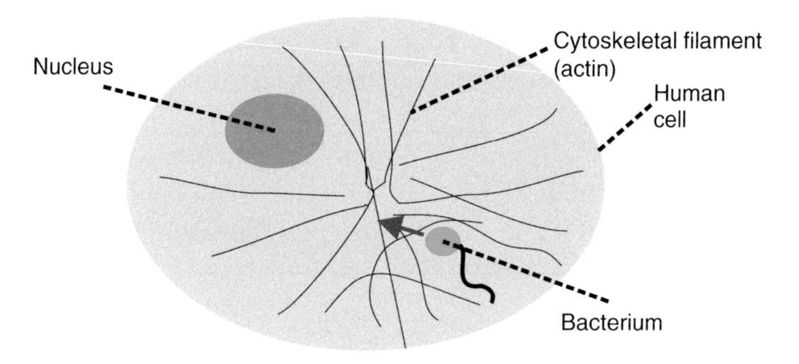

Figure 17.5 Schematic diagram of *Listeria monocytogenes* inside a human cell using the self-assembly of actin for propulsion. Also see Figure 19.1.

Type IV pili are appendages on the walls of bacteria that can perform a variety of functions including the exchange of genetic material (Figure 17.1c, Section 19.2). These pili can retract through the bacterial cell wall and synchronisation of these events can lead to twitching motility. For example, *P. aeruginosa* can alternate between flagellar and type IV pili-driven motility mechanisms when they explore surfaces,[26] whereas gonorrhoeae is restricted to type IV pili twitching. The pili-driven motion in *P. aeruginosa* can be further separated into vertical cell (walking) and horizontal cell (crawling) varieties that have distinct statistical descriptions e.g. the mean square displacements for the motion as a function of time can switch from subdiffusive to super-diffusive scaling (Chapter 2), dependent on the orientation of the bacteria.

Listeria monocytogenes can be pathogenic and these bacteria are sometimes transferred in badly prepared food e.g. cheese and uncooked vegetables. *L. monocytogenes* can move itself around inside human cells (Figure 17.5[27]). It does this by hijacking the self-assembling motors that are used for motility by the eukaryotic cells i.e. the bacterium is able to polymerise actin comets (Figure 19.1). *Shigella flexneri* (another pathogenic intracellular microorganism) is also able to use actin comets for motility.[28] It is found that *Shigella* remain firmly bound to actin filament ends as they elongate via self-assembly and it can be described with a lock, load and fire model. The motor is called autoclampin.[29] It is possible to examine the motility of both *Listeria* and *Shigella in vitro* with solutions containing the components required for actin self-assembly i.e. actin monomers, Arp213 complex, cofilin and capping proteins.[27] The small amplitudes of Brownian motion observed imply the bacteria are tightly bound to the actin and 5.4 nm actin substeps are seen for their motion (actin monomers are 8 nm in size, so this implies substeps during motion).[29]

Some work has been performed to make a census of the motility mechanisms of a wide range of microorganisms (plankton, larvae and bacteria) in marine environments.[30,31] Such exploratory studies are only starting to understand the extreme diversity of microorganisms in the oceans and they are crucial for an in-depth understanding of marine ecology.

The motion of bacteria in complex fluids is considered in Chapter 21.

17.2 Sedimentation of Bacteria

The *sedimentation rate* of non-motile synthetic colloidal spheres[32] can be calculated at low Reynolds number and provides a first approximation to understand the sedimentation of single immotile bacteria (Figure 17.6). The velocity of sedimentation is given by the ratio of the buoyancy force (due to gravity) to the friction coefficient (f),

$$v_{sed} = \frac{m'g}{f} = \frac{m'g}{6\pi\eta a} = 2a^2 \frac{(\rho_s - \rho)g}{9\eta}, \tag{17.2}$$

where $m'g$ is the weight of the bacterium (adjusted for buoyancy), ρ_s is the bacterial density, ρ is the density of water, g is gravity and a is the bacterial radius. This is a reasonably good approximation for spherical bacteria (e.g. the occasional example of unaggregated *S. aureus*) at low concentrations in water (Figure 17.6a). A key aspect of Equation (17.2) for the sedimentation rate is the proportionality with respect to a^2, the bacterial radius squared. A sphere of radius $a = 1$ μm, with density $\rho_s = 1.2$ gcm^{-3}, sediments at $v_{sed} = 4.4 \times 10^{-5}$ cms^{-1}. Thus, the majority of immotile bacteria will quickly drop to the bottom of a microfluidic container due to the buoyancy force (occasional examples of marine bacteria have gas-filled regions to modify their buoyancy, but they are unusual). Analytic expressions also exist for the friction coefficients of ellipsoids and rods (Figure 17.6b), which can be used to replace the friction coefficient (f) in Equation (17.2) for non-spherical bacteria. There are also numerical solvers for f that can be used for other less symmetrical shapes, including aggregates of bacteria, if it is assumed they are rigid.[33]

More sophisticated physical phenomena are expected during the *sedimentation* of bacteria based on comparative studies with synthetic colloidal suspensions,[34] although they still need to be rigorously demonstrated in experiments with bacteria. When multiple particles experience simultaneous sedimentation, their velocities increase beyond the values expected for isolated particles (Equation (17.2)) due to the hydrodynamic interaction between neighbouring particles, which causes the total collective hydrodynamic resistance to decrease. Sharp boundaries (that demonstrate clear demarcations between fluid, sedimenting fluid and sediment phases) can occur during sedimentation at high concentrations of synthetic colloids which can be described with sharp

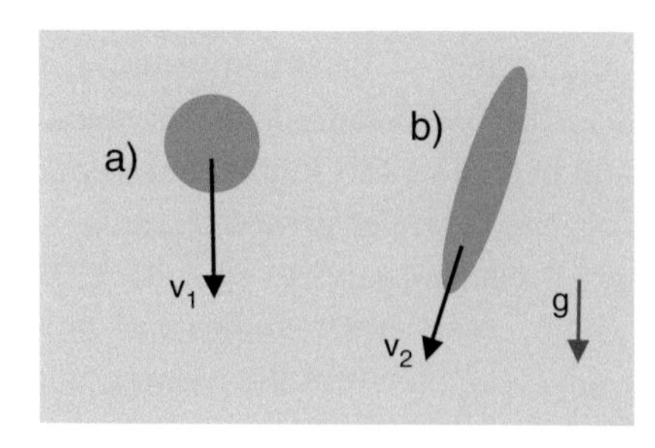

Figure 17.6 *Sedimentation* of (a) a spherical bacterium (velocity v_1) and (b) an elongated bacterium (velocity v_2). g is gravity.

shock solutions to a non-linear wave equation. Sedimentation rates also increase if the vessel in which the particles fall is inclined with respect to gravity and this is called the *Boycott effect*. Elongated objects (e.g. synthetic fibres, but it is also expected for filamentous bacteria) drift sideways as they sediment due to the hydrodynamic drag anisotropy (Figure 17.6b).

17.3 Swimming Near Surfaces

Bacteria are commonly found *swimming near surfaces* (Figure 17.7). This is a crucial process for the nucleation of biofilms,[35] but surfaces can also provide environments that are food rich and relatively secure for planktonic phases. The swimming behaviour of bacteria in Newtonian fluids (such as water) near to surfaces is strongly affected by the hydrodynamic boundary conditions. A crucial empirical factor used in conventional fluid mechanics calculations is that the velocity of fluids at a solid surface is zero (the boundary conditions). This is well established in experiments down to the nanoscale with many synthetic materials, although exceptions are known for superhydrophobic materials.[36] Exceptions are also known for viscoelastic complex fluids and they can have measurable slip lengths, which extend to micron scales in extreme cases e.g. entangled DNA solutions[23] and worm-like micelles.[37]

In general, the *hydrodynamic interaction* of a microorganism swimming in a Newtonian fluid in the vicinity of a surface can cause it to: (1) modify its motility and cause a change in the swimming speed of the organism, (2) change its trajectory due to the induced rotation and (3) drift towards the surface causing the accumulation of the microorganisms on the surface.[38] Motion of a microorganism near a single interface can lead to an (1) attraction to rigid walls, (2) attraction to both deforming and non-deforming interfaces, (3) swimming in circles, (4) scattering away from the rigid or free surface and (5) swimming at a fixed distance from the wall.

An effective *hydrodynamic attractive force* is often observed when microorganisms swim close to a solid/liquid interface created from Newtonian fluids.[39] Furthermore, bacteria will swim in circles when they are near to such interfaces. Specifically, *E. coli* in water swims in clockwise circles near solid/liquid interfaces.[40] Comparison of solid/

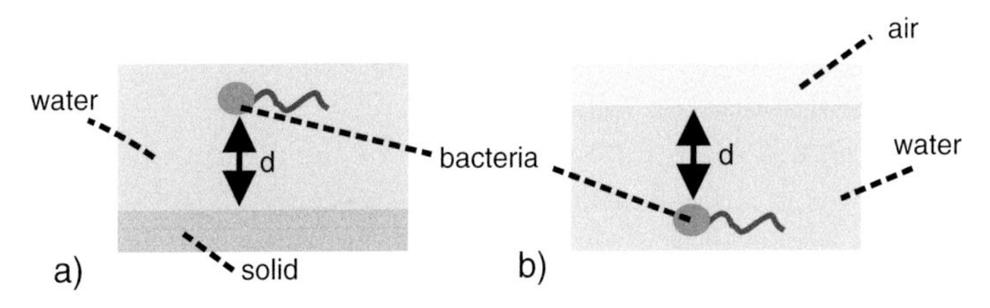

Figure 17.7 Bacteria swimming *near to surfaces*: (a) a solid/liquid interface and (b) a liquid/air interface. *d* is the perpendicular distance to the surface.

liquid and liquid/air interfaces shows they can cause a change of the sense of rotation for the circular motion i.e. the *E. coli* swim clockwise (solid/liquid) and counterclockwise (liquid/air), respectively.[41]

The residence times of bacterial cells that swim in the vicinity of a solid surface with an imposed external flow increase linearly with the externally imposed shear stress for *P. aeruginosa*.[42] These residence times reflect the decision-making processes of the bacteria as they modify their ability to associate with surfaces in response to mechanical cues from their environment.

Bacteria can release substances that assist their motility on surfaces. Thus, their swimming speed can increase as a power law of bacterial concentration on solid surfaces,[3] as the collective release process helps the motility of the community i.e. it is a common goods process in the terminology of game theory (Chapter 22). For example, *B. subtilis* can create a fluid layer on top of an agar gel that contains the surface-active wetting compound, surfactin, which helps bacterial motility (Section 23.2).

The pili in *P. aeruginosa* were found to preferentially follow slime trails of extracellular polymeric substance on surfaces created by either themselves or neighbouring bacteria.[43] This is broadly analogous to pheromone trails created by insects,[44] such as ants, but on much smaller length scales. Interestingly, this provides another memory mechanism for motility and creates a highly non-Markovian process for the stochastic motion of individual bacteria.

Holographic microscopy experiments were combined with optical tweezers to understand the entrapment of *E. coli* at water-air interfaces covered with surfactant.[45] The optical tweezer experiments allowed the forces of hydrodynamic coupling with the interface to be quantified.

Water in oil emulsions have been used to study the motility of *E. coli* confined inside $10-20\ \mu m$ sized aqueous droplets[46] (Figure 17.8). At low bacterial concentrations, the bacteria hugged the wall of the spherical droplets, similar to bacterial suspensions confined to two dimensions between two planar surfaces,[47] whereas at higher bacterial concentrations, surface scattering due to collisions of bacteria with one another populated the bulk of the droplets with bacteria.

Separate experiments with high-concentration bacterial suspensions using water in oil emulsion droplets found that vortices form with *B. subtilis* that have a continuous sense of rotation.[48] In contrast, droplets that contain *E. coli* have microscopic cycles in which the sense of the vortex periodically reversed in direction[49]; they resemble erratic stochastic washing machines. Additional experiments on *E. coli* suspensions in water in oil emulsion droplets explored the motion of the droplets themselves, which were driven by the bacterial motors they contained and could be described with a persistent random walk model.

There is often a large depletion of the concentration of bacteria from low shear regions in microfluidic geometries[47] (Figure 17.9) i.e. higher concentrations of bacteria accumulate at the walls. Theoretical predictions have been made for the average number density of *E. coli* confined between two glass plates when there is an external shear force.[39] At zero shear, the number density follows a parabolic profile with the bacteria enriched at the walls due to a hydrodynamic interaction.[47] Similar

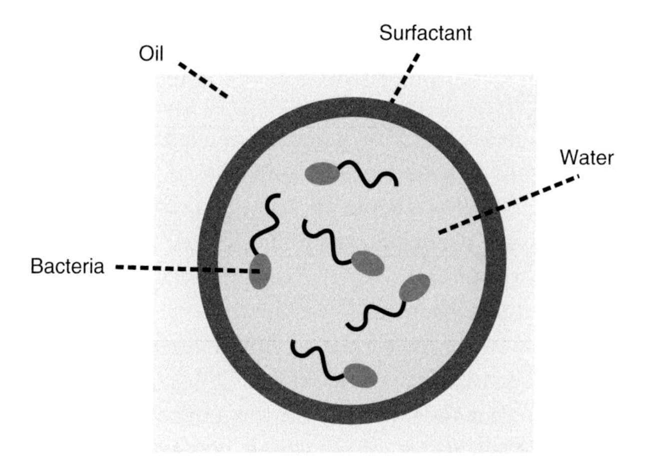

Figure 17.8 Motility of bacteria in oil and water emulsions stabilised by surface active molecules (surfactants).

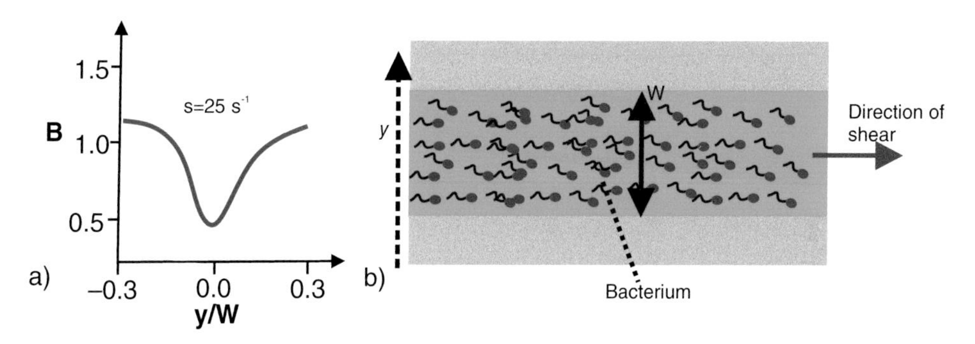

Figure 17.9 *B. subtilis* become depleted from low shear regions (the centre of the flow) and the effect is maximised at intermediate shear rates.[47] (a) Bacterial concentration (B) as a function of position (y) across the channel with a shear rate of $s = 25\,s^{-1}$. W is the width of the channel. (b) Schematic diagram of the pressure driven flow of bacteria in a channel. The axis y is perpendicular to the long axis of the channel.

experiments with *B. subtilis* show that depletion from high shear regions in microfluidic flows is optimised at intermediate shear rates (Figure 17.9).

A detailed model was given for the number density of *Caulobacter crescentus* cells, which also demonstrated a parabolic profile as a function of distance from the walls under shear.[50] Here, the accumulation of the bacteria at the interfaces was deduced to be due to collisions and the effects of rotational Brownian motion, rather than hydrodynamic interactions, which were calculated to be much smaller effects. This interpretation helps explain similar phenomena that occur with much larger motile cells, such as bull spermatozoa, and thus appears to be a more successful approach for modelling the phenomena.

Self-induced streaming instabilities occurred in growing films of *E. coli* that were restricted to move in two dimensions.[51] Both the swim speed of the *E. coli* and the amount of upstream motility were found to be independent of the bacterial cells' aspect ratios i.e. the ratio of the bacterial lengths to their diameters.

Active bacterial carpets can drive dust particles around if bacteria are attached to a synthetic solid surface (e.g. the bacteria are glued to a microscope slide or a colloidal sphere) and the bacteria have a motile phenotype.[52] This provides a mechanism to create a cyborg system in which living bacteria are combined with synthetic components to create active composites.[52] Specifically, the cell bodies of *Serratia marcescens* were adhered to two-dimensional PDMS surfaces and fluids in their vicinity were driven using the flagella. A similar methodology was also used with polystyrene spheres (10 μm diameter) that were driven around randomly by teams of around fifty bacteria.

Anomalous dispersion was found for suspensions of *E. coli* pushed through funnel-like constrictions in microfluidic experiments.[53] The bacterial concentration can build up in the flow after the point of constriction, which is opposite to the behaviour observed with passive particles. The phenomenon is thought to be due to hydrodynamic stresses that cause bacteria to desorb from the boundaries.

Suggested Reading

Aranson, I. S., Bacterial active matter. *Reports on Progress in Physics* **2022**, *85* (7), 1–43.

Lauga, E., *The Fluid Dynamics of Cell Motility*. Cambridge University Press: 2020.

Lauga, E.; Powers, T. R., The hydrodynamics of swimming microorganisms. *Reports on Progress in Physics* 2009, *72* (9), 1–36.

References

1. Berg, H. C., *E. coli in Motion*. Springer: 2004.
2. Luchsinger, R. H.; Bergersen, B.; Mitchell, J. G., Bacterial swimming strategies and turbulence. *Biophysical Journal* **1999**, *77* (5), 2377–2386.
3. Ben-Jacob, E.; Cohen, I.; Levine, H., Cooperative self-organization of microorganisms. *Advances in Physics* **2010**, *49* (4), 395–554.
4. Lauga, E., Bacterial hydrodynamics. *Annual Review of Fluid Mechanics* **2016**, *48* (1), 105–130.
5. Son, K.; Guasto, J. S.; Stocker, R., Bacteria can exploit a flagellar buckling instability to change direction. *Nature Physics* **2013**, *9* (8), 494–498.
6. Theves, M.; Taktikos, J.; Zaburdaev, V.; Stark, H.; Beta, C., A bacterial swimmer with two alternating speeds of propagation. *Biophysical Journal* **2013**, *105* (8), 1915–1924.
7. Turner, L.; Ryu, W. S.; Berg, H. C., Real-time imaging of fluorescent flagellar filaments. *Journal of Bacteriology* **2000**, *182* (10), 2793–2801.

8. Najafi, J.; Shaebani, M. R.; John, T.; Altegoer, F.; Bange, G.; Wagner, C., Flagellar number governs bacterial spreading and transport efficiency. *Science Advances* **2018**, *4* (9), eaar6425.
9. Levin, M. D.; Morton-Firth, C. J.; Abouhamad, W. N.; Bourret, R. B.; Bray, D., Origins of individual swimming behavior in bacteria. *Biophysical Journal* **1998**, *74* (1), 175–181.
10. Figueroa-Morales, N.; Soto, R.; Junat, G.; Darnige, T.; Douarche, C.; Martinez, V. A.; Lindner, A.; Clement, E., 3D spatial exploration by *E. coli* echoes motor temporal variability. *Physical Review X* **2020**, *10* (2), 021004.
11. Cluzel, P.; Surette, M.; Leibler, S., An ultrasensitive bacterial motor revealed by monitoring signaling proteins in single cells. *Science* **2000**, *287* (5458), 1652–1655.
12. Drescher, K.; Dunkel, J.; Cisneros, L. H.; Ganguly, S.; Goldstein, R. E., Fluid dynamics and noise in bacterial cell-cell and cell-surface scattering. *Proceedings of the National Academy of Sciences of the United States of America* **2011**, *108* (27), 10940–10945.
13. Antimicrobial peptide database. http://aps.unmc.edu/AP.
14. Pismen, L., *Active Matter Within and Around Us: From Self-propelled Particles to Flocks and Living Forms*. Springer: 2021.
15. Lele, P. P.; Roland, T.; Shrivastava, A.; Chen, Y.; Berg, H. C., The flagellar motor of *Caulobacter crescentus* generates more torque when a cell swims backward. *Nature Physics* **2016**, *12* (2), 175–178.
16. Spagnolie, S. E.; Lauga, E., Comparative hydrodynamics of bacterial polymorphism. *Physical Review Letters* **2011**, *106* (5), 058103.
17. Kaya, T.; Koser, H., Direct upstream motility in *E. coli*. *Biophysical Journal* **2012**, *102* (7), 1514–1523.
18. Jing, G.; Zoettl, A.; Clement, E.; Lindner, A., Chirality-induced bacterial rheotaxis in bulk shear flows. *Science Advances* **2020**, *6* (28), eabb2012.
19. Guyon, E.; Hulin, J. P.; Petit, L.; Mitescu, C. D., *Physical Hydrodynamics*, 2nd ed. Oxford University Press: 2015.
20. Berg, H. C.; Turner, L., Torque generated by the flagellar motor of *Escherichia coli*. *Biophysical Journal* **1993**, *65* (5), 2201–2216.
21. Gurung, J. P.; Gel, M.; Baker, M. A. B., Microfluidic techniques for separation of bacterial cells via taxis. *Microbial Cell* **2020**, *7* (3), 66–79.
22. Ahmed, T.; Stocker, R., Experimental verification of the behavioral foundation of bacterial transport parameters using microfluidics. *Biophysical Journal* **2008**, *95* (9), 4481–4493.
23. Malm, A. V.; Waigh, T. A., Elastic turbulence in entangled semi-dilute DNA solutions measured with optical coherence tomography velocimetry. *Scientific Reports* **2017**, *7* (1), 1186.
24. Tahara, H.; Takabe, K.; Sasaki, Y.; Kasuga, K.; Kawamoto, A.; Koizumi, N.; Nakamura, S., The mechanism of two-phase motility in the spirochete *Leptospira*: Swimming and crawling. *Science Advances* **2018**, *4* (5), eaar7975.
25. Yang, J.; Wolgemuth, C. W.; Huber, G., Kinematics of the swimming of spiroplasma. *Physical Review Letters* **2009**, *102* (21), 218102.

26. Conrad, J. C.; et al., Flagella and pili-mediated near-surface single-cell motility mechanisms in *P. aeruginosa*. *Biophysical Journal* **2011**, *100* (7), 1608–1616.
27. Lolsel, T. P.; Boujeman, R.; Pantaloni, D.; Carlier, M. F., Reconstitution of actin-based motility of Listeria and Shigella using pure proteins. *Nature* **1999**, *401* (6753), 613–616.
28. Pantaloni, D.; Le Clainche, C.; Carlier, M. F., Mechanism of actin-based motility. *Science* **2001**, *292* (5521), 1502–1506.
29. Dickinson, R. B.; Purich, D. L., Clamped-filament elongation model for actin-based motors. *Biophysical Journal* **2002**, *82* (2), 605–617.
30. Wheeler, J. D.; Secchi, E.; Rusconi, R.; Stocker, R., Not just going with the flow: The effects of fluid flow on bacteria and plankton. *Annual Review in Cellular Developmental Biology* **2019**, *35* (1), 213–237.
31. Guasto, J. S.; Rusconi, R.; Stocker, R., Fluid mechanics of planktonic microorganisms. *Annual Review of Fluid Mechanics* **2012**, *44*, 373–400.
32. Berg, H. C., *Random Walks in Biology*. Princeton University Press: 1993.
33. Carrasco, B.; de la Torre, J. G., Hydrodynamic properties of rigid particles: Comparison of different modeling and computational procedures. *Biophysical Journal* **1999**, *76* (6), 3044–3057.
34. Guazzelli, E.; Morris, J. F., *A Physical Introduction to Suspension Dynamics*. Cambridge University Press: 2012.
35. Harshey, R. M., Bacterial motility on a surface: Many ways to a common goal. *Annual Reviews of Microbiology* **2003**, *57*, 249–273.
36. Maali, A.; Bhushan, B., Measurement of slip length on superhydrophobic surfaces. *Philosophical Transactions of the Royal Society A* **2012**, *370* (1967), 2304–2320.
37. Lettinga, P.; Manneville, S., Competition between shear banding and wall slip in wormlike micelles. *Physical Review Letters* **2009**, *103* (24), 248302.
38. Desai, N.; Ardekani, A. M., Biofilms at interfaces: Microbial distribution in floating films. *Soft Matter* **2020**, *16* (7), 1731–1750.
39. Berke, A. P.; Turner, L.; Berg, H. C.; Lauga, E., Hydrodynamic attraction of swimming microorganisms by surfaces. *Physical Review Letters* **2008**, *101* (3), 038101.
40. Lauga, E.; DiLuzio, W. R.; Whitesides, G. M.; Stone, H. A., Swimming in circles: Motion of bacteria near solid boundaries. *Biophysical Journal* **2006**, *90* (2), 400–412.
41. Di Leonardo, R.; Dell'Arciprete, D.; Angelani, L.; Lebba, V., Swimming with an image. *Physical Review Letters* **2011**, *106* (3), 038101.
42. Lecuyer, S.; Rusconi, R.; Shen, Y.; Forsyth, A.; Vlamakis, H.; Kolter, R.; Stone, H. A., Shear stress increases the residence time of adhesion of *Pseudomonas aeruginosa*. *Biophysical Journal* **2011**, *100* (2), 341–350.
43. Gelimson, A.; Zhao, K.; Lee, C. K.; Kranz, W. T.; Wong, G. C. L.; Golestanian, R., Multicellular self-organization of *P. aeruginosa* due to interactions with secreted trails. *Physical Review Letters* **2016**, *117* (17), 178102.

44. Couzin, I. D.; Franks, N. R., Self-organized lane formation and optimized traffic flow in army ants. *Proceedings of Royal Society B – Biological Sciences* **2003**, *270* (1511), 139–146.
45. Bianchi, S.; Saglimbeni, F.; Frangipone, G.; Dell'Arciprete, D.; Di Lenarda, R., 3D dynamics of bacteria wall entrapment at a water-air interface. *Soft Matter* **2019**, *15* (16), 3397–3406.
46. Vladescu, I. D.; et al., Filling an emulsion drop with motile bacteria. *Physical Review Letters* **2014**, *113* (26), 268101.
47. Rusconi, R.; Guasto, J. S.; Stocker, R., Bacterial transport suppressed by fluid shear. *Nature Physics* **2014**, *10*, 212–217.
48. Hamby, A. E.; Vig, D. K.; Safonova, S.; Wolgemuth, C. W., Swimming bacteria power microscopic cycles. *Science Advances* **2018**, *4* (12), eaau0125.
49. Ramos, G.; Cordero, M. L.; Soto, R., Bacterial driving droplets. *Soft Matter* **2020**, *16* (5), 1359–1365.
50. Li, G.; Tang, J. X., Accumulation of microswimmers near a surface mediated by collision and rotational Brownian motion. *Physical Review Letters* **2009**, *103* (7), 078101.
51. Mather, W.; Mondragon-Palomino, O.; Danino, T.; Hasty, J.; Tsimring, L. S., Streaming instability in growing cell populations. *Physical Review Letters* **2010**, *104* (20), 208101.
52. Darnton, N.; Turner, L.; Breuer, K.; Berg, H. C., Moving fluid with bacterial carpets. *Biophysical Journal* **2004**, *86* (3), 1863–1870.
53. Altschuler, E.; Mino, G.; Perez-Penichet, C.; del Rio, L.; Lindner, A.; Rousselet, A.; Clement, E., Flow controlled densification and anomalous dispersion of *E. coli* through a constriction. *Soft Matter* **2013**, *9* (6), 1864–1870.

18 Chemotaxis and Detection

A wide variety of *sensory modules* exist in bacteria to detect physical stimuli, such as light (often bacteria are particularly sensitive to blue/UV wavelengths, since they create reactive oxygen species [ROS] that can damage them), temperature, chemoattractants, chemorepellants, magnetic fields and pH. Most bacteria have motile phases and respond to stimuli with altered forms of motility. Even a *Staphylococcus aureus* colony (which lacks any clear form of extracellular motility) can grow in a defined direction in response to a stimulus and thus, thermodynamically speaking, the bacteria are motile on large time scales. The direction of bacterial motility and other physiological responses (e.g. biofilm formation) is determined by sensory information from an array of molecular sensors on the surfaces of the bacteria. Due to their small size (μm), most bacteria need to compare variations of stimuli with time (e.g. by moving across concentration gradients), since insufficient variations occur in space across the organism's body.[1,2] Information from the sensory modules is combined using gene networks, so that decisions are made by the bacterial cells to optimise their survival with respect to their environments. Other biomolecular alternatives to protein sensors directly expressed by genes are also possible e.g. RNA thermometers can modify gene expression in a temperature-sensitive manner.[3]

Sensory transduction schemes in bacteria typically consist of a sensor module with a transmitter region that sends a signal to a receiver on a response modulator e.g. in the transduction modules of chemotaxis, osmoregulation and nitroregulation in *Escherichia coli*.[4]

A list of some chemicals that strongly affect the motility of *E. coli* is shown in Table 18.1. *E. coli* are sensitive to a wide range of molecules. The molecules can be simply classified as either attractants or repellents, although the responses of the bacteria to these stimuli are expected to be more nuanced in practice. A substantial amount of work was completed on the detection of simple sugars by bacteria in the 1960s, such as glucose, glycerol and galactose.[5,6]

Chemotactic sensors for *E. coli* form in clusters on one of the cellular poles. The use of multiple sensors in clustered arrays is due to the increase in amplification and thus sensitivity provided by allostery i.e. the coordinated activity of sensory receptors increases the sharpness of their responses.[7] With *E. coli*, adaptation of the sensors is possible due to the methylation of a sensory protein (CheR, Figure 18.1) i.e. the modulation of sensitivity in response to the varying availability of a chemoattractant.

Calculations can motivate the size of clusters of the sensors on the surface of the bacteria based on a balance of the external and internal noise on the sensory information e.g. the clustering can provide an optimal signal to noise for chemotaxis.[7]

Table 18.1 Chemicals that strongly affect the motility of *E. coli* can be placed into two broad classifications as *attractants* and *repellents*.[6]

Attractants	Amino acids, dipeptides, electron acceptors (oxygen, nitrate, fumarate), membrane-permeant bases, salts at low concentrations, sugar and sugar alcohols.
Repellents	Alcohols, amino acids, chemicals at high osmotic strength, divalent cations, glycerol or ethylene glycol at high concentrations, indole and membrane-permeant acids.

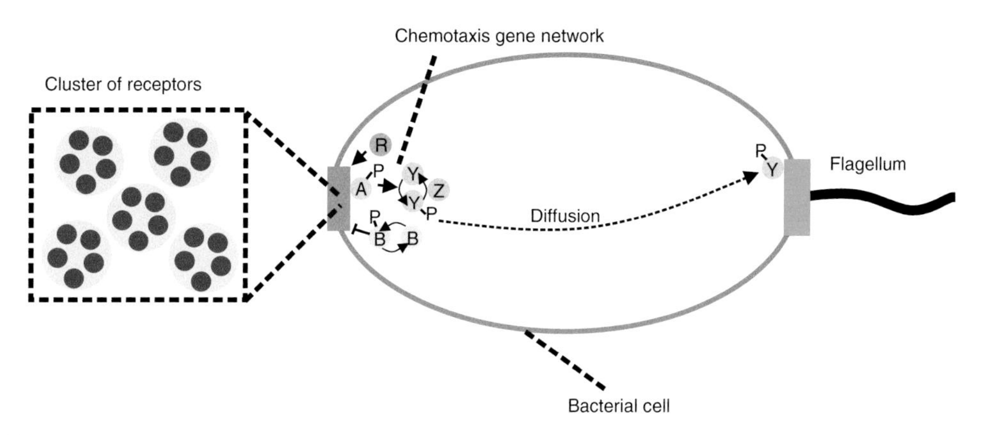

Figure 18.1 The cluster of receptors on the surface of *E. coli* interacts with a *chemotaxis gene network* causing the flagellum to rotate. *R* is CheR, *A* is CheA, *B* is CheB, *Y* is CheY and *Z* is CheZ. *P* is a phosphate group used in signalling.[7]

Typically, the array of sensors in *E. coli* are 0.2 – 0.4 μm in diameter and contain up to 5,000 receptors.

Chemotaxis becomes less efficient at higher bacterial concentrations due to collisions with other bacteria during run and tumble motions,[8] which randomise their motile responses. This is a possible motivation for the altered mechanisms observed for motility at high bacterial concentrations, such as swarming (Chapter 21).

Mechanical sensing of bacteria allows them to measure contact forces and fluid flows[9] e.g. via dedicated mechanically sensitive ion channels (Section 19.4). Such mechanical sensing can lead to feedback based on force-induced motility e.g. deep sea bacteria (piezophiles) are optimised for high pressure motility and increase the production of lateral flagella at high pressures, which facilitate high viscosity motility.

There is a fundamental question as to whether bacterial cells are too small to spatially measure concentration gradients.[1] Instead, they are thought to use temporal measurements as they move around to detect the concentration gradients.[10] However, there is some recent contradictory experimental evidence that some spatial information is available.[11] The time scale for rotational Brownian motion of microorganisms can also limit their ability to measure the direction of a source,[12] since it adds a noise term that randomises the microorganism's directional memory. There is some

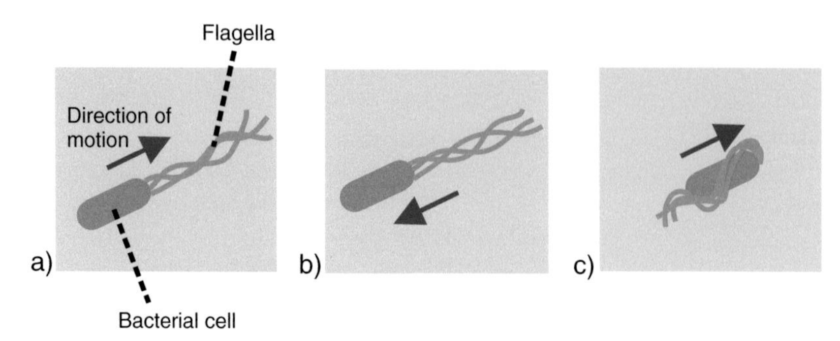

Figure 18.2 (a) *Push*, (b) *pull* and (c) *wrapped* modes for the swimming behaviour of *Pseudomonas putida*.[14] The wrapped mode is thought to be chemotactically sensitive, unlike the push and pull modes.

evidence that the chemostatic sensitivity of bacteria is very well optimised and close to its theoretical limit.[13]

Pseudomonas putida has three different swimming modes during chemotaxis: push, pull and wrapped[14] (Figure 18.2). The chemotactic behaviour could be satisfactory described by a model with the bacteria alternating between a wrapped mode that was chemotactically sensitive, combined with non-chemotactically sensitive push and stationary phases.

Detection of UV light is important for bacteria, since UV can produce ROS inside the cytoplasms of the bacterial cells, which in turn disrupt many biological molecules e.g. DNA can be damaged, which rapidly leads to death unless it is carefully repaired. The detection of UV light is often indirect via the sensing of ROS. For example, *Bacillus subtilis* swims away from UV light and then returns when it is switched off.[15] The swimming behaviour with *B. subtilis* could be modelled with reaction–diffusion equations (Chapter 5). Similar behaviour occurs with *Pseudomonas aeruginosa* and *E. coli*.[16,17]

Historically, the chemotaxis of large numbers of bacteria was measured using light scattering,[18] but current studies tend to prefer the increased information available from microscopy measurements on single bacteria. A continuum model for diffusion with advection can be used to describe light scattering data from the collective motility of large numbers of bacteria.

pH sensing of bacteria is an important component to allow the control of intracellular homeostasis.[19] Examples include *Helicobacter pylori* that in response to pH sensing releases urease to increase the pH of its microenvironment in the stomach (which is highly acidic) and extremophiles, such as *Acidithiobacillus ferrooxidans* (acidophile) and *Bacillus pseudofirmus* (alkaliphile) that are attracted to extreme environmental pHs.[20]

Two-component signalling (TCS) systems are common in archaea and bacteria[21] (Figure 18.3). Classic examples of TCS are CheY in the chemotaxis circuit of *E. coli* (Figure 18.1), sporulation in *Bacillus anthracis*, temperature and osmotic pressure sensing. TCS systems demonstrate bistability and hysteresis in sensitivity (common non-linear phenomena, Chapter 12). More sophisticated signalling gene networks also occur in bacteria e.g. mitogen-activated protein kinase (MAPK) in *Myxococcus xanthus*. MAPK allows signal filtering and amplification. It is thus more sophisticated

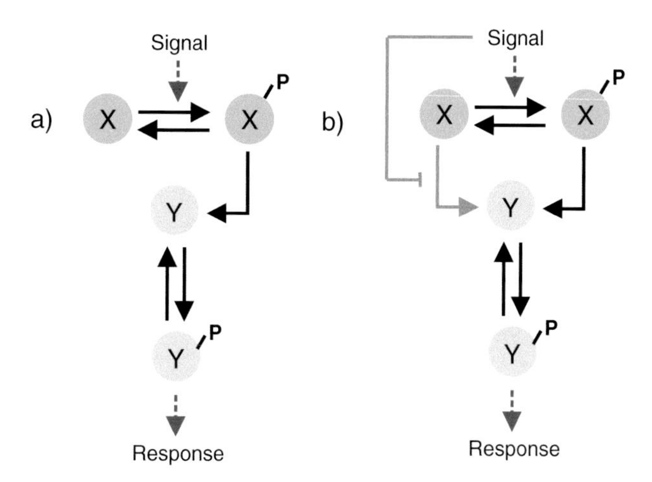

Figure 18.3 Two varieties of *two-component signalling* (TCS) systems are commonly observed with bacteria (a) and (b). *X* is the sensor and *Y* is the receiver. *P* is a phosphate group. The signal causes phosphorylation of the sensor, which then leads to phosphorylation of the receiver.[21]

than TCS and is commonly found in eukaryotes. Other gene transcription networks can also lead to signalling and chemotaxis. Circadian oscillations are driven during the elongation of the cyanobacterium *Synechococcus* by three proteins and a source of energy [adenosine triphosphate (ATP)].[22]

18.1 General Models of Chemotaxis

The statistics and hydrodynamics of *bacterial motility* were considered in Chapters 2, 9 and 11. Such considerations restrict a bacterium's physically permitted responses during chemotaxis. Purcell created a minimal model for the motility of microorganims[1] via the Scallop theorem (Chapter 9). However, the Scallop theorem is only rigorously true for dilute microorganisms in bulk Newtonian fluids and exceptions are possible e.g. near surfaces, in non-Newtonian fluids or for high-concentration bacterial suspensions.

One of the simplest models for chemotaxis is based on a *biased random walk*[23] (Chapter 2) and its origins are in Ed Purcell's famous articles on 'Life at Low Reynolds Number'[1] and 'The Physics of Chemotaxis'.[2] For a perfectly absorbing spherical microorganism, the sensing uncertainty for the concentration of a chemical is

$$\frac{\left\langle (\delta c)^2 \right\rangle_\tau}{\bar{c}^2} = \frac{1}{4\pi D a \bar{c} \tau}, \tag{18.1}$$

where D is the diffusion coefficient of the chemical, a is the radius of the sphere, δc is the variation of the number of absorbed molecules that are sensed, $\left\langle (\delta c)^2 \right\rangle_\tau$ is the average square variation over time τ, τ is the duration of sensing and $\bar{c}$ is the average

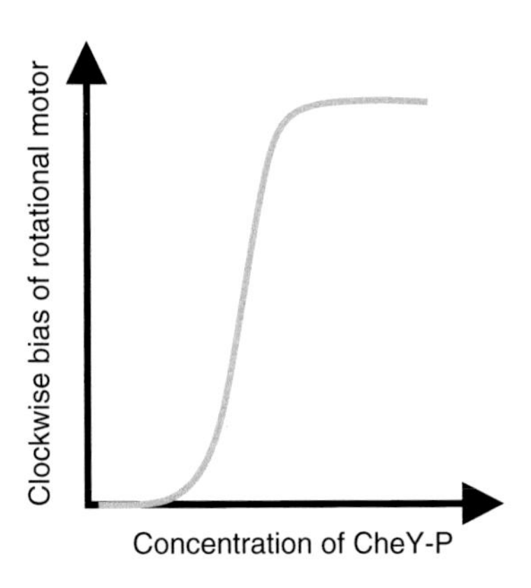

Figure 18.4 The CheY-P system in *E. coli* provides *ultrasensitive detection.*[24] Individual flagellar motors in *E. coli* are very sensitive to the CheY-P concentration in terms of the clockwise bias of the rotational motor.

concentration of chemicals away from the sphere. Thus, higher chemical diffusion coefficients, larger bacteria, higher chemical concentrations and longer durations of sensing lead to more accurate detection of chemical signals by bacteria via diffusion.

Ultrasensitive detection of extracellular chemicals by *E. coli* is achieved by the CheY-P system[24] (Figures 18.1 and 18.4) i.e. phosphorylation of the CheY protein. Very small changes in CheY-P cause a large change in the bias for clockwise/counter-clockwise rotation of the bacterial flagella causing persistent motion. The frequency of switching of the sense of motility (persistent motion switching to tumbling motion) is also very sensitive to the CheY-P concentration.

Strong heterogeneity is observed in the chemical sensitivity of individual marine bacteria during chemotaxis.[25] This can explain part of the heterogeneity of the motility of individual bacteria i.e. their large dispersion. Chemotaxis data was described using an *advection–diffusion model*,

$$\frac{\partial c}{\partial t} = D\nabla^2 c - \nabla.(vc), \tag{18.2}$$

where c is the concentration of the bacteria, D is the long-time effective diffusion coefficient of the swimming bacteria (Chapter 2) and v is the chemotactic drift velocity. Similar models are used to describe the transport of antibiotics across lawns of bacteria (Chapter 24).

The cells of *Cyanobacterium synechocystis* act as microlenses and allow the bacteria to detect the direction of light sources.[26] Interbacterial communication via photons has not yet been demonstrated.

Weber's law states that

$$p = k \ln \frac{S}{S_0}, \tag{18.3}$$

where p is the strength of perception in the organism, S is the signal strength, S_0 is the threshold signal strength and k is a constant. This law holds very generally for

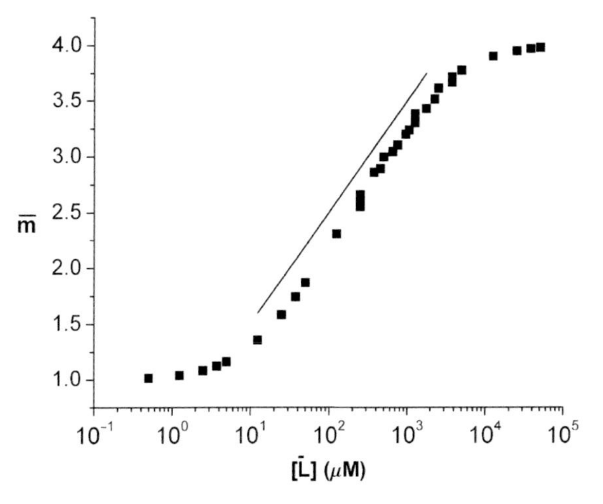

Figure 18.5 *Weber's law* is observed for chemosensing of *E. coli*.[27] The mean methylation level $\overline{m}$ of CheR is shown as a function of ligand concentration $\left[\overline{L}\right]$ on a semi-log plot. The linear region occurs over three decades of ligand concentration and follows Weber's law i.e. logarithmic sensing. Reprinted from Kalinin, Y. V.; Jiang, L.; Tu, Y.; Wu, M., Logarithmic sensing in *E. coli* bacterial chemotaxis. *Biophysical Journal* **2009**, *96* (6), 2439–2448 with permission from Elsevier.

sensory systems in biology. In general, Weber's law is thought to provide an optimal manner to compress information from a sensory system that is required to explore a wide range of signal strengths i.e. the law reflects the logarithmic compression of signals needed to transmit signals that occur over many orders of magnitude that are restricted by the finite information-carrying capacity of the process of information transfer (the channel capacity). It is commonly observed in human hearing where it leads to the decibel scale for loudness, but it is also thought to occur in touch, taste, sight and smell with humans. *E. coli* chemotaxis follows a logarithmic (Weber's law) for its sensitivity over three orders of magnitude for chemoattractant concentrations (MeAsp or L-serine, Figure 18.5).[27]

18.2 Pattern Formation during Chemotaxis

The motion of bacteria in a temperature gradient is affected by intracellular communication.[28] A microfluidic device was used to demonstrate motility waves due to chemical communication between *E. coli* that was initiated by careful control of the temperature.[28] The mode of motility was described using a modified Keller–Segel model.

P. aeruginosa can preferentially follow trails of its own extracellular polymeric substance on surfaces using the motility of type IV pili.[29] Thus, social trail following occurs and bacterial tracks will demonstrate non-Markovian properties as a result i.e. they have a memory.

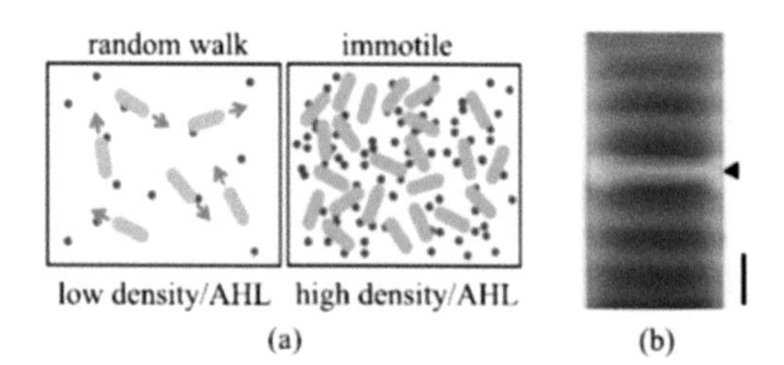

Figure 18.6 Periodic stripes are observed in suspensions of engineered *E. coli*, which release lactone molecules.[30] (a) High densities of lactone molecules cause the bacteria to become immotile, whereas low densities allow motility. (b) Stripes occur in one dimension. Image contrast is provided by higher densities of bacteria. Reprinted figure with permission from Fu, X.; Tang, L. H.; Liu, C.; Huang, J. D.; Hwa, T.; Lenz, P., Stripe formation in bacterial systems with density-suppressed motility. *Physical Review Letters* **2012**, *108*, 198102. Copyright (2012) by the American Physical Society.

Periodic stripes are observed in engineered *E. coli* bacteria that have a density-suppressed mechanism for their motility in response to lactone (Figure 18.6).[30] High densities of lactone molecules caused the bacteria to become immotile. The behaviour could be motivated using two coupled partial differential equations. The first equation describes the lactone concentration (h) and the second the density of the bacterial cells (ρ),

$$\frac{\partial h}{\partial t} = D_h \frac{\partial^2 h}{\partial x^2} + \alpha\rho - \beta h, \tag{18.4}$$

$$\frac{\partial \rho}{\partial t} = \frac{\partial^2}{\partial x^2}\left[\mu(h)\rho\right] + \gamma\rho\left(1 - \frac{\rho}{\rho_s}\right), \tag{18.5}$$

where t is the time, x is the distance in 1D perpendicular to the stripes, D_h is the diffusion coefficient of the lactone molecules, β is the rate at which lactone is removed from the system, α is the rate of lactone production, γ is the bacterial growth rate, $\mu(h)$ is a motility function of lactone concentration and ρ_s is the saturation density of the bacteria (for logistic growth, Equation (16.7)).

Salmonella typhimurium forms periodic arrays of continuous or perforated rings in response to different concentrations of the chemoattractant,[31] potassium succinate (Figure 18.7). Two coupled reaction–diffusion equations were used to predict the 2D cell density (n) as a function of time and space,

$$\frac{\partial n}{\partial t} = D_n \nabla^2 n - \nabla.\left(\frac{k_1 n}{(k_2 + c)^2}\nabla c\right) + f(n,c,s), \tag{18.6}$$

$$\frac{\partial c}{\partial t} = D_c \nabla^2 c + g(n,c,s) - h(n,c,s), \tag{18.7}$$

where D_n is the diffusion coefficient of whole bacterial cells, k_1 is the constant of proportionality which describes the chemotactic response of the cells, k_2 is the receptor dissociation constant, c is the concentration of chemoattractant (potassium succinate), s is the substrate concentration and $f(n,c,s)$ is the proliferation function. Furthermore,

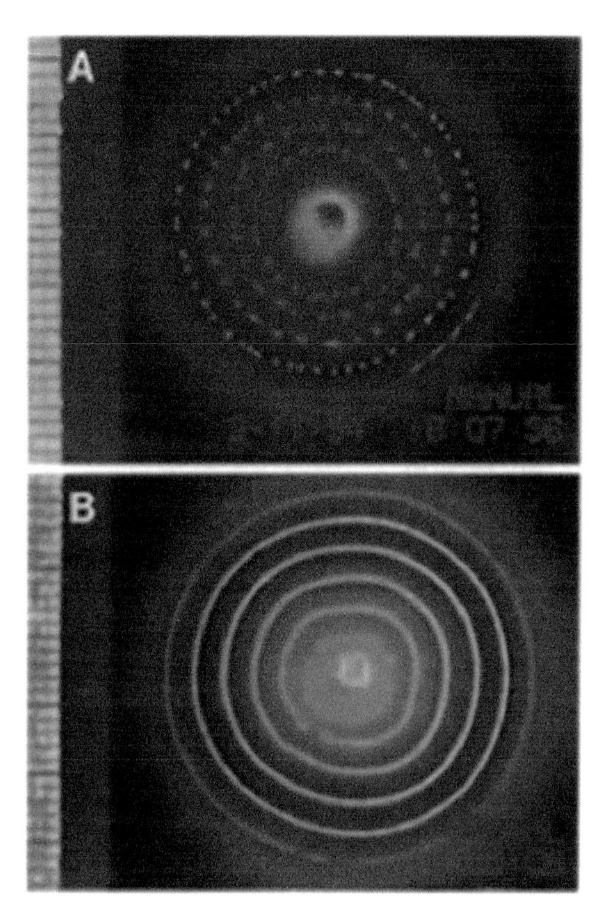

Figure 18.7 Pattern formation with *Salmonella typhimurium* grown on soft agar.[31] Different concentrations of potassium succinate create (a) perforated rings and (b) continuous rings. Reprinted from Woodward, D. E.; Tyson, R.; Myerscough, M. R.; Murray, J. D.; Budrene, E. O.; Berg, H. C., Spatio-temporal patterns generated by *S. typhimurium*. *Biophysical Journal* 1995, *68* (5), 2181–2189 with permission from Elsevier.

D_c is the chemoattractant diffusion coefficient, $g(n,c,s)$ is the production rate of chemoattractants and $h(n,c,s)$ is the consumption rate of the chemoattractants. Periodic solutions are found for the equation in reasonable agreement with the Salmonella data. Thus, gradients of chemoattractant amino acids can induce spatiotemporal patterns in *S. typhimurium* grown on agar gels.[31]

Keller-Segel models are also commonly used to describe chemotactic pattern formation by bacteria.[32,33] They can be used to describe swarm rings,

$$\frac{\partial \rho}{\partial t} = D_b \nabla^2 \rho - \nabla.\left(k\rho\nabla c\right) + a\rho, \tag{18.8}$$

$$\frac{\partial c}{\partial t} = D_c \nabla^2 c + \alpha\rho, \tag{18.9}$$

where ρ is the bacterial density, D_b is the bacterial diffusion coefficient, k is the chemoattractant coefficient, a is the rate of bacterial division, D_c is the diffusion coefficient of the chemoattractant and α is the rate that the chemoattractant is produced (it can be negative). Such reaction–diffusion models have been revisited with an agent-based modelling (ABM) approach (Chapter 27) to include more physical constraints[34] and this is expected to be an important direction for future research.

Simple experiments with *E. coli* chemotaxis involve using a capillary closed at either end with a galactose energy source. Two bands form spontaneously of concentrated bacteria (one band is aerobic and the other is anaerobic).[6] A simple scaling model was used to describe this behaviour.[35] Sharp gradients in oxygen concentration caused a motility transition. Oxygen-driven motility was also found in *E. coli* when the local oxygen concentration was measured using fluorescent probes.[35]

The introduction of microfluidic geometries has improved the accuracy with which chemotactic transport properties can be measured e.g. chemotactic sensitivity and random motility parameters.[36] Corrections to the Keller-Segel model were proposed to understand spatiotemporal-varying environments created in microfluidic devices with *E. coli*.[37] This extended model was better able to explain the phase shift between the chemical signal and the response of the bacteria.

18.3 Models of Quorum Sensing

Quorum sensing in bacteria is often performed with a two-component detection system due to its robustness[38] (Figure 18.3). A simplified model for quorum sensing in *Vibrio fischeri*[39] (the original bacterium in which quorum sensing was discovered, which has a fascinating symbiotic relationship with Bobtailed squid) is

$$\frac{dA}{dt} = k_0 I - r\left(A - A_{ext}\right) - 2k_1 A^2 \left(R_T - 2R\right)^2 + 2k_2 R, \tag{18.10}$$

$$\frac{dR}{dt} = k_1 A^2 \left(R_T - 2R\right)^2 - k_2 R, \tag{18.11}$$

$$\frac{dI}{dt} = a_0 + \frac{aR}{K_m + R} - bI, \tag{18.12}$$

$$\frac{dA_{ext}}{dt} = pr\left(A - A_{ext}\right) - dA_{ext}, \tag{18.13}$$

where k_0, k_1, k_2, a_0, a and b are positive constants, A is the intracellular concentration of the autoinducer (AHL), A_{ext} is the extracellular concentration of AHL, R is the concentration of the active LuxR-AHL complex, I is the concentration of LuxI, $r\left(A - A_{ext}\right)$ is the rate of diffusion of AHL into the cell, p is the population density and d is the rate of loss of AHL. Such systems biology models for quorum sensing are reasonably complex and numerical solutions with computers are needed to interpret their behaviour e.g. to provide a sensitivity analysis to the different parameters to understand which are critical for quorum sensing.

18.4 Magnetotaxis

Magnetotactic bacteria are found in marine environments and in salt marshes.[40] Magnetosomes are organelles inside bacteria that are used to sense magnetic fields and they are carefully arranged in chains by the cells using an actin analogue (MamK).[41,42] It is thought the chains enhance the effect of dipolar coupling between the magnetosomes and thus optimise their sensitivity for the detection of magnetic fields. The magnetosomes are known to contain magnetite (Fe_3O_4) in freshwater bacteria from Mossbauer spectroscopy.[43] Changes in the magnetic field of ~0.5 Gauss (5×10^{-5} T, on the order of the Earth's magnetic field) can lead to magnetotactic responses.[40] Figure 18.8 shows a microscopy image of magnetosomes that are aligned in bacteria.[41]

The motility of *Magnetospirillum gryphiswaldense* was studied on surfaces with magnetic patterns.[44] The magnetic sensing ability is found to be a balance between the motility and the magnetic moment of the magnetosome chain.

Six different magneto-aerotactic behaviours were observed for bacteria based on a study of 12 species.[45] The magnetotactic behaviour allows bacteria to swim in straight lines without losing their bearings due to thermal diffusion (both translational and rotational diffusion). Magnetotaxis is mixed with aerotaxis in bacteria from marine/brackish habitats.[46] Magnetotactic bacteria demonstrate unusual circular trajectories when placed in rotating magnetic fields.[47]

Magnetically induced birefringence, super-conducting quantum interference devices (SQUIDS), spin defects in diamonds and spin valves have been used to measure the magnetic moments of bacteria (Section 13.4.6).[48]

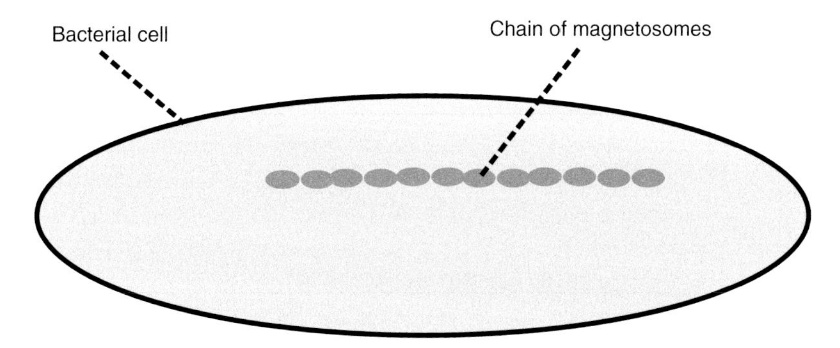

Figure 18.8 Schematic diagram of *magnetosomes* (blue) inside the cell of a magnetotactic bacterium.[41] The cytoskeletal protein Mamk is required to properly organise the chain of magnetosomes.

Suggested Reading

Berg, H. C.; Purcell, E. M., Physics of chemoreception. *Biophysical Journal* **1977**, *20* (2), 193–219.

Endres, R. G., *Physical Principles in Sensing and Signalling: With an Introduction to Modelling in Biology*. Oxford University Press: 2013.

Purcell, E. M., Life at low Reynolds number. *American Journal of Physics* **1977**, *45* (3), 3–11.

References

1. Purcell, E. M., Life at low Reynolds number. *American Journal of Physics* **1977**, *45* (3), 3–11.
2. Berg, H. C.; Purcell, E. M., Physics of chemoreception. *Biophysical Journal* **1977**, *20* (2), 193–219.
3. Kortmann, J.; Narberhaus, F., Bacterial RNA thermometers: Molecular zippers and switches. *Nature Reviews Microbiology* **2012**, *10* (4), 255–265.
4. Parkinson, J. S., Signal transduction schemes of bacteria. *Cell* **1993**, *73* (3), 857–871.
5. Adler, J., Chemoreceptors in bacteria. *Science* **1969**, *166* (3913), 1588–1597.
6. Adler, J., Chemotaxis in bacteria. *Science* **1966**, *153* (3737), 708–716.
7. Endres, R. G., *Physical Principles in Sensing and Signalling*. Oxford University Press: 2013.
8. Colin, R.; Drescher, K.; Sourjik, V., Chemotactic behaviour of *E. coli* at high cell density. *Nature Communications* **2019**, *10* (1), 5329.
9. Dufrene, Y. F.; Dersat, A., Mechanomicrobiology: How bacteria sense and respond to forces. *Nature Reviews Microbiology* **2020**, *18* (4), 227–240.
10. Hu, B.; Chen, W.; Rappel, W. J.; Levine, H., Physical limits on cellular sensing of spatial gradients. *Physical Review Letters* **2010**, *105* (4), 048104.
11. Thar, R.; Kuhl, M., Bacteria are not too small for spatial sensing of chemical gradients: An experimental evidence. *Proceedings of the National Academy of Sciences of the United States of America* **2003**, *100* (10), 5748–5753.
12. Dusenberg, P. B., *Living at Microscale*. Harvard: 2009.
13. Brumley, D. R.; Carrara, F.; Hein, A. M.; Yawata, Y.; Levin, S. A.; Stocker, R., Bacteria push the limits of chemotactic precision to navigate dynamic chemical gradients. *Proceedings of the National Academy of Sciences of the United States of America* **2019**, *116* (22), 10792–10797.
14. Alirezaeizanjani, Z.; Grossmann, R.; Pfeifer, V.; Hintsche, M., Chemotaxis strategies of bacteria with multiple run modes. *Science Advances* **2020**, *6* (22), eaaz6153.

15. Delprato, A. M.; Samadani, A.; Kudrolli, A.; Tsimring, L. S., Swarming ring patterns in bacterial colonies exposed to ultraviolet radiation. *Physical Review Letters* **2001**, *87* (15), 158102.
16. Blee, J. A.; Roberts, I. S.; Waigh, T. A., Membrane potentials, oxidative stress and the dispersal response of bacterial biofilms to 405 nm light. *Physical Biology* **2020**, *17* (3), 036001.
17. Akabuogu, E. U.; Martorelli, V.; Krasovec, R.; Roberts, I. S.; Waigh, T. A., Emergence of ion-channel mediated electrical oscillations in *E. coli* biofilms. *eLife* **2023**, to appear.
18. Berg, H. C.; Turner, L., Chemotaxis of bacteria in glass capillary arrays. *Biophysical Journal* **1990**, *58* (4), 919–930.
19. Krulwich, T. A.; Sachs, G.; Padan, E., Molecular aspects of bacterial pH sensing and homeostasis. *Nature Reviews Microbiology* **2011**, *9* (5), 330–343.
20. MacNab, R. M.; Castle, A. M., A variable stoichiometry model for pH homeostasis in bacteria. *Biophysical Journal* **1987**, *52* (4), 637–647.
21. Voigt, E. O., *A First Course in Systems Biology*, 2nd ed. Garland Science: 2017.
22. Rust, M. J.; Markson, J. S.; Lane, W. S.; Fisher, D. S.; O'Shea, E. K., Ordered phosphorylation governs oscillation of a three-protein circadian clock. *Science* **2007**, *318* (5851), 809–812.
23. Berg, H. C., *Random Walks in Biology*. Princeton University Press: 1993.
24. Cluzel, P.; Surette, M.; Leibler, S., An ultrasensitive bacterial motor revealed by monitoring signaling proteins in single cells. *Science* **2000**, *287* (5458), 1652–1655.
25. Salek, M. M.; Carrara, F.; Fernandez, V.; Guasto, J. S.; Stocker, R., Bacterial chemotaxis in a microfluidic T-maze reveals strong phenotypic heterogeneity in chemotactic sensitivity. *Nature Communications* **2019**, *10* (1), 1877.
26. Schuergers, N.; et al., Cyanobacteria use micro-optics to sense light direction. *eLife* **2016**, *5*, e12620.
27. Kalinin, Y. V.; Jiang, L.; Tu, Y.; Wu, M., Logarithmic sensing in *Escherichia coli* bacterial chemotaxis. *Biophysical Journal* **2009**, *96* (6), 2439–2448.
28. Salman, H.; Zilman, A.; Loverdo, C.; Jefroy, M.; Libchaber, A., Solitary modes of bacterial culture in a temperature gradient. *Physical Review Letters* **2006**, *97* (11), 118101.
29. Gelimson, A.; Zhao, K.; Lee, C. K.; Kranz, W. T.; Wong, G. C. L.; Golestanian, R., Multicellular self-organization of *P. aeruginosa* due to interactions with secreted trails. *Physical Review Letters* **2016**, *117* (17), 178102.
30. Fu, X.; Tang, L. H.; Liu, C.; Huang, J. D.; Hwa, T.; Lenz, P., Stripe formation in bacterial systems with density-suppressed motility. *Physical Review Letters* **2012**, *108* (19), 198102.
31. Woodward, D. E.; Tyson, R.; Myerscough, M. R.; Murray, J. D.; Budrene, E. O.; Berg, H. C., Spatio-temporal patterns generated by *Salmonella typhimurium*. *Biophysical Journal* **1995**, *68* (5), 2181–2189.
32. Brenner, M. P.; Levitov, L. S.; Budrene, E. O., Physical mechanisms for chemotactic pattern formation by bacteria. *Biophysical Journal* **1998**, *74* (4), 1677–1693.

33. Lambert, G.; Bergman, A.; Zhang, Q.; Bortz, D.; Austin, R., Physics of biofilms: The initial stages of biofilm formation and dynamics. *New Journal of Physics* **2014**, *16* (4), 045005.
34. Farrell, F.; Hallatschek, O.; Marenduzzo, D.; Waclaw, B., Mechanically driven growth of quasi-two-dimensional microbial colonies. *Physical Review Letters* **2013**, *111* (16), 168101.
35. Douarche, C.; Buguin, A.; Salman, H.; Libchaber, A., *E. coli* and oxygen: A motility transition. *Physical Review Letters* **2009**, *102* (19), 198101.
36. Ahmed, T.; Stocker, R., Experimental verification of the behavioral foundation of bacterial transport parameters using microfluidics. *Biophysical Journal* **2008**, *95* (9), 4481–4493.
37. Zhu, X.; et al., Frequency-dependent *Escherichia coli* chemotaxis behaviors. *Physical Review Letters* **2012**, *108* (12), 128101.
38. Alon, U., *An Introduction to Systems Biology: Design Principles of Biological Circuits*, 2nd ed. CRC Press: 2020.
39. Ingalls, B. P., *Mathematical Modeling in Systems Biology: An Introduction*. MIT Press: 2013.
40. Blakemore, R., Magnetotactic bacteria. *Science* **1975**, *190* (4212), 377–379.
41. Komeili, A.; Li, Z.; Newman, D. K.; Jensen, G. J., Magnetosomes are cell membrane invaginations organized by the actin-like protein MamK. *Science* **2006**, *311* (5758), 242–245.
42. Klumpp, S.; Lefevre, C. T.; Bennet, M.; Faivre, D., Swimming with magnets: From biological organisms to synthetic devices. *Physics Reports* **2019**, *789* (2), 1–54.
43. Frankel, R. B.; Blakemore, R. P.; Wolfe, R. S., Magnetite in freshwater magnetotactic bacteria. *Science* **1979**, *203* (4387), 1355–1356.
44. Loehr, J.; Pfeiffer, D.; Schuler, D.; Fischer, T. M., Magnetic guidance of the magnetotactic bacterium *Magnetospirillum gryphiswaldense*. *Soft Matter* **2016**, *12* (15), 3631–3635.
45. Lefevre, C. T.; et al., Diversity of magneto-aerotactic behaviors and oxygen sensing mechanisms in cultured magnetotactic bacteria. *Biophysical Journal* **2014**, *107* (2), 527–538.
46. Frankel, R. B.; Bazylinski, D. A.; Johnston, M. S.; Taylor, B. L., Magneto-aerotaxis in marine coccoid bacteria. *Biophysical Journal* **1997**, *73* (2), 994–1000.
47. Erglis, K.; Wen, Q.; Ose, V.; Zeltins, A.; Sharipo, A.; Janmey, P. A.; Cebers, A., Dynamics of magnetotactic bacteria in a rotating magnetic field. *Biophysical Journal* **2007**, *93* (4), 1402–1412.
48. Rosenblatt, C.; Torres de Araujo, F. F.; Frankel, R. B., Birefringence determination of magnetic moments of magnetotactic bacteria. *Biophysical Journal* **1982**, *40* (1), 83–85.

19 Molecular Machines

Molecular machines in bacteria can do work in the thermodynamic sense (they create a force that moves a load a distance in the direction of the force) and are thus able to act as motors. Standard functions fulfilled by molecular machines include: enzymes in catalysis, signalling and regulatory mechanisms, DNA replication, DNA repair, DNA recombination and transposition, RNA transcription, protein synthesis and folding, proteolysis, transport, connectivity and communication, chemotaxis, photosynthesis, membrane channels, membrane transporters and the immune system.[1]

The *intracellular structure* of bacteria is very well organised e.g. molecular machines are carefully positioned by the bacterial cytoskeleton, such as MamK, which are analogues of more well-known cytoskeletal components found in eukaryotic cells, such as actins and microtubules.[2] Bacteria are not just disordered bags of biochemicals. In many cases, the organisation is crucial for the activity of the molecular machines bacteria contain. Furthermore, the molecular concentrations inside bacteria are so high that intermolecular interactions play an important role e.g. the kinetics and self-assembly of molecular machines are strongly affected by intermolecular interactions with neighbouring molecules. There is currently a large interest in the statistical physics community in the study of intracellular machines where it is branded *active matter*.[3] The challenge is to effectively describe the driven non-equilibrium ensembles of interacting particles in motors, whether they are machines inside cells or the cells themselves, in highly organised, highly concentrated systems.

The first job in research on molecular machines is to define their *structure*. Tools such as X-ray crystallography, nuclear magnetic resonance and cryoelectron microscopy are used to provide molecular-scale information. A secondary task is to understand the dynamics of the machines and then explore their dynamics *in vivo* to understand their function. Standard tools include single-molecule fluorescence microscopy, atomic force microscopy, optical and magnetic tweezers (Chapter 13). Thus, individual molecular motors can be investigated and the minute pN forces they exert can be quantified both *in vitro* and *in vivo*.

19.1 Self-assembling Fibres

Intracellular bacteria that infect human cells, such as *Listeria* and *Shigella*, are unusual in that they hijack the self-assembly of actin filaments inside the human cells to move. In human cells, actin monomers self-assemble into actin filaments, which

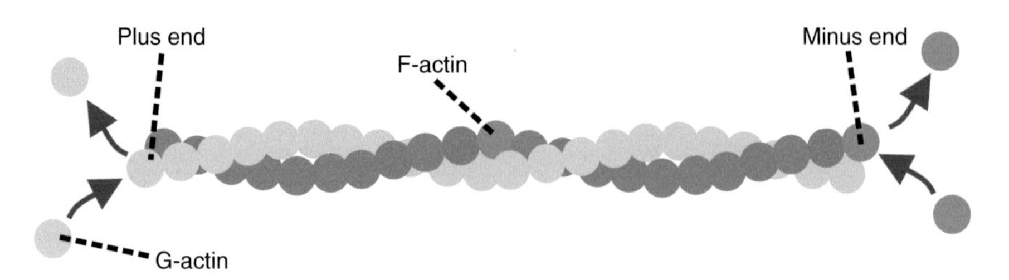

Figure 19.1 *Self-assembling cytoskeletal fibres* (e.g. actins or microtubules) are used by eukaryotic cells for motility. A schematic diagram of the self-assembly of F-actin fibres from G-actin monomers is shown. Self-assembly occurs at both the plus and minus ends of the F-actin fibres. The motility of eukaryotic self-assembled fibres can be hijacked by microorganisms e.g. *Listeria* or *Shigella*. Analogous self-assembling cytoskeletal proteins are also made by bacteria themselves e.g. MamK.

push on the cell's membrane and lead to amoeboid-type movements of entire cells (Figure 19.1). *Listeria* and *Shigella* hijack the self-assembly of actin to move themselves around inside the infected human cells. Brownian motion provides significant forces on the scale of individual molecular motors, such as actin fibres, and the challenge is for such self-assembling motors to make progress when experiencing these thermal forces. A paradigm to understand this behaviour is that of a *Brownian ratchet* (BR).[4] In the BR model, the motor provides an asymmetric potential that rectifies Brownian fluctuations of the displacement due to thermal energy (the fluctuations follow symmetric Gaussian distributions) and can lead to directed transport.

Thus, *BRs* are a standard paradigm to understand stochastic molecular motors, such as the actin polymerisation induced by *Listeria*.[4] Passive random diffusion described by Fick's second law (Equation (2.4)) is rectified and the process of rectification requires a source of energy, which in turn supplies the work done for the motor (otherwise the process would violate the second law of thermodynamics[5]). A simple model for self-assembling motors in one dimension is therefore

$$\frac{\partial c}{\partial t} = D\frac{\partial^2 c}{\partial x^2} + \left(\frac{fD}{kT}\right)\frac{\partial c}{\partial x} + \alpha\left[c(x+\delta,t) - \theta(x-\delta)c(x,t)\right] + \beta\left[H(x-\delta)c(x-\delta,t) - c(x,t)\right], \tag{19.1}$$

where c is the actin concentration, D is the diffusion coefficient, kT is the thermal energy, t is the time, x is the displacement in one dimension, f is the load force, α is the polymerisation rate, θ is the step function, β is the depolymerisation rate and δ is the size of a monomer. The solution for the velocity of the self-assembling fibre is

$$v = \delta\left[\alpha e^{-\omega} - \beta\right], \tag{19.2}$$

where ω is the dimensionless work done in adding a monomer. This model was subsequently extended to consider the effects of elasticity.[6] Many of the cytoskeletal proteins inside bacteria are also thought to self-assemble and can be described by similar one-dimensional (1D) BR models.[2,7] BR models provide credible paradigms for molecular motors, but the simplest models underestimate the efficiency by a large margin. They could be improved by more accurate modelling of mesoscopic forces and the inclusion of non-linear effects e.g. due to molecular elasticity.

19.2 Pili

Pili are relatively short, hair-like structures that are found on the surface of bacteria and archaea. Currently, five different types of bacterial pili are defined in Gram-negative bacteria: chaperone-usher, type IV, conjugative type IV secretion, type V and curli fibres[9] (Figure 19.2). The naming scheme is a historical anomaly, since there are now no type II or III pili that are recognised. Conjugative pili are used as a conduit for DNA translocation and are thus part of a translocation machine. Curli pili (Figure 19.2b) are functional amyloids and they are often shed into bacterial biofilms providing them with important mechanical components.

Type IV pili (Figure 19.2d) are found to drive the motility of some bacteria[10] e.g. twitching motility.[11] Optical tweezer experiments demonstrate that type IV pili in *Neisseria gonorrhoeae* can retract and exert substantial 100 pN forces.[12]

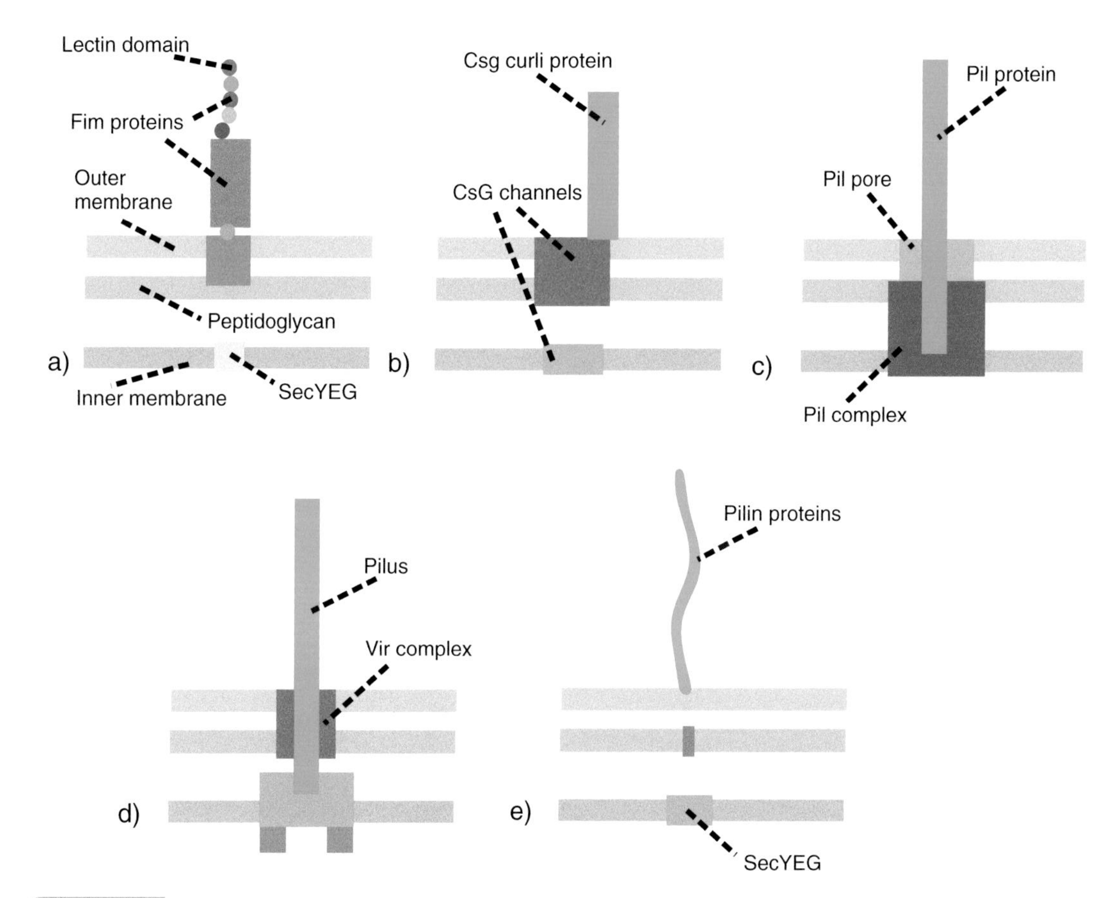

Figure 19.2 The structure of *five types of pili* from Gram-negative bacteria: (a) *chaperone-usher pili*, (b) *curli*, (c) *type IV pili*, (d) *type IV secretion pili* and (e) *type V pili*.[8]

Dynamic force spectroscopy using an atomic force microscope has been used to study the pili of *Escherichia coli*.[13] A sticky model was used to describe the mechanics of the extension of the pili. Adhesive pili on uropathogenic bacteria are 6.8 nm in diameter and ~1 μm in length.[14]

19.3 Flagellar Motors

The *flagellar motor* in *E. coli* is constructed from 24 different proteins and is a marvel of nanotechnology (Figure 19.3).[16] In most bacteria, the motor is driven by proton gradients, rather than adenosine triphosphate (ATP), which is the standard currency of energy in eukaryotes e.g. ATP is used to drive the contraction of mammalian muscle. Sodium gradients are used with marine bacteria that live in alkaline conditions where protons are scarce. BR models (Section 19.1) have been extended to describe flagellar motors and are reasonably successful, although questions on efficiency also exist, similar to models of actin (Section 19.1). The flagellar motor is closely related to the Fo-ATPase enzymes used to construct ATP in eukaryotic mitochondria, indicating the evolutionary origin of mitochondria as symbiotic bacteria that exist as living fossils (endosymbionts) in modern-day eukaryotic cells.

Flagellar motors are driven by gradients of charged ions and electrorotation by external electric fields can be used to calculate the torque on flagellar motors.[17] The flagella are created by self-assembly, which occurs by the diffusion of subunits down

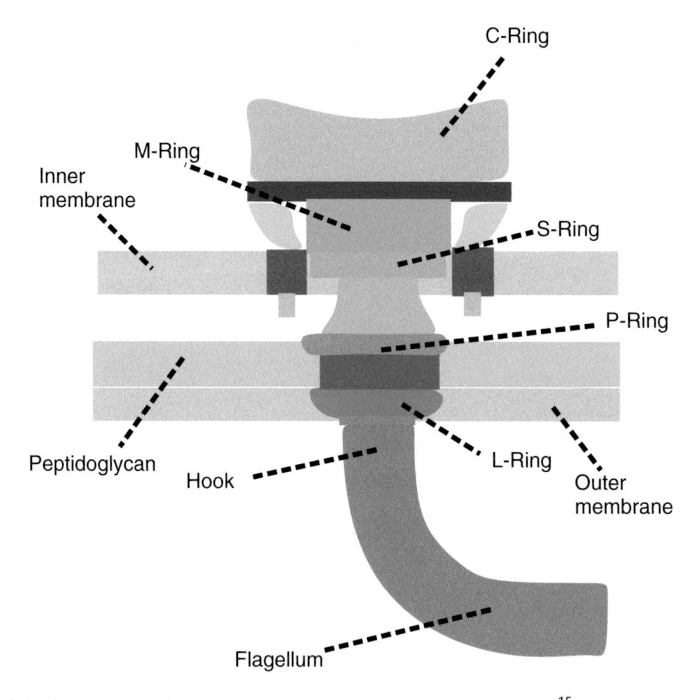

Figure 19.3 The structure of the *flagellar motor* from *E. coli* which consists of 24 different proteins.[15]

each flagellum. *E. coli* motors can rotate at up to 300 rev s^{-1}. The times over which *E. coli* flagellar motors rotate were found to follow a gamma distribution.[18] This provided evidence for multiple hidden states that were interpreted using a hidden Markov model.[18] The statistics of single bacterial motors from *E. coli* have been studied using fluorescence microscopy and clockwise versus counterclockwise switching has been quantified,[18] which in turn affects the frequency of runs and tumbles during cellular motility (Chapter 2). A wide range of mechanical performances have been measured for different varieties of rotational bacterial motors.[19]

19.4 Ion Channels

The first crystallographic structures obtained for ion channels were from bacteria e.g. K^+ channels from *Streptomyces lividans*.[21] A large variety of prokaryote ion channels are now known using comparative genetics studies with their eukaryotic counterparts, but their exact physiological roles can be less clear due to the small size of bacteria, which makes them challenging to patch clamp and thus perform detailed electrophysiological measurements. For example, a potassium ion channel Kch was found in *E. coli* by Milkman (1994).[22] There is some partial evidence from mutagenesis studies that it is required to modulate the membrane potential of the bacteria,[23] whereas other potassium ion channels are needed for the simple import/export of the potassium ions. Further recent evidence from our group demonstrates that Kch is involved in modulating the membrane potential in response to light stress, but more research is required to rigorously understand its activity.[24] There is also extensive evidence for mechanically sensitive channels in bacteria (Figure 19.4) and they play a role in

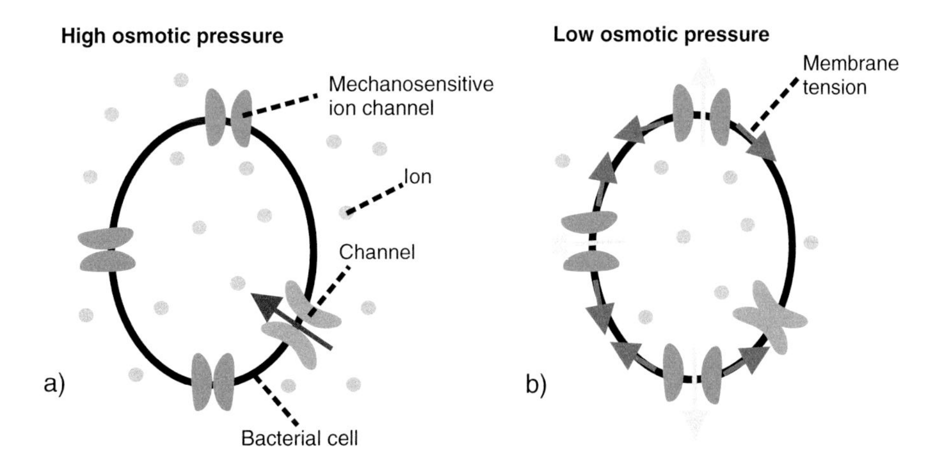

Figure 19.4 *Mechanosensitive ion channels* in bacterial membranes allow them to resist osmotic shocks.[20] (a) With high external osmotic pressure, the mechanosensitive ion channels are closed. (b) With low external osmotic pressure, the mechanosensitive ion channels are open.

osmoregulation.[20] Gap junctions have also been discovered in bacteria, presumably for the transfer of ions between cells.[25]

Prokaryote ion channels can be active (e.g. ATP-fuelled) or passive (transport occurs down concentration gradients). Voltage-gated ion channels occur and they are associated with action potentials in eukaryotic cells. Standard prokaryotic ion channels include: porins, mechanosensitive channels, channel-forming toxins and bacterial homologs of mammalian channels.[26] Larger molecules can be transported through membranes using secretion machines (Section 19.6).

19.5 Nucleic Acid Associating Machines

The *central dogma of molecular biology* is shown in Figure 19.5, which is presented in slightly more detail than when it was introduced in Figure 11.1. Molecular machines are required to perform all of the key steps e.g. DNA polymerase is needed to replicate DNA.

DNA in bacteria is often stored in tightly wound circles (plectonemic structures), so they fit inside bacterial cells whose diameter is much smaller than the length of DNA. The circular DNA chains are tightly wound by enzymes that control their topology. There are type I and II topoisomerases, which act on chain topologies, cleaving one strand or two strands at a time, respectively. Changes in the topology and the accessibility of the DNA chains are required during both replication and transcription.

DNA polymerase is an enzyme capable of replicating DNA and was first discovered in *E. coli* (Figure 19.6). DNA Pol III incorporates base pairs into a DNA chain in *E. coli* (it is a replicative polymerase). The process of DNA replication also requires helicase, primase, single-stranded DNA, binding proteins, a clamp and a clamp loader. The complete complement of molecules for DNA replication is called the *replisome*. DNA Pol I (DNA ligase) in *E. coli* is required to join up the pieces of DNA created by Pol III. Single-molecule experiments have been performed on DNA machinery *in vivo*

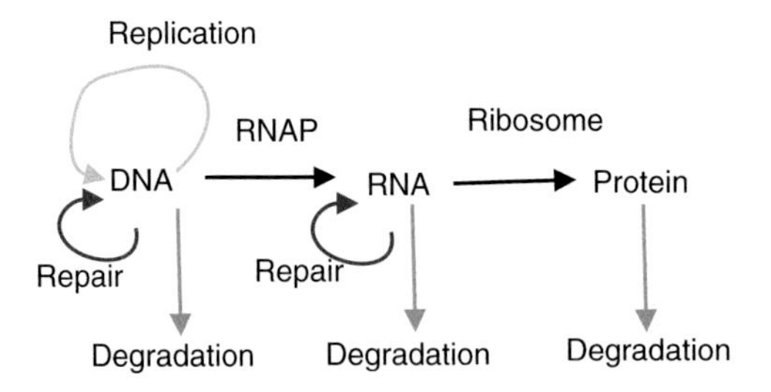

Figure 19.5 The *central dogma of molecular biology* states that information is stored in DNA. Transcription then transfers this information to RNA (via RNAP) and translation in turn converts the information into proteins (via ribosomes). DNA and RNA molecules can be repaired if they contain defects. Replication produces identical copies of the DNA during cell division.

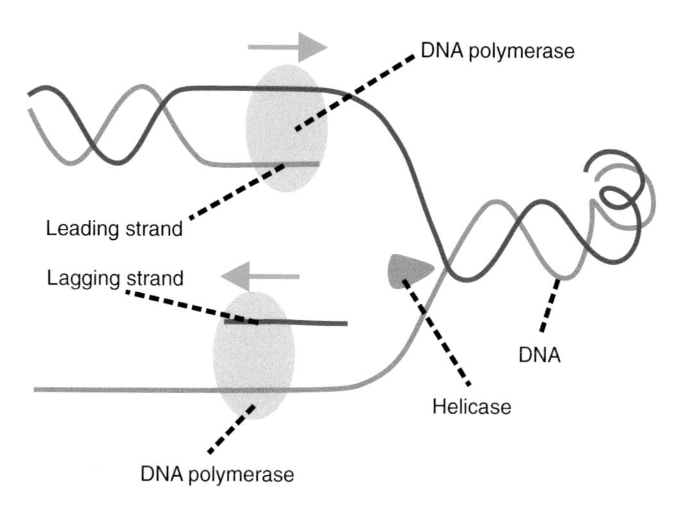

Figure 19.6 *DNA polymerase* is a molecular machine used for the replication of DNA in bacteria.

inside bacteria e.g. with the DNA polymerase of *Bacillus subtilis*, the DNA is moved processively through the machine and not vice versa (the machine is fixed in place).[27]

Damage to DNA is handled by a separate set of repair proteins.[1] The nucleotide-excision repair pathway handles DNA lesions by UV light, reactive oxygen species (ROS) and other risk factors. Mismatch repair (when the DNA bases are not with their matched partners i.e. As with Ts and Gs with Cs) involves a different pathway in *E. coli* and so too does double-strand repair.

RNA is synthesised from DNA via *RNA polymerase* (RNAP) and there is only a single variety of RNAP in both bacteria and archaea that produces mRNA, rRNA and tRNA. RNA polymerases from bacteria and archaea have 5 and 11 subunits, respectively. In bacteria, an additional sigma (σ) factor is needed for transcription initiation. Exosome complexes degrade RNA and are needed for quality control and to adjust levels of expression. RNA polymerases were found to be arranged in banded clusters in *E. coli* using PALM super-resolution fluorescence microscopy.[28] Two to eight bands were observed in nutrient-rich media, which reduced to 1–2 bands in nutrient-poor media.

CRISPR (clustered regularly interspaced short palindromic repeats) represents part of an adaptive immune system for bacteria that primarily acts against viruses i.e. bacteriophages.[29] The CRISPR mechanism is based on sequences of DNA that create small guide RNA transcripts targeted at the phages[30,31] (Figure 19.7). As a form of acquired immunity, CRISPR allows the bacterium to learn to react to new viral threats.[31] Both foreign RNA and DNA are targeted by different CRISPR machinery (bacteriophages can be based on either RNA or DNA). Standard model laboratory strains of bacteria, such as *E. coli* and *S. thermophiles* are known to have four different CRISPR systems[30] and it is expected that more will be discovered. Type II CRISPR/Cas has been used for RNA-guided editing of human cells[32] and has thus attracted a huge amount of interest as a precise tool for gene editing. CRISPR interference has also been shown to limit horizontal gene transfer in *S. aureus* and can thus limit the

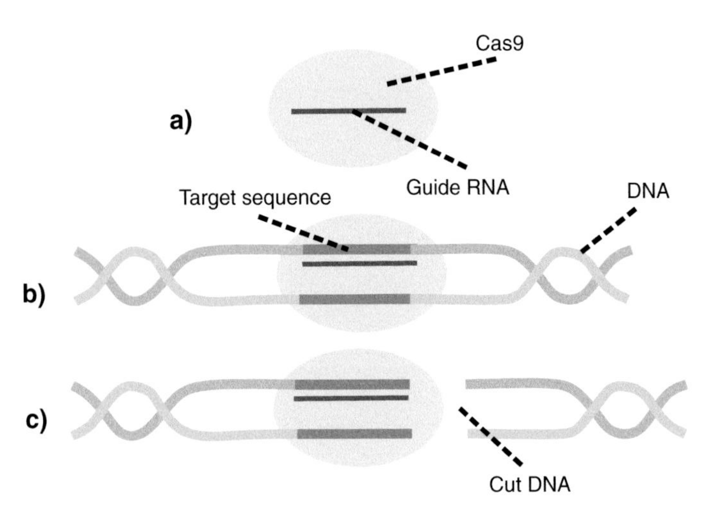

Figure 19.7 The *CRISPR* (clustered regularly interspaced short palindromic repeats) part of the adaptive immune system of bacteria consists of three stages. (a) The Cas9 protein creates a complex with the guide RNA. (b) The complex attaches to the matching DNA sequence in a cell. (c) The complex cuts the double-stranded DNA. DNA can then be inserted into the break if the Cas9 protein is used for *in vitro* synthetic biology applications.

rate that antibiotic-resistant genes are acquired.[33] CRISPR is only one of the components of the *bacteriophage resistome* of bacteria.[34]

19.6 Secretion Complexes

Proteins embedded in the one or two external membranes of bacteria are required to help the export and import of molecules.[35] *Secretion systems* in Gram-negative bacteria are classified as types I, II (Tat, Sec), III, IV, V and VI. Gram-positive bacteria have different protein secretion systems to Gram negatives due to the different structures of their cell walls e.g. type VII systems occur with *Mycobacterium corynebacterium* to export proteins through a dense waxy layer of lipid in their cell walls.

A range of *secretion machines* are used by bacteria to inject proteins into other prokaryotic (type VI) and eukaryotic (types III and IV) cells[36] e.g. R-type bacteriocin from *Pseudomonas aeruginosa* is injected by molecular syringes. The proteins injected into eukaryotic cells are predominantly cell modulatory rather than cell destructive. In comparison with simpler bacterial exotoxin systems that involve simple poisons, it is thought that the main evolutionary pressure has been to allow the transfer of multiple types of protein at the same time to provide more nuanced interactions. However, bacteria do use type V secretion machines as weapons against other bacteria and they then need to store immunity proteins (antitoxins), so they are not killed by their own poisons.

Type III secretion machines can deliver proteins into host cells.[37] They are evolutionary related to flagella and are present in both animal and plant pathogens. They

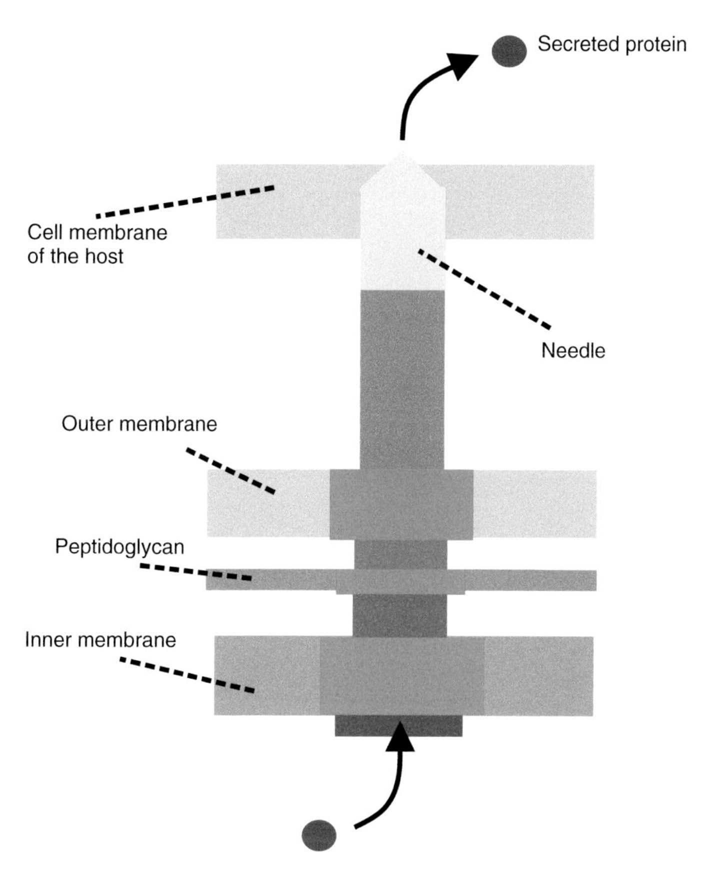

Figure 19.8 A *type III secretion complex* embedded in a bacterial membrane. Effector proteins can be secreted into host cells.

are sophisticated machines and are often made up of >20 proteins and form a needle complex that is ~120 nm long (Figure 19.8). Such type III machines occur in lots of disease-causing bacteria, including *Enteropathogenic E. coli*, *Enterohemorrhagic E. coli*, *P. aeruginosa* and *Shigella*, among others. It has been possible to follow the type III secretion of effectors into host cells in real time using fluorescence microscopy e.g. with *Salmonella*.[38,39]

Other *non-ionic channels* can also be important for the transport of cargoes across bacterial membranes. *ABC transporters* are important for both the import and export of molecules in bacteria.[40] Such transporters are required to transfer slime, capsular and biofilm macromolecules.

19.7 Other Machines

A wide range of other machines occur in bacteria and many eukaryotic machines find their evolutionary origins in bacteria.[1] For example, there is a dynamin

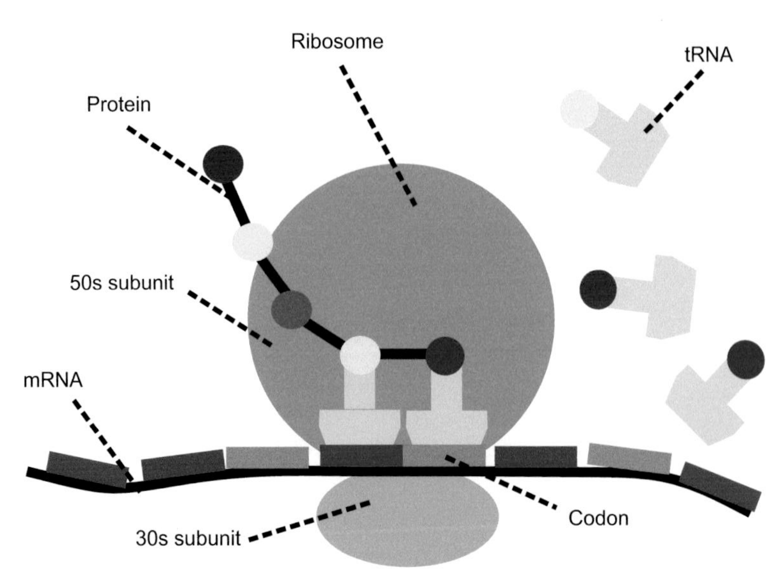

Figure 19.9 *Ribosomes* are required in the central dogma of molecular biology (Figure 19.5) where they translate the information from mRNA into the sequences of proteins.

analogue that pinches off and tubulates lipid membranes driven by guanosine triphosphate (GTP).

Ribosomes are crucial machines in the central dogma of molecular biology (Figure 19.9). They link amino acids together to form proteins based on the sequence of the messenger RNA (mRNA) that is input. Ribosomes look similar for all life forms on planet Earth, indicating their common evolutionary origins, but the exact sequence depends on the species. The exact sequencing of ribosomes is thus used to classify species of bacteria and determine evolutionary relationships e.g. it was used in the original discovery of archaea.[41] Ribosomes contain two subunits that are described by their sedimentation coefficients, 30S and 50S.

Photosynthetic units (PSUs) are constructed from light-harvesting complexes (LHCs) and reaction centers[1] (Section 31.3). LHCs perform exciton transfer to transmit energy from photons to the reaction centres. The LHCs increase the rate that photons are absorbed by the reaction centres by two orders of magnitude. There are typically about 200–250 chlorophyll molecules in a PSU.

Suggested Reading

Bensimon, D.; Croquette, V.; Allemand, J. F.; Michalet, X.; Strick, T., *Single-molecule Studies of Nucleic Acids and Their Proteins*. Oxford University Press: 2018.

Steven, A.; Baumeister, W.; Johnson, L. N.; Perham, R. N., *Molecular Biology of Assemblies and Machines*. Garland Science: 2016.

References

1. Steven, A. C.; Baumeister, W.; Johnson, L. N.; Perham, R. N., *Molecular Biology of Assemblies and Machines*. Garland Science: 2016.
2. Ramos-Leon, F.; Ramamurthi, K. S., Cytoskeletal proteins: Lessons learned from bacteria. *Physical Biology* **2022**, *19* (2), 021005.
3. Pismen, L., *Active Matter Within and Around Us: From Self-propelled Particles to Flocks and Living Forms*. Springer: 2021.
4. Peskin, C. S.; Odell, G. M.; Oster, G. F., Cellular motions and thermal fluctuations: The Brownian ratchet. *Biophysical Journal* **1993**, *65* (1), 316–324.
5. Feynman, R., *Feynman Lectures on Physics*. Basic Books: 2010.
6. Mogliner, A.; Oster, G. F., Cell motility driven by actin polymerization. *Biophysical Journal* **1996**, *71* (6), 3030–3045.
7. Philips, R.; Kondev, J.; Theriot, J.; Garcia, H.; Kondev, J., *Physical Biology of the Cell*. Garland Science: 2012.
8. Fronzes, R.; Remaut, H.; Waksman, G., Architectures and biogenesis of non-flagellar protein appendages in Gram-negative bacteria. *The EMBO Journal* **2008**, *27* (17), 2271–2280.
9. Hospenthal, M. K.; Costa, T. R. D.; Waksman, G., A comprehensive guide to pilus biogenesis in Gram-negative bacteria. *Nature Reviews Microbiology* **2017**, *15* (6), 365–379.
10. Maier, B., The bacterial type IV pilus system – a tunable molecular motor. *Soft Matter* **2013**, *9* (24), 5667–5671.
11. Skerker, J. M.; Berg, H. C., Direct observation of extension and retraction of type IV pili. *Proceedings of the National Academy of Sciences of the United States of America* **2001**, *98* (12), 6901–6904.
12. Merz, A. J.; So, M.; Sheetz, M. P., Pilus retraction powers bacterial twitching motility. *Nature* **2000**, *407* (6800), 98–102.
13. Andersson, M.; Fallman, E.; Uhlin, B. E.; Axner, O., Dynamic force spectroscopy of *E. coli* P Pili. *Biophysical Journal* **2006**, *91* (7), 2717–2725.
14. Bullitt, E.; Makowski, L., Bacterial adhesion pili are heterologous assemblies of similar subunits. *Biophysical Journal* **1998**, *74* (1), 623–632.
15. Xing, J.; Bai, F.; Berry, R.; Oster, G., Torque-speed relationship of the bacterial flagellar motor. *Proceedings of the National Academy of Sciences of the United States of America* **2006**, *103* (5), 1260–1265.
16. Berg, H. C., *E. coli in Motion*. Springer: 2004.
17. Berg, H. C.; Turner, L., Torque generated by the flagellar motor of *Escherichia coli*. *Biophysical Journal* **1993**, *65* (5), 2201–2216.
18. Korobkova, E. A.; Emonet, T.; Park, H.; Cluzel, P., Hidden stochastic nature of a single bacterial motor. *Physical Review Letters* **2006**, *96* (5), 058105.
19. Li, G.; Tang, J. X., Low flagellar motor torque and high swimming efficiency of *Caulobacter crescentus* swarmer cells. *Biophysical Journal* **2006**, *91* (7), 2726–2734.

20. Blount, P.; Iscia, I., Life with bacterial mechanosensitive channels, from discovery to physiology to pharmacological target. *Microbiology and Molecular Biology Reviews* **2020**, *84* (1), e00055-19.
21. MacKinnon, R., Potassium channels and the atomic basis of selective ion conduction. *Angewandte Chemie* **2004**, *43* (33), 4265–4277.
22. Milkman, R., An *Escherichia coli* homologue of eukaryotic potassium channel proteins. *Proceedings of the National Academy of Sciences of the United States of America* **1994**, *91* (9), 3510–3514.
23. Kuo, M. M. C.; Saimi, Y.; Kung, C., Gain-of-function mutations indicate that *Escherichia coli* Kch forms a functional K+ conduit in vivo. *The EMBO Journal* **2003**, *22* (16), 4049–4058.
24. Akabuogu, E. U.; Martorelli, V.; Krasovec, R.; Roberts, I. S.; Waigh, T. A., Emergence of ion-channel mediated electrical oscillations in *Escherichia coli* biofilms. *eLife* 2023, to appear.
25. Weiss, G. L.; Kieninger, A. K.; Maldener, I.; Forchhammer, K.; Pilhofer, M., Structure and function of a bacterial gap junction analog. *Cell* **2019**, *178* (2), 374–384.
26. Compton, E. L. R.; Misdell, J. A., Bacterial ion channels. *EcoSalPlus* **2010**, *4* (1).
27. Lemon, K. P.; Grossman, A. D., Localization of bacterial DNA polymerase: Evidence for a factory model of replication. *Science* **1998**, *282*, 1516–1519.
28. Endesfelder, U.; Finan, J.; Holden, S. J.; Cook, P. R.; Kapanidis, A. N., Multiscale spatial organization of RNA polymerase in *Escherichia coli*. *Biophysical Journal* **2013**, *105* (1), 172–181.
29. Wiedenheft, B.; Sternberg, S. H.; Doudna, J. A., RNA-guided genetic silencing systems in bacteria and archaea. *Nature* **2012**, *482* (7385), 331–338.
30. Horvath, P.; Barrangou, R., CRISPR/Cas, the immune system of bacteria and archaea. *Science* **2010**, *327* (5962), 167–170.
31. Barrangou, R.; Fremaux, C.; Deveau, H.; Richards, M.; Boyaval, P.; Moineau, S.; Romero, D. A.; Horvath, P., CRISPR provides acquired resistance against viruses in prokaryotes. *Science* **2007**, *315* (5819), 1709–1712.
32. Cong, L.; et al., Multiplex genome engineering using CRISPR/Cas systems. *Science* **2013**, *339* (6121), 819–823.
33. Marraffini, L. A.; Sontheimer, E. J., CRISPR interference limits horizontal gene transfer in Staphylococci by targeting DNA. *Science* **2008**, *322* (5909), 1843–1845.
34. Hyman, P.; Abedon, S. T., Bacteriophage host range and bacterial resistance. *Advances in Applied Microbiology* **2010**, *70*, 217–248.
35. Green, E. R.; Mecsas, J., Bacterial secretion systems: An overview. In *Virulence Mechanisms of Bacterial Pathogens*, Kudva, I. T. et al., Eds. American Society for Microbiology: 2016; pp. 1–19.
36. Galan, J. E.; Waksman, G., Protein-injection machines in bacteria. *Cell* **2018**, *172* (6), 1306–1318.
37. Galan, J. E.; Collmer, A., Type III secretion machines: Bacterial devices for protein delivery into host cells. *Science* **1999**, *284* (5418), 1322–1328.

38. Enninga, J.; Mounier, J.; Sansonetti, P.; van Nhein, G. T., Secretion of type III effectors into host cells in real time. *Nature Methods* **2005**, *2* (12), 959–965.
39. van Engelenburg, S. B.; Palmer, A. E., Imaging type-III secretion reveals dynamics and spatial segregation of Salmonella effectors. *Nature Methods* **2010**, *7* (4), 325–330.
40. Locher, K. P.; Lee, A. T.; Rees, D. C., The *E. coli* BtuCD structure: A framework for ABC transporter architecture and mechanism. *Science* **2002**, *296* (5570), 1091–1098.
41. Woese, C. R.; Kandler, O.; Wheelis, M. L., Towards a natural system of organisms: Proposal for the domains archaea, bacteria, and eucarya. *Proceedings of the National Academy of Sciences of the United States of America* **1990**, *87* (12), 4576–4579.

20 Capsules and Slimes

Emphasis is given in this book to *biofilms*, *slimes* and *capsules*, since they are all virulence factors for human infections. Capsules can be considered evolutionary intermediates to biofilms, since lipopolysaccharide (LPS) capsules (also lipopeptide and other polymer chemistries occur) can be shed to form a viscoelastic fluid called a *slime layer*. Slime layers superficially resemble primitive biofilms, but biofilms tend to be more structured gels (with transport channels and intricate cross-links between the polymers) driven by intercellular signalling (quorum sensing) and are created by much more complex genetic programmes. The word *capsule* is a bit of a misnomer, since on the molecular scale they resemble biopolymer brushes and appear to follow the physical laws developed separately by physical chemists for synthetic polymer brushes[1] (Section 4.1.2). Thus, *capsular brushes* might be a better nomenclature. An image of a capsular brush on a bacterium is shown in Figure 20.1.

Why do bacteria form capsules? Many of the advantages and disadvantages of capsules mirror those of biofilms (compare Tables 20.1 and 23.1). The capsules provide a soft semi-permeable shield for the bacteria. They offer some protection against antimicrobials (e.g. cationic surfactants), viruses and phagocytosis by immune cells. The protection against phagocytosis is due to both the steric repulsive potential of the brushes and the choice of capsular chemistry, since the capsular brushes can mimic naturally occurring chemistries in the host and can thus be invisible to the host's immune system. Bacteria can also bind the wrong complement components to confuse the host's immune system. Disadvantages of the expression of capsules include the metabolic cost for their production, increased hydrodynamic drag, reduction in sensitivity during chemotaxis, barriers to adhesion to host cells and the reduced availability of food. Since there are some disadvantages to the expression of capsules, bet hedging occurs with the expression of capsules in the case of *E. coli* (Chapter 22)[3] and it is probably a general phenomenon. A small fraction of the encapsulated bacteria switch off their soft armoured coatings, which allows them to adhere to host cells more effectively and thus allows them to enter cells and propagate the infections *in vivo* where they can then switch the capsules' genetic programmes back on and regrow their soft armour.[3] The analogous list of the advantages and disadvantages with biofilms (Table 23.1) is much larger than with capsules and many more analogous functions may be established with capsules.

Some of the most dangerous pathogenic strains of bacteria possess capsules e.g. *E. coli* that cause urinary tract infections, *Klebsiella pneumoniae* that cause lung infections or *Bacillus anthracis* that cause anthrax when spores enter the body. The glucan capsule of *Streptococcus mutans* and other species forms the matrix of dental

Table 20.1 Advantages and disadvantages of the expression of *capsules* on bacteria.

Advantages	Disadvantages
Repel the host's phagocytes due to the steric potential of the brushes.	Reduced availability of food.
Adsorb antibiotics.	Present a barrier for adhesion to host cells.
Capsular chemistry can make bacteria invisible to the host's immune system.	Require energy for their production.
Upregulation can lead to slime layers, which have similar advantages as a primitive biofilm.	Increased hydrodynamic drag.
Barrier to adhesion in unfavourable environments.	Reduce the signal to noise ratios during chemotaxis.
Bind to the wrong complement to confuse the host's immune system.	
Can block bacteriophage penetration.	

The analogous list for biofilms (Table 23.1) implies more entries are possible, but they need to be rigorously established in the context of capsules.

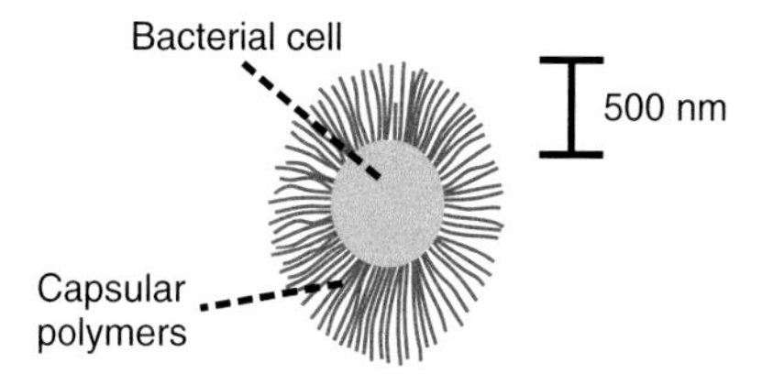

Figure 20.1 Schematic diagram of *bacterial capsules* e.g. on *Streptococcus pneumoniae*.[2] The capsular brush is ~500 nm in thickness.

plaque.[4] However, there are some relatively benign capsular bacteria that make reasonable model experimental systems with less challenging health and safety issues for experimentalists e.g. *E. coli* Nissle 1917 (K5 capsule) that are used in food supplements as probiotics.

Most non-capsular Gram-negative bacteria have short polysaccharide chains adhered to their surface (e.g. LPS), which are on the order of 10–200 nm in length. *Rough* (~10 nm size) and *smooth* (~100 nm size) varieties of LPS are classified (Figure 20.2). The O polysaccharide is missing or truncated in the rough form. LPS contains a highly hydrophobic lipid in the outer membrane leaflet (often lipid A), a core polysaccharide and an O polysaccharide. Thus, most Gram-negative bacteria have relatively short polymer brushes on their surfaces due to LPS. However, capsular polysaccharides (CPS) can be much longer, with lengths >200 nm, and this increase in length has dramatic implications for their physical properties. Synthetic variants of polymeric brushes can create non-stick coatings on nanoparticles and were invented separately by physical chemists to stabilise the phases of colloidal

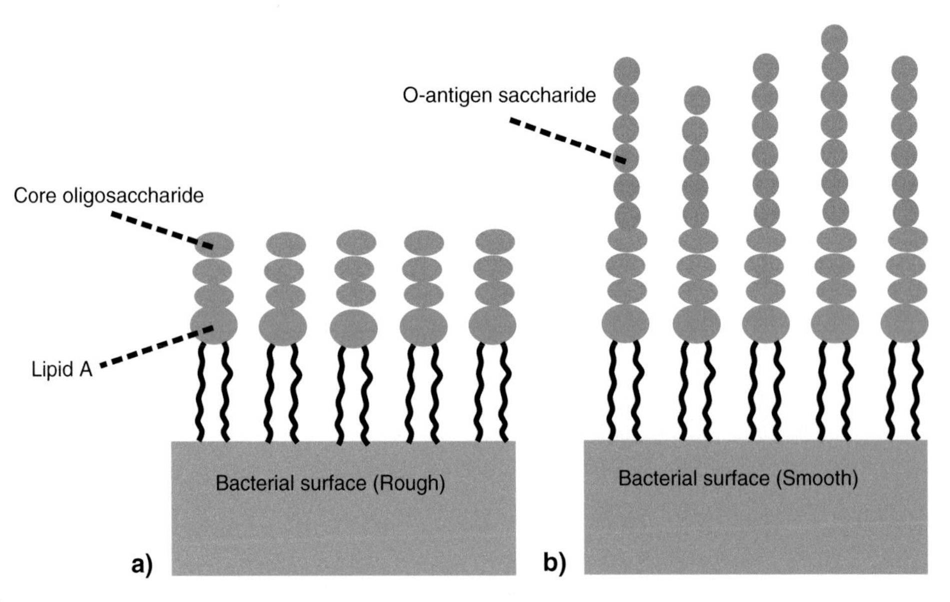

Figure 20.2 (a) *Rough* (short) and (b) *smooth* (longer) lipopolysaccharides (LPS) can occur on the surfaces of non-encapsulated bacteria.

dispersions[5] e.g. polymeric hairs can be added to the surfaces of paint nanoparticles to stop them aggregating (Section 4.1.2). The polymeric hairs significantly modify the intracellular potential of the colloidal particles on the mesoscale and provide a repulsive potential against most other nanoparticles (the only exceptions are oppositely charged colloidal particles). Quantitative models can be adapted from colloidal physics for the steric potentials of polymeric brushes and used to describe bacteria.[6,7] As the size of polymer brushes increases, so too does the length scale and magnitude of the steric repulsive potential experienced by the bacteria.

In evolutionary terms, many bacteria have been in an arms race with single-celled eukaryotic microorganisms, such as amoeba, for many millions of years. Due to the preponderance of such competitive interactions, capsules are thought to have originated from them i.e. they are a protective mechanism against phagocytosis by single-celled organisms. The capsules were then subsequently successful in promoting infections in multicellular eukaryotic organisms, such as humans.

Historically, capsular bacteria were used to establish the phenomenon of genetic transformation for the first time i.e. bacteria can take up DNA from other bacteria.[8] Capsular strains of *S. pneumoniae* cause a violent infection followed by death in mice, whereas non-capsular strains do not. Mixing the non-capsular bacteria with DNA from heat-killed capsular strains was found to re-establish the virulence of the infections.

Unusually, *B. anthracis* capsules are made of amino acids (poly-γ – D-glutamic acid) rather than carbohydrates. *B. anthracis* has both an S-layer and a capsule.

One origin of *slime layers* is the continual shedding of capsular glycolipids. Another is the simple upregulation of the extracellular polymeric substance (EPS) in biofilms. More rheological data is needed to develop a broad overview of the physical properties of slime layers and their composites with capsules and biofilms. The phenomena observed are qualitatively in agreement with those expected for viscoelastic fluids, but their viscoelasticity has not been extensively explored. CPS can be expressed in parallel with EPS in biofilms, although CPS is downregulated in some quiescent biofilms.[9] Capsules can play an important role in biofilm formation.[10] A sensitive interaction is found among pili, CPS and biofilm fluidity in AFM experiments.

Fungi can also express capsules on their membranes. The main example in which the presence of a capsule is clearly associated with pathogenicity is *Cryptococcus neoformans*,[11] which can cause lung disease and meningitis.

Almost 80 different varieties of CPS (*serotypes*) are known to exist in *E. coli*, whereas 93 serotypes are found with *S. pneumoniae*.[12] Each serotype will provide a different challenge for the immune response of the host organism e.g. the production of antibodies targeted at the bacteria due to their different chemistries. An analogy is made with confectionary via M and Ms; the chocolate (the bacteria) is always the same inside the sweets, but the sugar coating varies between the different serotypes.

Lipid rafts have been observed in bacteria and occur due to the phase separation of membrane components.[13] Capsular components appear to grow in rafts with *E. coli*, although the raft motility is subdiffusive and fairly restricted.[7]

Capsules can block the adsorption of bacteriophage, which is a crucial initial step in the life cycle of the virus.[14] Some phages have evolved enzymes to cleave specific capsular polymers to facilitate their entry.

Rough LPS has been modelled using Hartree–Fock self-consistent field calculations[15] (Figure 20.2a). However, high charge fractions make accurate simulations challenging due to the long-range nature of the charge interaction (an intractable many-body problem). Polyelectrolyte scaling theories have many advantages in this respect[16] (Section 4.1.2).

The original studies of *capsular forces* were performed with *K. pneumoniae* using atomic force microscopy. Studies also measured the effect of antimicrobial peptides on *K. pneumoniae* capsules (Chapter 24).[17,18]

Experiments comparing capsular and non-capsular strains of *Staphylococcus aureus* in laminar flow show two orders of magnitude increases in the attachment rate for non-capsular bacteria. This again indicates the anti-adhesion nature of capsular coatings.[19]

In the mucoidal strain of Australian *K. pneumoniae*, CPS is shed and forms a thick viscoelastic *slime layer*. There is evidence that capsules provide resistance to antimicrobial peptides in *K. pneumoniae*.[20] However, our own group saw no significant changes in resistance for capsular *E. coli*[21] with antimicrobial peptide surfactants, although large self-assembled peptide aggregates could be imaged stuck on the capsular wall and the membrane-induced process of peptide self-assembly is a complicating factor. It is expected that this mechanism of capsular adsorption of peptides depletes the available concentration of the peptides, but no significant changes were seen in the minimum inhibitory concentrations (MICs) of the *E. coli* in response to the peptides with capsules during our experiments (Chapter 24). However, there is good evidence for resistance against bacteriophages provided by capsules,[14] although

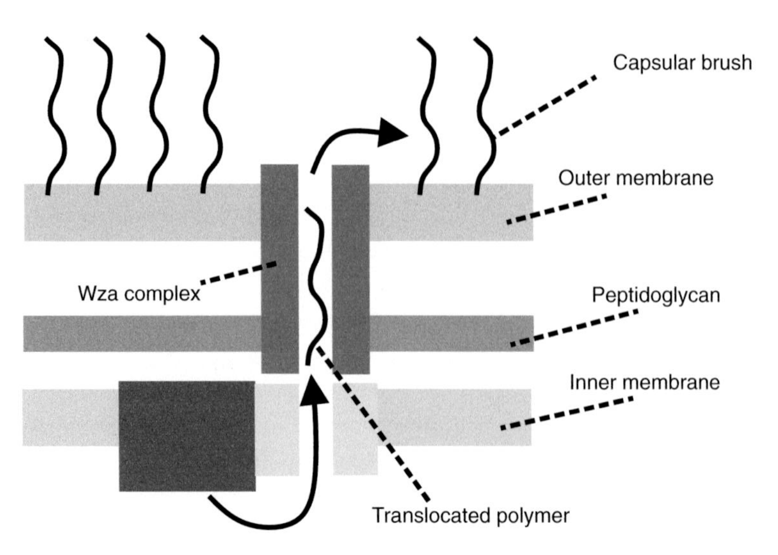

Figure 20.3 *Translocation* of CPS through the external membranes of Gram-negative bacteria. A mechanism is needed to thread the large polymeric molecules through a hole in each membrane. There will be a substantial free energy barrier to these processes due to the entropy of the polymer chains.

here again the story is complex e.g. bacteriophages can produce specific enzymes for capsule degradation and are thus strain-specific.

Translocation machines, such as ABC transporters, are required for the creation of *E. coli* capsules and they still have a number of outstanding physical questions.[22] The translocation of capsular LPS has some unique features.[23] How the length of capsule brushes is determined is still not completely understood. For *E. coli*, there are *molecular clock*, *molecular ruler* and *variable geometry* models.[24] The variable geometry model is the current front-runner. Different molecular challenges are presented to the process of translocation in Gram-positive bacteria versus Gram-negatives (Figure 20.3). Translocation is more challenging with double membrane barriers (Gram negative), since two consecutive translocation events (or three if the peptidoglycan wall is also considered) are required, each with a substantial free energy barrier due to the entropy of the polymer chains.

Super-resolution fluorescence microscopy images (STORM, Section 13.2.4) show that capsules of lyso-PG nucleate at all points on the external surface of *E. coli*, although the brushes are longer at the poles[7] (Figure 20.4). Thus, geometrical effects seem to play a role in the growth of capsular brushes.

Microelectrophoresis has been used to measure the charge and softness of capsular coatings on *Shewanella* species.[25] Zeta potentials (measured with Zeta sizers, Section 13.3.5) can also be used to characterise *Shewanella* capsular bacteria, since their values change with the expression of the capsule.[26]

In terms of physical processes, the *translocation mechanisms* across bacterial membranes can be separated between small and large molecules (Figure 20.3). Large polymeric molecules have a significant entropic penalty to be squeezed through a small

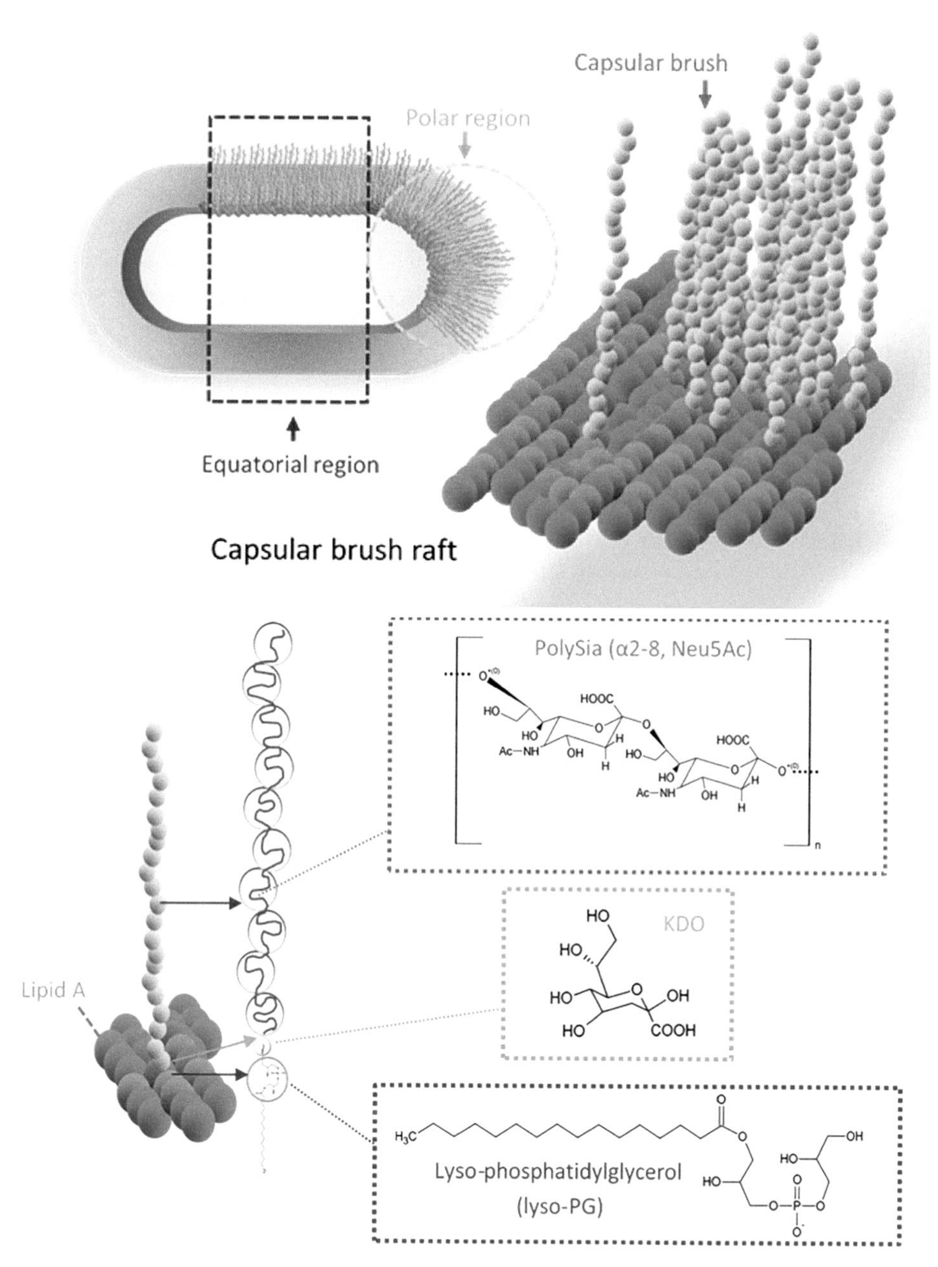

Figure 20.4 *Lyso PG polymeric brushes* occur in the capsules of a serotype of *E. coli*.[7] The brushes are longer on the polar regions of the *E. coli* when observed with super-resolution fluorescence microscopy.

hole and a motor is required to work against the free energy barrier,[27] since the process will not occur spontaneously. Osmotic flow can help drag particles through the hole during translocation facilitating transport. The retraction of the polymer chains back into the bacterial cell needs to be prohibited once they are threaded into the pore e.g. a specific binding event with a chaperone is required. Anomalous scaling exponents (β) are expected for the dynamics of translocation events,[27]

$$\langle \Delta m^2(t) \rangle \sim t^{\beta}, \tag{20.1}$$

where $\langle \Delta m^2(t) \rangle$ is the mean square displacement of a polymer chain through the channel in one dimension. Such scaling laws are an example of anomalous diffusion, Equation (1.8). Some bacteria use the translocation process to kill other cells e.g. membrane pore haemolysis produced by *S. aureus* can kill human cells by inducing them to leak their contents. Wza is the translocon for *E. coli* synthesis of type I capsules[28] i.e. the collection of genes required to synthesise and translocate the capsular polymers (Figure 20.3). There are still gaps in our understanding of how the translocation proteins function with *E. coli.*

Attempts have been made with atomic force microscopy measurements to understand how S-layers under the capsular brushes reinforce the mechanics of the cell walls. It is expected that the S-layers make the bacteria more rigid to compression, but initial results were inconclusive.[29]

Phagocytosis of bacteria by eukaryotic cells is an important part of the host's immune response to clear infections.[30] Some progress has been made with modelling the physics of phagocytosis, but extensions are still required for the presence of capsular brushes[30] e.g. to model the dependence of phagocytosised particles on their mesoscopic interparticle potentials.

Continuum mechanics calculations provide some useful models to understand the deformation of bacteria e.g. when they are deformed during an atomic force microscopy experiment[31] (Section 4.1.2). Among them is the collapse pressure needed to deform a water-filled elastic shell. For spherical elastic shells filled with water, the collapse pressure (P_c) is given by[32]

$$P_c = \frac{4E}{\sqrt{12(1-v^2)}}\left(\frac{h}{R}\right)^2, \tag{20.2}$$

where E is the modulus of the elastic shell, h is the shell thickness, R is the shell radius and v is the Poisson ratio of the shell's material.

The indentation of ellipsoidal and cylindrical elastic shells has also been investigated.[33] Such calculations are useful for comparison with atomic force microscopy measurements and some studies have been performed with bacteria,[34] although there is still some discussion in the literature as to the exact prefactors and the effects of prestresses. For pressurised ellipsoids and spheres (and generalisations to other smooth Gaussian surfaces),[33] it is found that the force experienced by elastic shells follows

$$\begin{aligned} F &= \pi\left(\frac{\kappa_G}{\kappa_M^2}\right) p\kappa_M^{-1}\delta \quad \delta << t \\ F &= \pi\, p\kappa_M^{-1}\delta \qquad \delta >> t \end{aligned} \tag{20.3}$$

where two regimes are defined depending on whether the deformation of the shells (δ) is greater than the shell thickness (t). κ_G is the Gaussian curvature $\left(\kappa_G = (R_x R_y)^{-1}\right)$, R_x

and R_y are the minor and major axes of the ellipsoid, respectively, κ_M is the mean curvature $\left(\kappa_M = \left(R_x^{-1} + R_y^{-1}\right)/2\right)$, τ is a dimensionless measure of the stress due to the pressure and p is the turgor pressure. The turgor pressure of bacteria can thus be calculated from indentation experiments. For intact *E. coli* cells, the wall is stiffer in the circumferential direction (Young's modulus 49 ± 20 MPa) than the axial direction (Young's modulus 23 ± 8 MPa).[35] The turgor pressure was 29 ± 3 kPa. The original AFM study with *Magnetospirillum gryphiswaldense* by Arnoldi et al.[34] found the effective spring constant of the cell wall to be ~0.05 N/m and the turgor pressure to be ~100 kPa.

Molecular dynamics simulations indicate that the attachment of large polymeric chains on one side of a lipid membrane gives rise to a substantial entropic force that will cause the membrane to increase its curvature.[36] No sudden change in curvature of bacterial walls has been observed during the growth of capsules on bacteria, so the entropic force of the polymer chains must be resisted by structures in the bacterial walls[7] e.g. peptidoglycan or adhesions with cytoskeletal filaments.

20.1 Slime Layers

From a historical perspective, a challenge has been to differentiate between a *slime layer* and a *biofilm* with bacteria. Slime layers were originally used to denote both slime layers and biofilms. It is now known that slime layers and biofilms are due to different genetic programmes and robust classification depends on genetic studies. Slime layers are thought to be due to simple upregulation of the production of capsular material or extracellular polymeric material. Slime layers are relatively simple viscoelastic fluids or weak physical gels and often can be simply washed off a surface by diluting their concentration. Biofilms in contrast are much more structured gels, driven by intercellular communication via quorum sensing and correspond to more sophisticated genetic programmes. There is thus a continuum of states expected between the extremes of pure biofilm and pure slime layer. Mixtures can also occur with slimes incorporated into more structured biofilms and the bacteria can simultaneously express capsules with slimes and biofilms.

Slime layers can contain both proteins and carbohydrates e.g. M proteins in *Streptococcus pyogenes* impair complement function in the host's immune system, protein A in *S. aureus* binds immunoglobulin to interfere with phagocytosis and LPS of Gram-negative bacteria delay or blunt acute inflammation responses.[37]

Suggested Reading

King, J. S.; Roberts, I. S. Bacterial surfaces: Front lines in host-pathogen cell-surface interactions. In the *Biophysics of infection*, Leake, M. C., Ed., 2016; pp. 129–156. Considers capsular *E. coli* in urinary tract infections.

Muthukumar, M., *Polymer Translocation*. CRC Press: 2016. It is an interesting challenge to understand the translocation of lipopolysaccharides through the membranes of bacterial cells. This book describes some theoretical progress, predominantly from the perspective of driven translocation of DNA through apertures during electrophoresis.

Palyulin, V. V.; Aln-Nissila, T.; Metzler, R., Polymer translocation: The first two decades and the recent diversification. *Soft Matter* **2014**, *10* (45), 9016–9037. Considers anomalous transport during translocation.

Pincus, P., Colloid stabilization with grafted polyelectrolyte. *Macromolecules* **1991**, *24* (10), 2912–2919. Scaling calculations of polyelectrolyte brushes.

Richards, D. M.; Endres, R. G., How cells engulf: A review of theoretical approaches to phagocytosis. *Reports on Progress in Physics* **2017**, *80* (12), 126601.

Rubinstein, M.; Colby, R., *Polymer Physics*. Oxford University Press: 2003. Excellent modern introduction to polymer scaling theories.

Zhulina, E. B.; Borisov, O. V., Polyelectrolytes grafted to curved surfaces. *Macromolecules* **1996**, *29*, 2618–2626. Scaling calculations of polyelectrolyte brushes on curved surfaces.

References

1. Rubinstein, M.; Colby, R. H., *Polymer Physics*. Oxford University Press: 2003.
2. Wen, Z.; Zhang, J. R., Bacterial capsules. In *Molecular Medical Microbiology*, Tang, Y. W., Sussman, M., et al., Eds., Academic Press: 2015; Vol. 1, pp. 33–53.
3. King, J. E.; Owaif, H. A. A.; Jia, J.; Roberts, I. S., Phenotypic heterogeneity in expression of the K1 polysaccharide capsule of uropathogenic *Escherichia coli* and downregulation of the capsule genes during growth in urine. *Infection and Immunity* **2015**, *83* (7), 2605–2613.
4. Jakubovics, N. S.; Goodman, S. D.; Mashburn-Warren, L.; Stafford, G. P.; Cieplik, F., The dental plaque biofilm matrix. *Periodontology 2000* **2021**, *86* (1), 32–56.
5. Milner, S. S., Polymer brushes. *Science (New York, N. Y.)* **1991**, *251* (4996), 905–914.
6. Wang, H.; Wilksch, J. J.; Lithgow, T.; Strugnell, R. A.; Gee, M. L., Nanomechanics measurements of live bacteria reveal a mechanism for bacterial cell protection: The polysaccharide capsule in *Klebsiella* is a responsive polymer hydrogel that adapts to osmotic stress. *Soft Matter* **2013**, *9* (31), 7560–7567.
7. Phanphak, S.; Georgiades, P.; Li, R.; King, J.; Roberts, I. S.; Waigh, T. A., Super-resolution fluorescence microscopy study of the production of K1 capsules by *Escherichia coli*: Evidence for the differential distribution of the capsule at the poles and the equator of the cell. *Langmuir: The ACS Journal of Surfaces and Colloids* **2019**, *35* (16), 5635–5646.
8. Gladwin, M. T.; Trattler, W.; Mahan, C. S., *Clinical Microbiology Made Ridiculously Simple*, 7th ed. MedMaster: 2019.

9. Limoli, D. H.; Jones, C. J.; Wozniak, D. J., Bacterial extracellular polysaccharides in biofilm formation and function. *Microbiology Spectrum* **2015**, *3* (3), 10.1128.
10. Wang, H.; Wilksch, J. J.; Strugnell, R. A.; Gee, M. L., Role of capsular polysaccharides in biofilm formation: An AFM nanomechanics study. *ACS Applied Materials & Interfaces* **2015**, *7* (23), 13007–13013.
11. Hogan, L. H.; Klein, B. S.; Levitz, S. M., Virulence factors of medically important fungi. *Clinical Microbiology Reviews* **1996**, *9* (4), 469–488.
12. Whitfield, C., Biosynthesis and assembly of capsular polysaccharides in *Escherichia coli*. *Annual Review of Biochemistry* **2006**, *75*, 39–68.
13. Bramkamp, M.; Lopez, D., Exploring the existence of lipid rafts in bacteria. *Microbiology and Molecular Biology Reviews: MMBR* **2015**, *79* (1), 81–100.
14. Hyman, P.; Abedon, S. T., Bacteriophage host range and bacterial resistance. *Advances in Applied Microbiology* **2010**, *70*, 217–248.
15. Lins, R. D.; Straatsma, T. P., Computer simulation of the rough lipopolysaccharide membrane of *Pseudomonas aeruginosa*. *Biophysical Journal* **2001**, *81* (2), 1037–1046.
16. Dobrynin, A. V.; Rubinstein, M., Theory of polyelectrolytes in solutions and at surfaces. *Progress in Polymer Science* **2005**, *30* (11), 1049–1118.
17. Mularski, A.; Wilksch, J. J.; Wang, H.; Hossain, M. A.; Wade, J. D.; Separovic, F.; Strugnell, R. A.; Gee, M. L., Atomic force microscopy reveals the mechanobiology of lytic peptide action on bacteria. *Langmuir: The ACS Journal of Surfaces and Colloids* **2015**, *31* (22), 6164–6171.
18. Mularski, A.; Wilksch, J. J.; Hanssen, E.; Strugnell, R. A.; Separovic, F., Atomic force microscopy of bacteria reveals the mechanobiology of pore forming peptide action. *Biochimica et Biophysica Acta* **2016**, *1858* (6), 1091–1098.
19. Prince, J. L.; Dickinson, R. B., Kinetics and forces of adhesion for a pair of capsular/unencapsulated *Staphylococcus* mutant strains. *Langmuir: The ACS Journal of Surfaces and Colloids* **2003**, *19* (1), 154–159.
20. Campos, M. A.; Vargas, M. A.; Regueiro, V.; Llompart, C. M.; Alberti, S.; Bengoechea, J. A., Capsule polysaccharide mediates bacterial resistance to antimicrobial peptides. *Infection and Immunity* **2004**, *72* (12), 7107–7114.
21. Phanphak, S. Investigation of K1 bacterial capsular morphology and single molecules in capsular biosynthesis using super-resolution fluorescence microscopy (dSTORM), PhD Thesis. University of Manchester: 2019.
22. Muthukumar, M., *Polymer Translocation*. CRC Press: 2019.
23. Woodward, L.; Naismith, J. H., Bacterial polysaccharide synthesis and export. *Current Opinion in Structural Biology* **2016**, *40*, 81–88.
24. King, J. D.; Berry, S.; Clarke, B. R.; Morris, R. J.; Whitfield, C., Lipopolysaccharide O antigen size distribution is determined by a chain extension complex of variable stoichiometry in *Escherichia coli* O9a. *Proceedings of the National Academy of Sciences of the United States of America* **2014**, *111* (17), 6407–6412.
25. Dague, E.; Duval, J.; Jorand, F.; Thomas, F.; Gaboriaud, F., Probing surface structures of *Shewanella spp.* by microelectrophoresis. *Biophysical Journal* **2006**, *90* (7), 2612–2621.

26. Gaboriaud, F.; Gee, M. L.; Strugnell, R. A.; Duval, J. F. L., Coupled electrostatic, hydrodynamic and mechanical properties of bacterial interfaces in aqueous media. *Langmuir: The ACS Journal of Surfaces and Colloids* **2008**, *24* (19), 10988–10995.
27. Palyulin, V. V.; Ala-Nissila, T.; Metzler, R., Polymer translocation: The first two decades and the recent diversification. *Soft Matter* **2014**, *10* (45), 9016–9037.
28. Dong, C.; Beis, K.; Nesper, J.; Brunkan-LaMontagne, A. L.; Clarke, B. R.; Whitfield, C.; Naismith, J. H., Wza the translocon for *E. coli* capsular polysaccharides defines a new class of membrane protein. *Nature* **2006**, *444* (7116), 226–229.
29. Schar-Zammaretti, P.; Ubbink, J., The cell wall of lactic acid bacteria: Surface constituents and macromolecular conformations. *Biophysical Journal* **2003**, *85* (6), 4076–4092.
30. Richards, D. M.; Endres, R. G., How cells engulf: A review of theoretical approaches to phagocytosis. *Reports on Progress in Physics* **2017**, *80* (12), 126601.
31. Audoly, B.; Pomeau, Y., *Elasticity and Geometry*. Oxford University Press: 2010.
32. Munglani, G.; Wittel, F. K.; Vetter, R.; Bianchi, F.; Herrmann, H. J., Collapse of orthotropic spherical shells. *Physical Review Letters* **2019**, *123* (5), 058002.
33. Vella, D.; Ajdari, A.; Vaziri, A.; Boudaoud, A., Indentation of ellipsoidal and cylindrical elastic shells. *Physical Review Letters* **2012**, *109* (14), 144302.
34. Arnoldi, M.; Fitz, M.; Bauerlein, E.; Fritz, M.; Radmacher, M.; Sackmann, E.; Boulbitch, A., Bacterial turgor pressure can be measured by atomic force microscopy. *Physical Review E* **2000**, *62* (1 Pt B), 1034–1044.
35. Deng, Y.; Sun, M.; Shaevitz, J. W., Direct measurement of cell wall stress stiffening and turgor pressure in live bacterial cells. *Physical Review Letters* **2011**, *107* (15), 158101.
36. Lipowsky, R., In *Physics of Biological Membranes*. Bassereau, P.; Sens, P., Eds. Springer: 2019; pp. 3–44.
37. Torok, E.; Moran, E.; Cooke, F., *Oxford Handbook of Infectious Diseases and Microbiology*. Oxford University Press: 2016.

PART III

INTERACTING BACTERIA AND BIOFILMS

The final section considers the phenomena involved with interacting bacteria in suspensions and biofilms. Several challenging physical problems related to interacting bacteria are presented and some progress is highlighted with quantitative treatments for bacterial viscoelasticity, population dynamics, biofilm formation, action of antibiotics, disease, systems biology, simulations, mechanisms of communication, bioelectricity and some applications.

21 Dynamics and Viscoelasticity of Suspensions and Biofilms

The dynamics of high-concentration bacteria at surfaces and in suspensions are strongly influenced by intercellular interactions, such as hydrodynamics, mesoscopic forces, synchronised active motility and chemical signalling. The flow properties of swarming and concentrated bulk bacterial suspensions are first considered. Then the motion of bacteria through viscoelastic solutions is presented e.g. *Helicobacter pylori*, an ulcer-forming bacterium, which moves through mucus in the stomach. Finally, the viscoelasticity of biofilms is covered.

21.1 Swarming and Higher Concentration Bacterial Suspensions

In general, colloidal suspensions have viscosities that increase with colloid concentration, since more colloids increase the total amount of dissipation and interparticle interactions also increase dissipative effects. For dilute concentrations of colloidal suspensions, neglecting interparticle interactions, a linear increase of viscosity with colloid concentration is expected following the *Einstein relationship* (Figure 21.1),

$$\eta = \eta_0 \left(1 + \frac{5}{2}\phi\right), \tag{21.1}$$

where η_0 is the viscosity of the solvent and ϕ is the volume fraction of the colloids. At higher concentrations, more rapid increases in viscosity can occur, with a sudden jump in the value of the viscosity at the jamming point. Moderate reductions of viscosity can also occur when the colloid concentration is increased, if elongated colloids are considered and they experience liquid crystalline alignment transitions (Chapter 8). Aggregation phenomena due to sticky interactions are common with bacteria and they will also cause large deviations from Equation (21.1) i.e. much larger increases in viscosity will occur than predicted based on the volume fraction (ϕ) of single bacteria.

Dramatic signatures of *active viscoelasticity* are observed for bacterial suspensions in concentrated solutions due to the motor-driven motility of the bacteria. For example, active suspensions of bacteria can behave as *room temperature super-fluids* where zero and negative viscosities have been observed[1] (the bacterial suspensions move in the opposite direction to that in which they are sheared, Figure 21.2). A reduction in viscosity is also seen in microrheology measurements of free-standing liquid films that contain *Escherichia coli* due to collective hydrodynamic effects.[2,3] The viscosity of the films was measured with magnetic tweezer experiments and it decreased by up to a factor of seven because of the bacterial motility.

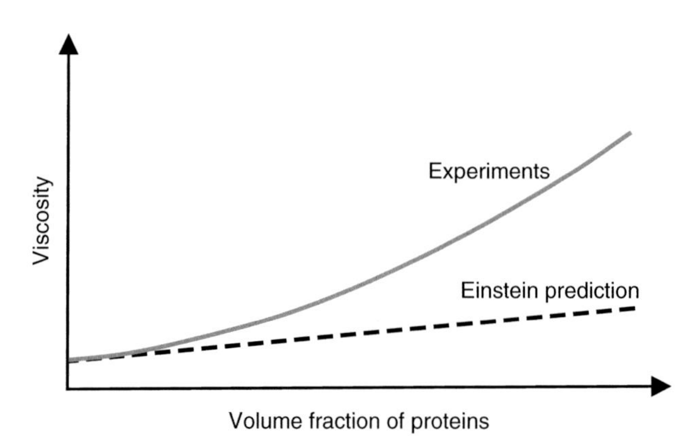

Figure 21.1 Schematic diagram of the viscosity (η) as a function of volume fraction (ϕ) e.g. for small protein colloids. The continuous line indicates the experiments. The Einstein prediction (Equation (21.1), dashed line) provides a lower limit for the dependence of the viscosity on the protein volume fraction (ϕ).

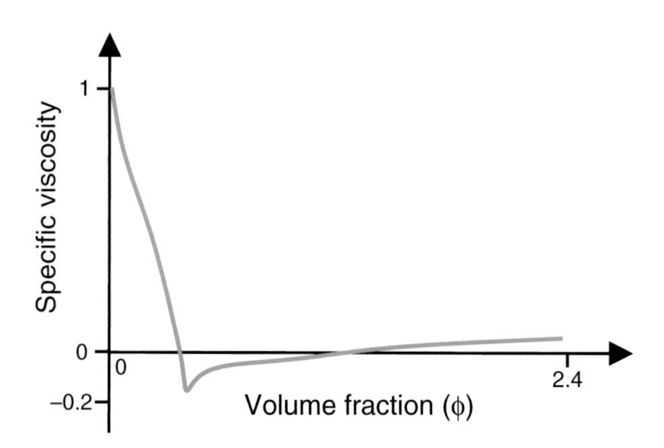

Figure 21.2 Specific viscosity (η_p / η_0) as a function of volume fraction (ϕ) for motile *Escherichia coli* strains (ATCC9637 and RP437) from rheology experiments using a shear rate ($\dot{\gamma}$) of 0.04 s^{-1}[1]. Superfluid behaviour with zero or negative viscosities is observed over a range of concentrations. Lower viscosities and higher motilities are observed under oxygenated conditions.

The *rheology* of strains of *E. coli* was examined in a low shear Couette rheometer[1] (Figure 21.2). Two of the strains had zero or negative viscosities for bacterial volume fractions of >0.4%. The motility of the *E. coli* was sensitive to both oxygen concentration and L-serine availability. A simple theory was developed for the viscosity of the suspension (η). A linear decrease of η with the volume fraction (ϕ) of the bacteria was predicted and observed (the opposite trend to that expected from the Einstein relationship, Equation (21.1)),

$$\eta = \eta_0 \left(1 - K \left(\frac{\tau}{t_c} \right) \phi \right), \tag{21.2}$$

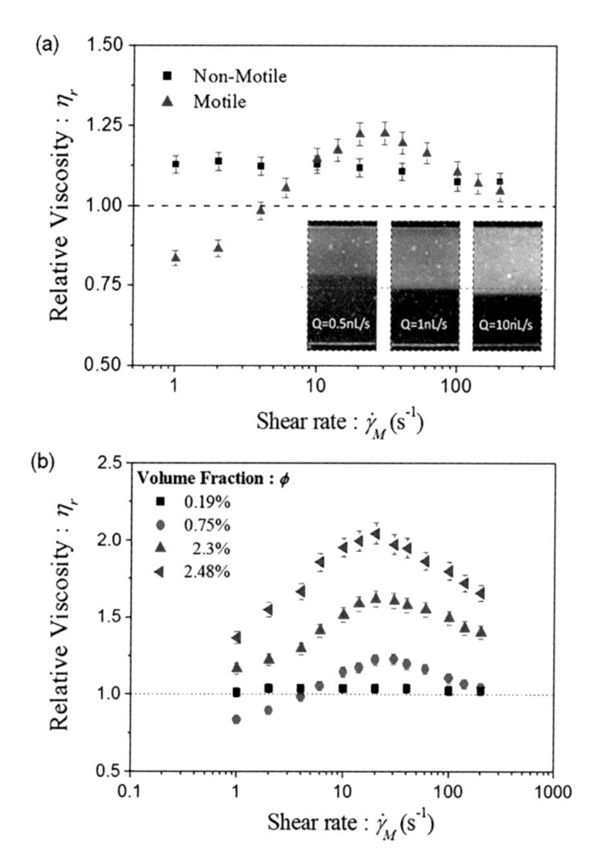

Figure 21.3 *Shear thickening* and *shear thinning* are observed with suspensions of *E. coli* in Y-shaped microfluidic channels.[4] (a) The relative viscosity (η_r) as a function of shear rate for a volume fraction of 0.35%, comparing motile and non-motile bacteria. (b) Relative viscosity (η_r) as a function of shear rate for *E. coli* for different volume fractions. Shear thickening is observed followed by shear thinning. Reprinted figure with permission from Gachelin, J.; Mino, G.; Berthet, H.; Lindner, A.; Rousselet, A.; Clement, E., Non-Newtonian viscosity of E. coli suspensions. *Physical Review Letters* 2013, *110*, 268103. Copyright (2013) by the American Physical Society.

where η_0 is the solvent viscosity, K is a constant, $K \propto \left[B(\tau/t_c)-A\right]$, τ characterises the directional persistence of an average swimming trajectory of the bacteria and t_c is the time taken for the bacteria to drag the solvent over a distance equal to their size. A and B are constants that solely depend on the bacterial shape and K is found to be -120 ± 10 experimentally.

The *non-linear rheology* of *E. coli* suspensions has also been explored. Measurements on a Y-shaped microfluidic channel with concentrated suspensions of *E. coli* showed that both *shear thickening* and *shear thinning* occurred as a function of shear rate, which depended on the bacterial concentration[4] (Figure 21.3). Shear thickening is where the viscosity increases with the shear rate, whereas shear thinning is the reverse. At low shear rates, the viscosity of the *E. coli* suspensions was lower than

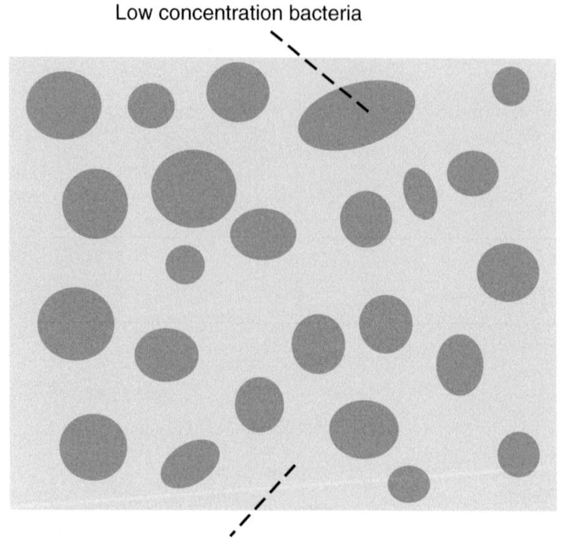

Figure 21.4 Schematic diagram of *motility-induced phase separation* predicted for suspensions of motile particles using simulations.[7] The simulations used a lattice model that incorporates run and tumble motility for the bacteria. The local density of the bacteria was predicted as a function of position across the two-dimensional surface.

that of the surrounding solvent, in qualitative agreement with other linear rheology experiments that demonstrate the existence of superfluidity.

A *jamming transition* is observed with *Bacillus subtilis* suspensions under shear flows as their concentration is increased and the transition is very sensitive to the levels of hydration.[5] Jamming is a common viscoelastic phenomenon for high-concentration colloidal suspensions e.g. it is observed in dense suspensions of custard power.[6]

The concept of *active phase separation* provides a new framework to understand pattern formation in concentrated bacterial suspensions[8,9] (Figure 21.4). For example, motility can drive phase separation in suspensions of active colloidal particles.[7] Experimentally, synthetic biology tools have been used with *E. coli* suspensions to study active phase separation. Using a genetically modified strain of *E. coli*,[10] stripes were found to form in suspensions due to a motility mechanism that was suppressed at higher bacterial densities (Section 18.2). The stripes could be modelled using two coupled reaction–diffusion equations, but active phase separation provides an alternative paradigm to model the phenomenon e.g. an effective fluid theory with a Landau–Ginzburg formalism could be used.

Many models for the motility of nano/micro-sized active particles in complex and crowded environments have been developed as a result of experiments on synthetic non-biological systems.[11] *de novo* swimmers created by synthetic chemists provide simple robust systems to explore the physics of active motile suspensions.

The distinctive motility of bacteria at high concentrations on a solid surface is called *swarming*[13] (Figure 21.5a) and it is associated with a distinct genetic

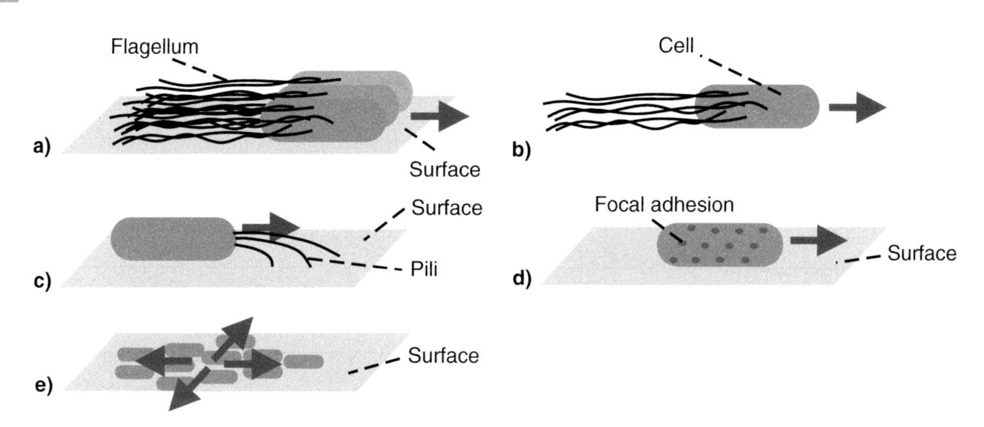

Figure 21.5 Different motility mechanisms used by bacteria at surfaces.[12] (a) *Swarming* and (b) *swimming* are powered by flagella. (c) *Twitching* is due to the extension of pili. (d) *Gliding* uses focal adhesion complexes on the surfaces of the bacteria. (e) *Sliding* occurs due to the translocation of surfactants through the bacterial membranes.

programme. Swarming is thus closely associated with the interactions of bacteria with surfaces that were covered in Chapter 17, but high-concentration bacterial suspensions (both at surfaces and in the bulk) have some separate phenomena that will be highlighted in the current section. *Swarming* and *rafting* (coherent motions of groups of bacteria) can both occur, and there is a lag time for the occurrence of swarming as the bacteria switch their genetic programmes.[12] Swarming has thus both genetic and physical origins.

The statistics of swarming has been studied for *E. coli* moving on an agar surface.[13] The standard run and tumble motion of *E. coli* is modified in these concentrated films due to randomisation caused by collisions with other bacteria (Chapter 2). Furthermore, there is often local alignment of cells induced by collisions with neighbours and short-range hydrodynamic interactions, which become more substantial for bacterial cells with higher aspect ratios i.e. nematic liquid crystalline ordering occurs for the swarming of bacteria with elongated cell shapes (Chapter 8). During swarming, *E. coli* were found to multi-nucleate, synthesise large numbers of flagella, produce wetting agents and coordinate their motion.[13] Thus, a complex pattern of gene expression is associated with this swarming phenotype.

The *Vicsek model*[14] is a very simple agent-based technique (Chapter 27) used to describe the collective movement of motile particles, e.g. swarming of bacteria, and it has become a standard tool in the physics simulation community.[15] The model incorporates a directional order parameter to describe the flocking motion i.e. the averaged normalized velocity in which the flock of motile particles is moving. The model can exhibit a dynamic phase transition, which is observed in experiments with a wide range of motile organisms e.g. coherent flocks can form that have a common direction of rotation, due to a modest tendency for mutual alignment superposed on the random motion of individual organisms (also observed in murmurations of starlings, shoals of tropical fish, etc.). Self-generated vortices from a state of randomly moving swarms

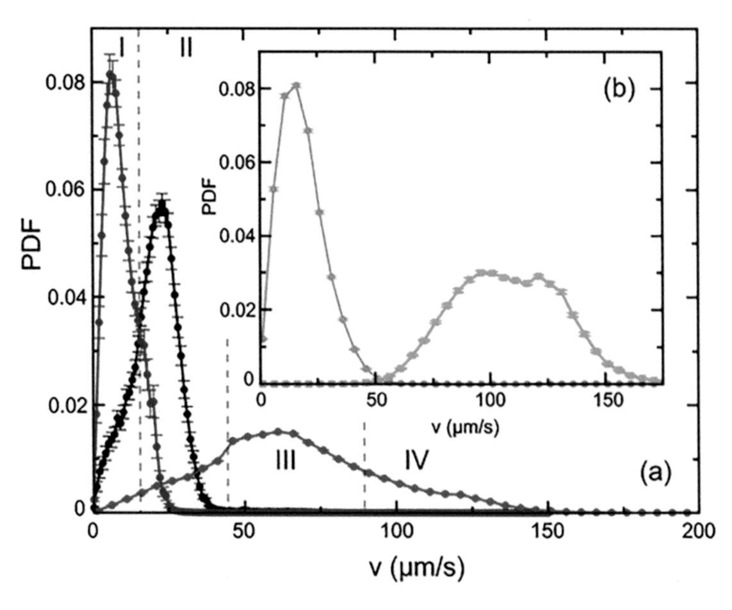

Figure 21.6 A transition to a swimming liquid crystalline phase is observed in experiments with *B. subtilis* suspensions.[17] (a) A probability distribution is shown for the speeds (v) adopted by the bacteria. Black – dilute swimmers, blue – semi-dilute swimmers and red – concentrated swimmers. (b) Fast (green) and slow (yellow) subphases are observed in the suspensions of concentrated swimmers. Reprinted figure with permission from Cisneros, L. H.; Kessler, J. O.; Ganguly, S.; Goldstein, R. E., Dynamics of swimming bacteria: transition to directional order at high concentration. *Physical Review E* **2011,** *83*, 061907. Copyright (2011) by the American Physical Society.

can occur with bacteria and the Vicsek model has been used to describe their creation.[9,16] The Vicsek model describes a separate phenomenon from that of nematic liquid crystallinity, since it considers velocity alignment rather than directional alignment of mesogens. Both coherent flocking and liquid crystallinity are possible with bacterial suspensions, so careful experiments are needed to differentiate between the two possible mechanisms.

The transition to a swimming nematic phase in suspensions of *B. subtilis* has been studied as a function of bacterial concentration[17] (Figure 21.6). The average bacterial velocity first decreases and then increases as the liquid crystalline phase transition occurs. Steric forces of the elongated cells play an important role, similar to that seen in the equilibrium phases of liquid crystals (Chapter 8) e.g. rods spontaneously organise in solution to form a nematic phase if they have sufficiently large aspect ratios. Specifically, length/diameter ratios above a limit given by the Onsager threshold $\frac{1}{4}c\pi L^2 D_{eff} = 4$ are predicted to cause nematic phases in suspensions, where L is the length and D_{eff} is the effective diameter adjusted by the charge interaction and c is the concentration of bacteria.[18] More accurate calculations need to incorporate the effects of the polydispersity of the bacteria and their exact interparticle potential on the positions of the liquid crystalline phase boundaries.[19] Motor activity will also alter the behaviour e.g. flagella will provide active fluctuations in steric potential.

The *swarming statistics* of *B. subtilis* on agar were found to sensitively depend on the length of mutant bacteria.[20] A universal behaviour was found for the correlation of the dynamics of *B. subtilis* as a function of dynamic flock size in microscopy experiments.[21] Active sedimentation flows occur in sessile droplets of concentrated bacterial solutions and can also lead to the large-scale coherence of the dynamics.[22]

Swarming bacteria are able to make collective decisions with regard to their navigation in *Paenibacillus vortex*.[23] Experiments have studied the transport of synthetic colloids due to collective swarming.

Swarming bacteria (*B. subtilis*) were treated with photosensitizers and exposed to white light, which caused the creation of reactive oxygen species (ROS, Chapter 24). The ROS subsequently caused a loss of collective motion of the bacteria, since it damaged molecules inside them i.e. an antibiotic stress was created.[24] This was analysed with an agent-based model for the motility with an alignment rule similar to a Vicsek model. The effect of antibiotics on swarming behaviour was also quantified.[24]

Swarming colonies of the light-responsive bacteria *Serratia marcescens* were exposed to a wide spectrum of ultraviolet (UV)/visible light.[25] Medium-light dosages temporarily suppressed collective motion and large dosages led to paralysed jammed regions of the bacterial suspensions.

Colloidal beads were placed in quasi-two-dimensional solutions of *E. coli* (suspended liquid films) and the motion of the beads was tracked[26] (Figure 21.7).

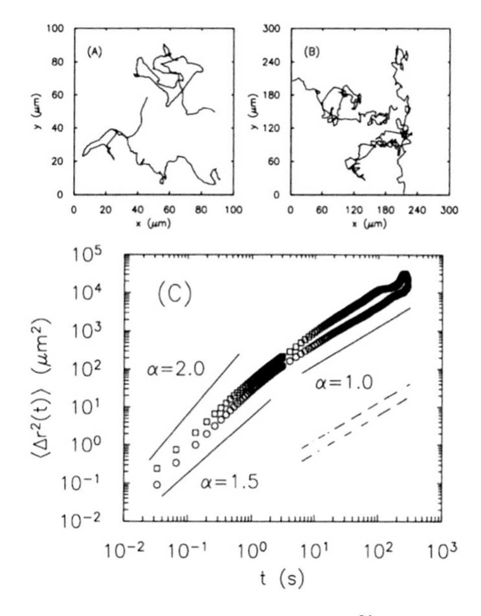

Figure 21.7 *Particle tracking* experiments with colloidal beads in suspensions of *E. coli*.[26] (a) and (b) show particle tracks for shorter and longer times, respectively. (c) The time averaged mean square displacement $\left(\left\langle \Delta r^2(t) \right\rangle\right)$ as a function of lag time for 4.5 μm (circles) and 10 μm (squares) colloidal beads. Power law trend lines are shown with different exponents: $\alpha = 2$ is ballistic, $\alpha = 1.5$ is super-diffusive and $\alpha = 1$ is diffusive. Reprinted figure with permission from Wu, X. L.; Libchaber, A., Particle diffusion in a quasi-2D bacterial bath. *Physical Review Letters* **2000,** *84* (13), 3017–3020. Copyright (2000) by the American Physical Society.

Clear evidence was found for active super-diffusive motion at short times in the scaling of the mean square displacement of the beads as a function of time interval $\left(\langle \Delta r^2(\tau) \rangle \sim \tau^{\alpha}, \alpha > 1 \text{ for } \tau < 2\text{ s}\right)$, that crossed over to diffusive behaviour at longer times $(\alpha = 1 \text{ for } \tau > 2\text{ s})$. The response of colloidal beads is commonly used in passive particle tracking microrheology experiments (Section 13.4.1). Such experiments can help bridge the gap from the microscopic properties of the bacteria to the macroscopic viscoelasticity of the solutions observed in bulk experiments.[27,28] These experiments with *E. coli* thus present clear evidence for the active viscoelasticity of the solutions i.e. the flagella of the bacteria are driving the motion of the colloidal beads in their local environments, which is observed as active viscoelasticity. Similar experiments on the displacements of colloidal spheres placed in high concentration suspensions of *E. coli* showed an active noise power spectrum with a characteristic $\omega^{-1/2}$ frequency scaling[29] (ω is the frequency). Thus, the linear viscoelasticity of the bacterial suspensions has a well-defined signature characteristic of the active noise spectrum driven by the flagella of the bacteria.

Separate particle tracking studies also used 10 μm diameter colloidal spheres as probes in *E. coli* suspensions and focused on the collisions of the bacteria with the spheres to explain the anomalous super-diffusive statistics of the spheres' motion[30] (Chapter 2) i.e. the study went beyond the standard Gaussian fluctuations assumed in particle tracking microrheology that can be analysed with the generalised fluctuation-dissipation theorem.[31] Fluorescence and bright-field microscopy were used to track the motion of the spheres.

During swimming on a surface, *Pseudomonas aeruginosa* can produce rhamnolipids, which are controlled by a quorum sensing mechanism. These cause the bacteria to form branched tendril patterns during swarming motility. Simulations indicate the pattern formation is driven by a Marangoni effect i.e. the flow is driven by gradients in the surface tension caused by the rhamnolipids.[32]

Bacterial turbulence can occur at low Reynolds numbers (Chapter 9, opposite to that expected for conventional turbulence) with high concentrations of bacteria in suspensions of *B. subtilis* (Figure 21.8). Such high concentration of bacterial suspensions may in turn lead to the formation of biofilms,[33,34] so biofilms can be nucleated from a turbulent phase. The correlations of bacterial velocity and orientation are found to be scale invariant for the dynamics of bacterial clusters.[21,22,35] A vector field theory was reasonably successful in describing the long-wavelength dynamics of bacterial turbulence in both two and three dimensions.[33,36,37]

Intermittency (erratic increases and decreases in the velocity of the flows) is a standard signature of classical turbulence.[38] It is also found for the velocity of *B. subtilis* cells during bacterial turbulence in microchannels.[39]

Coarse-grained hydrodynamic theories indicate that in general self-propelled colloidal particles (including planktonic bacteria) have *giant number density fluctuations* and their flows are unstable.[40] This provides a theoretical route to motivate the phenomenon of *bacterial turbulence*. Giant number fluctuations were observed in experiments on the collective gliding motion of *Myxococcus xanthus* mutants[35]

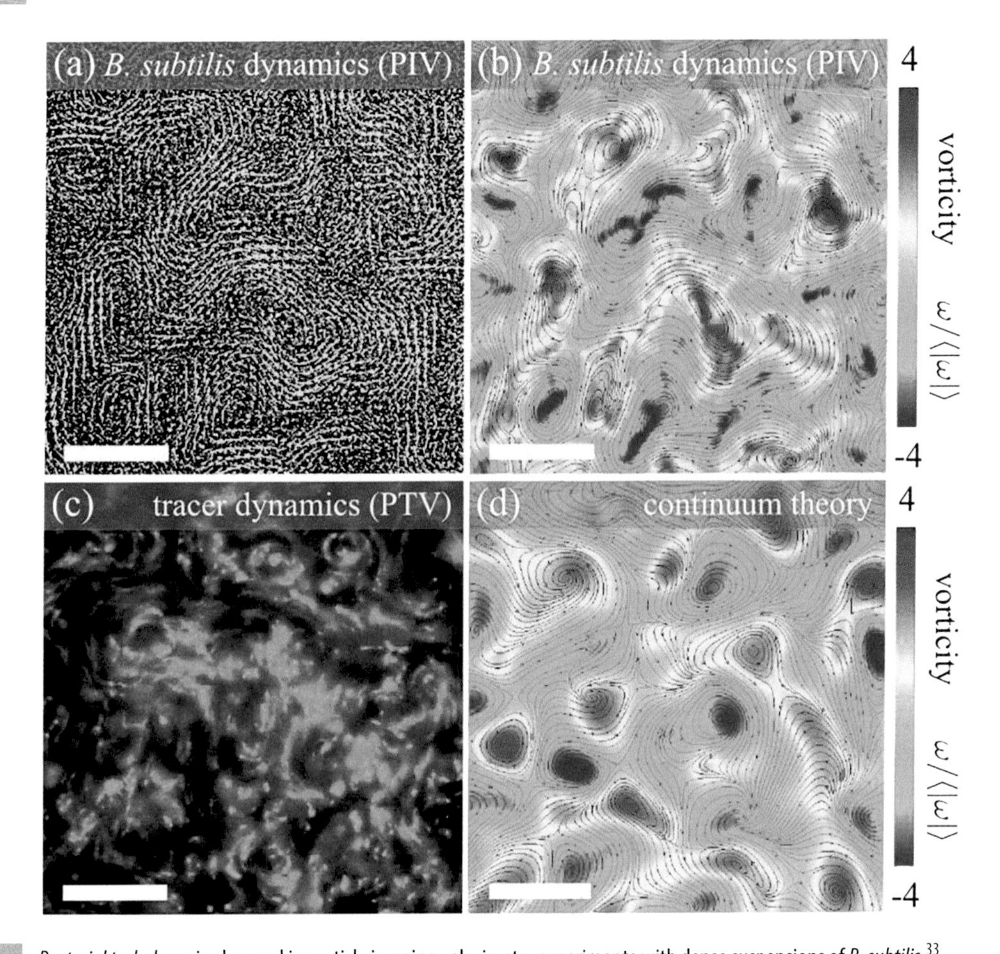

Figure 21.8 *Bacterial turbulence* is observed in particle imaging velocimetry experiments with dense suspensions of *B. subtilis*.[33] (a) The direction of motion of the bacteria. (b) The vorticity of the bacteria. (c) The dynamics of the tracer particles. (d) Results of a continuum theory that predicts the vorticity.

(Figure 21.9). The average number fluctuations $(\langle \Delta n(l) \rangle)$ were found to scale with the number of bacterial cells $(n(l))$ in a box of side l,

$$\langle \Delta n(l) \rangle \propto \langle n(l) \rangle^{\beta}, \tag{21.3}$$

where $\langle \Delta n(l) \rangle = \left(\langle n(l)^2 \rangle - \langle n(l) \rangle^2 \right)^{1/2}$ and β is a constant. With thermal fluctuations for the number of particles in a non-motile colloidal system, it is expected that $\beta = \frac{1}{2}$, whereas giant fluctuations were observed for *M. xanthus* with $\beta > \frac{1}{2}$. The collective motion was driven by cluster formation and occurred above a critical cell packing fraction (>17%). The interplay between the rod-like shape of the bacteria and their mechanisms of self-propulsion was the main factor that drove the type of collective motion observed.

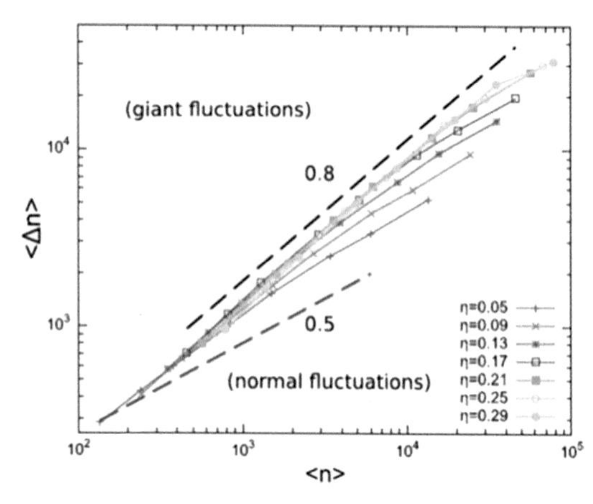

Figure 21.9 *Giant number fluctuations* are observed in the gliding motion of *Myxococcus xanthus* mutants.[35] η is the volume fraction. $\langle \Delta n \rangle$ is the average number of fluctuations and $\langle n \rangle = n(l)$ is the number of bacteria in a box of size l. Reprinted figure with permission from Peruani, F.; Starrus, J.; Jakovljevic, V.; Sogaard-Andersen, L.; Deutsch, A.; Baer, M., Collective motion and non-equilibrium cluster formation in colonies of gliding bacteria. *Physical Review Letters* **2012,** *108*, 098102. Copyright (2012) by the American Physical Society.

21.2 Motion of Bacteria in Viscoelastic Fluids and Porous Materials

The study of *viscoelasticity* (fluids that demonstrate intermediate behaviour between solids and fluids) is a huge endeavour, with many outstanding technical challenges in both theory and experiment[41] (Chapter 10). For example, important problems exist to extend the Navier–Stokes equation to viscoelastic materials (Equation (9.1)), which is particularly relevant for the study of non-linear viscoelasticity e.g. with biofilms under flow.

Some progress has been made with hydrodynamic calculations for the motions of single bacteria through viscoelastic fluids e.g. oscillating filaments in viscoelastic fluids.[42] Surprisingly, with some bacteria, the addition of viscoelastic polymers to suspensions can enhance their motility[43] i.e. they swim faster. A range of theoretical explanations have been provided, including polymer density fluctuations, chirality and colloidal effects.[44,45]

H. pylori are a major cause of stomach ulcers in humans and they appear to have coevolved over many thousands of years with their hosts e.g. the historical dispersion of humans from Africa can be mapped using *H. pylori* genetics.[47] It is a challenging calculation to understand how *Helicobacter* propels itself through the mucin gel in the stomach, since mucin is highly viscoelastic.[48] Mucin glycoproteins experience a pH switchable gelation transition at low pHs in the stomach. The challenge to understand the mechanism of bacterial motility is compounded by the use of urea by the

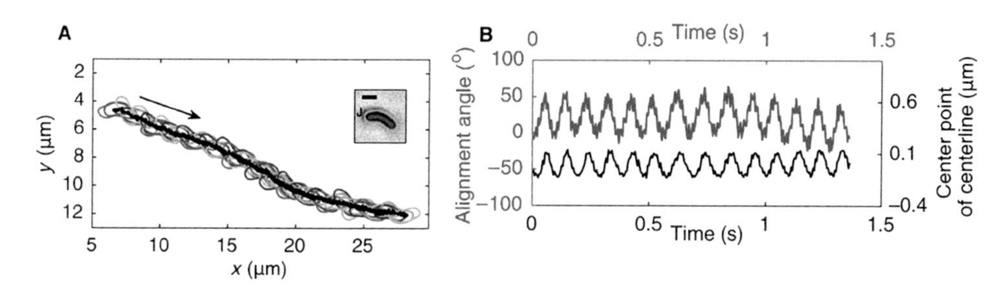

Figures 21.10 The motions of *H. pylori* through mucin gels follow helical trajectories.[46] (a) Trajectories of bacteria through porcine gastric mucin from particle tracking experiments. (b) The rotation of bacteria as they move through the mucin, where the alignment angle of the helix is plotted as a function of time.

bacteria to reduce the viscoelasticity of the mucin i.e. the bacteria modify their local viscoelasticity to facilitate their motility (Figure 21.10). The helical trajectories of the bacteria swimming in mucin have been examined in detail.[46]

The *extracellular polymeric substances* (EPSs) found in biofilms are known to be viscoelastic[49,50] (Section 21.3). Some predatory bacteria infiltrate biofilms of other species and rehouse themselves in the biofilms.[51] The viscoelasticities of the biofilms present barriers for motility during the process of infiltration.

It is also a challenge to understand the motility of bacteria in porous media e.g. soil.[52] Although porous media is viscoelastic, the emphasis is often on geometrical effects to determine the response of the bacteria (which are often much softer than the porous materials enclosing them). The form of bacterial motility observed depends on the geometry of the pores, which can increase the tortuosity of their motion e.g. for *Pseudomonas putida* moving in sand columns. Microscopy experiments were used with particle imaging velocimetry (PIV) to study *B. subtilis* motility in porous microfluidic environments as a model for soil.[53]

The fragmentation of linear aggregates of *Lactobacillus rhamnosus* under shear is industrially important for the creation of dairy products (Section 31.1). Shear flows from spray drying can break chains of bacteria into much smaller aggregates at high industrially relevant shear rates, reducing the viscosity of the bacterial suspensions (a shear thinning phenomenon).[54]

The motility of *P. mirabilis* has been studied in suspensions of synthetic lyotropic liquid crystals.[55] The liquid crystal orients the motion of the bacteria and encourages the formation of multi-cellular aggregates.

Relatively little work has been performed on *capsule viscoelasticity*. It is expected that measurable viscoelastic effects will occur for the motion of water through the polymer brushes by reference to the literature on synthetic surface-attached polymer chains.[56] Viscoelastic measurements could be performed using atomic force microscopy.

Slimes are widely used as food additives e.g. gellan and xanthan. They are found to be weak physical gels[57,58] (Chapter 6).

21.3 Biofilms

Polymer gels, such as the EPS of biofilms, are classic examples of viscoelastic materials and concentrated bacterial suspensions without the EPS are classified as *colloidal suspensions*, which are again viscoelastic (Figure 21.11). Thus, bacterial biofilms can demonstrate viscoelastic behaviours that are characteristic of both polymeric and colloidal suspensions (Figure 6.6, Chapters 6 and 7). Furthermore, fluid flows have a series of effects on the viscoelasticity of bacterial biofilms as the bacteria adapt to their hydrodynamic environments. Perhaps the best studied phenomenon is when shear flows harden up biofilms i.e. the shear moduli of biofilms increase with shear flow rate. This was demonstrated in particle tracking microrheology experiments with *Staphylococcus aureus*.[49] The phenomenon has also been observed with bulk rheology, microfluidics and magnetic microrheology experiments.[60–62] Shear flow hardening has thus been seen with a variety of standard bacterial species (*E. coli*, *S. aureus*, *P. aeruginosa*, mixed species from water sludge, etc.), so it is a generic phenomenon. A clear evolutionary advantage is offered for bacteria with the shear-hardening phenotype, since they can remain in a favourable environmental niche over a wide range of flow velocities. Biofilms with extremely large viscoelastic moduli have the disadvantage that they are much more challenging for the bacteria to disrupt when the bacteria reach the phase of growth in which dispersal is favoured. Thus, it is expected and observed that the mechanical moduli of biofilms (e.g. their shear moduli) take on a limited range of relatively soft values (kPa–MPa) due to the conflicting requirements for a mechanically protective environment that can be quickly disrupted to allow subsequent dispersal.

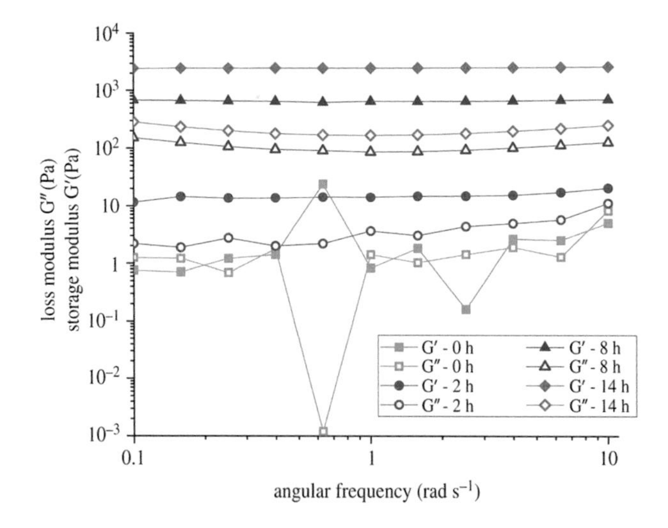

Figure 21.11 *Complex shear moduli* $(G^* = G' + iG'')$ from a *B. subtilis* biofilm as a function of angular frequency measured in a rheometer.[59] A weak power law dependence for both G' and G'' occurs with frequency, where $G' > G''$ and this is characteristic of gels.

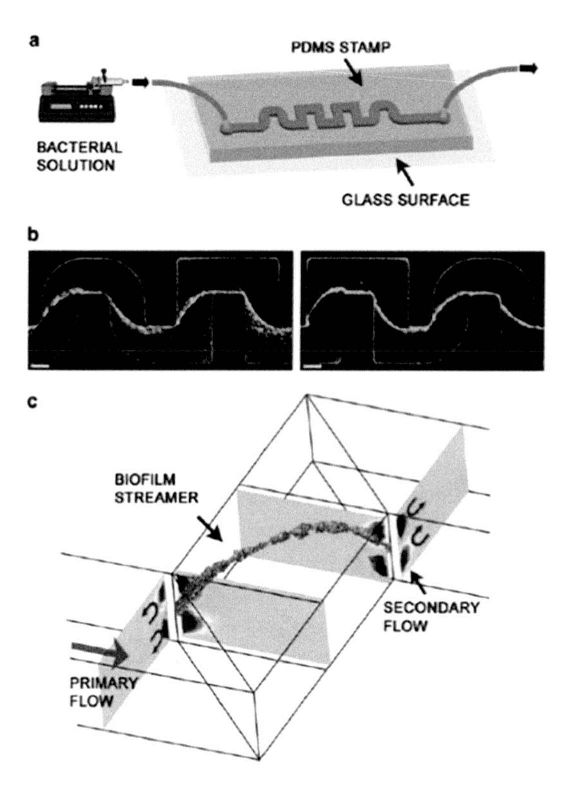

Figure 21.12 Streamers of *P. aeruginosa* were created by shear flows across biofilms grown in microfluidic geometries.[65] (a) Microfluidic set up. (b) Confocal microscopy image of a biofilm. (c) Suggested mechanism for streamer formation. Reprinted from Rusconi, R.; Lecuyer, S.; Autrusson, N.; Guglielmini, L.; Stone, H. A., Secondary flow as a mechanism for the formation of biofilm streamers. *Biophysical Journal* **2011,** *100* (6), P1392–1399 with permission from Elsevier.

Microrheology has also been used to gauge the efficacy of anti-biofilm[49] and antibiotic molecules in disrupting the viscoelasticity of biofilms.[63] Microrheology could thus be used as a general method for screening treatments targeted at biofilm mechanics and the treatments could then be formulated with antibiotics to provide effective methods to combat biofilm-forming infections.

Bulk rheology measurements were used to explore the viscoelasticity of *P. aeruginosa*, *Pseudomonas fluorescens* and *P. putida* biofilms when exposed to a wide range of small molecules (non-enzymatic) that could potentially perturb the biofilm mechanics.[64] It was found that the biofilms were remarkably robust to the chemical perturbations. The most effective treatments were $FeCl_3$ and $Al_2(SO_4)_3$, which increased the shear moduli of the biofilms, and citric acid, which decreased the shear moduli.

Streamers are formed by long threads of biofilm in shear flows (Figure 21.12) that extend away from the point of surface attachment. The formation of streamers is found to be strongly influenced by secondary flows.[65] Streamer formation was examined in zigzag microfluidic geometries.[65] It was found to depend sensitively on the curvature of the zigzags and thus the secondary flows.

Some progress has been made in calculating the elastic modulus for streamers.[66] The Young's moduli are found to have values of 10–6,000 kPa, reasonably close to values observed for surface attached biofilms.[67]

The colloidal properties of bacteria in biofilms were highlighted in a recent bulk rheology study of the non-linear viscoelasticity of *B. subtilis*, *Comamonas denitrificans*, *P. fluorescens* and *P. aeruginosa*.[68] Strain hardening was observed, which is a classic signature of colloidal suspensions.

Suggested Reading

Lauga, E., Bacterial hydrodynamics. *Annual Review of Fluid Mechanics* **2016**, *48*, 105–130. Excellent theoretical overview of bacterial motility.

Lauga, E. *The Fluid Mechanics of Cellular Motility*. Cambridge University Press: 2020. Some interesting calculations on bacteria in viscoelastic fluids are considered.

Pismen, L. *Active Matter Within and Around Us*. Springer: 2021. Excellent introduction to the active motility of cells.

References

1. Lopez, H. M.; Gachelin, J.; Douarche, C.; Auradou, H.; Clement, E., Turning bacteria suspensions into superfluids. *Physical Review Letters* **2015**, *115* (2), 28301.
2. Sokolov, A.; Aranson, I. S.; Kessler, J. O.; Goldstein, R. E., Concentration dependence of the collective dynamics of swimming bacteria. *Physical Review Letters* **2007**, *98* (15), 158102.
3. Sokolov, A.; Aranson, I. S., Reduction of viscosity in suspension of swimming bacteria. *Physical Review Letters* **2009**, *103* (14), 148101.
4. Gachelin, J.; Mino, G.; Berthet, H.; Lindner, A.; Rousselet, A.; Clement, E., Non-Newtonian viscosity of *Escherichia coli* suspensions. *Physical Review Letters* **2013**, *110* (26), 268103.
5. Rhodeland, B.; Hoeger, K.; Ursell, T., Bacterial surface motility is modulated by colony-scale flow and granular jamming. *Journal of Royal Society Interface* **2020**, *17* (167), 20200147.
6. Waitukaitis, S. R.; Jaeger, H. M., Impact-activated solidification of dense suspensions via dynamic jamming fronts. *Nature* **2012**, *487* (7406), 205–209.
7. Cates, M. E.; Tailleur, J., Motility-induced phase separation. *Annual Review of Condensed Matter Physics* **2015**, *6*, 219–244.
8. Stenhammer, J.; Tiribocchi, A.; Allen, R. J.; Marenduzzo, D.; Cates, M. E., Continuum theory of phase separation kinetics for active Brownian particles. *Physical Review Letters* **2013**, *111* (14), 145702.

9. Pismen, L., *Active Matter Within and Around Us: From Self-propelled Particles to Flocks and Living Forms*. Springer: 2021.
10. Fu, X.; Tang, L. H.; Liu, C.; Huang, J. D.; Hwa, T.; Lenz, P., Stripe formation in bacterial systems with density-suppressed motility. *Physical Review Letters* **2012**, *108* (19), 198102.
11. Bechinger, C.; Di Leonardo, R.; Lowen, H.; Reichardt, C.; Volpe, G.; Volpe, G., Active particles in complex and crowded environments. *Reviews of Modern Physics* **2016**, *88* (4), 045006.
12. Kearns, D. B., A field guide to bacterial swarming motility. *Nature Reviews Microbiology* **2010**, *8* (9), 634–644.
13. Darnton, N. C.; Turner, L.; Rojevsky, S.; Berg, H. C., Dynamics of bacterial swarming. *Biophysical Journal* **2010**, *98* (10), 2082–2090.
14. Vicsek, T.; Czirok, A.; Ben-Jacob, E.; Cohen, I.; Shochet, O., Novel type of phase transition in a system of self-driven particles. *Physical Review Letters* **1995**, *75* (6), 1226–1229.
15. Mounfield, C. C., *The Handbook of Agent Based Modelling*. Independent Publishing: 2020.
16. Czirok, A.; Ben-Jacob, E.; Cohen, I.; Vicsek, T., Formation of complex bacterial colonies via self-generated vortices. *Physical Review E* **1996**, *54* (2), 1791.
17. Cisneros, L. H.; Kessler, J. O.; Ganguly, S.; Goldstein, R. E., Dynamics of swimming bacteria: Transition to directional order at high concentration. *Physical Review E* **2011**, *83* (6 Pt 1), 061907.
18. Dogic, Z.; Fraden, S., Development of model colloidal liquid crystals and the kinetics of the isotropic-smectic transition. *Philosophical Transactions of the Royal Society A* **2001**, *359* (1782), 997–1015.
19. Vroege, G. J.; Lekkerkerker, H. N. W., Phase transitions in lyotropic colloidal and polymeric liquid crystals. *Reports on Progress in Physics* **1992**, *55* (8), 1241–1309.
20. Ilkanaiv, B.; Kearns, D. B.; Ariel, G.; Be'er, A., Effect of cell aspect ratio on swarming bacteria. *Physical Review Letters* **2017**, *118* (15), 158002.
21. Chen, X.; Dong, X.; Be'er, A.; Swinney, H. L.; Zhang, H. P., Scale-invariant correlation in dynamic bacterial clusters. *Physical Review Letters* **2012**, *108*, 148101.
22. Dombrowski, C.; Cisneros, L.; Chatkaew, S.; Goldstein, R. E.; Kessler, J. O., Self-concentration and large-scale coherence in bacterial dynamics. *Physical Review Letters* **2004**, *93* (9), 098103.
23. Shklarsh, A.; Finkelshtein, A.; Ariel, G.; Kalisman, O.; Ingham, C.; Ben-Jacob, E., Collective navigation of cargo-carrying swarms. *Interface Focus* **2012**, *2* (6), 786–798.
24. Lu, S.; Bi, W.; Liu, F.; Wu, X., Loss of collective motion of bacteria undergoing stress. *Physical Review Letters* **2013**, *111* (20), 208101.
25. Yang, J.; Arratia, P. E.; Patterson, A. E.; Gopinath, A., Quenching active swarms: Effects of light exposure on collective motility in swarming *Serratia marcescens*. *Journal of Royal Society Interface* **2019**, *16* (156), 20180960.

26. Wu, X. L.; Libchaber, A., Particle diffusion in a quasi-two-dimensional bacterial bath. *Physical Review Letters* **2000**, *84* (13), 3017–3020.
27. Waigh, T. A., Advances in the microrheology of complex fluids. *Reports on Progress in Physics* **2016**, *79* (7), 074601.
28. Waigh, T. A., Microrheology of complex fluids. *Reports on Progress in Physics* **2005**, *68* (3), 685.
29. Chen, D. T. N.; Lau, A. W. C.; Hough, L. A.; Islam, M. F.; Goulian, M.; Lubensky, T. C.; Yodh, A. G., Fluctuations and rheology in active bacterial suspensions. *Physical Review Letters* **2007**, *99* (14), 148302.
30. Lagarde, A.; Dages, N.; Nemoto, T.; Demery, V.; Bartolo, D.; Gibaud, T., Colloidal transport in bacteria suspensions: From bacteria collision to anomalous and enhanced diffusion. *Soft Matter* **2020**, *16* (32), 7503.
31. Mason, T. G.; Weitz, D. A., Optical measurements of frequency-dependent linear viscoelastic moduli of complex fluids. *Physical Review Letters* **1995**, *74* (7), 1250.
32. Du, H.; Xu, Z.; Anyan, M.; Kim, O.; Leevy, W. M.; Shrout, J. D.; Alber, M., High density waves of the bacterium *Pseudomonas aeruginosa* in propagating swarms result in efficient colonization of surfaces. *Biophysical Journal* **2012**, *103* (3), 601–609.
33. Dunkel, J.; Heidenreich, S.; Drescher, K.; Wensick, H. H.; Bar, M.; Goldstein, R. E., Fluid mechanics of bacterial turbulence. *Physical Review Letters* **2013**, *110*, 228102.
34. Wolgemuth, C. W., Collective swimming and the dynamics of bacterial turbulence. *Biophysical Journal* **2008**, *95* (4), 1564–1574.
35. Peruani, F.; Starrus, J.; Jakovljevic, V.; Sogaard-Andersen, L.; Deutsch, A.; Baer, M., Collective motion and nonequilibrium cluster formation in colonies of gliding bacteria. *Physical Review Letters* **2012**, *108* (9), 098102.
36. Wensink, H. H.; Dunkel, J.; Heidenreich, S.; Drescher, K.; Goldstein, R. E.; Lowen, H.; Yeomans, J. M., Meso-scale turbulence in living fluids. *PNAS* **2012**, *109* (36), 14308–14313.
37. Dunkel, J.; Heidenreich, S.; Bar, M.; Goldstein, R. E., Minimal continuum theories of structure formation in dense active fluids. *New Journal of Physics* **2013**, *15* (4), 045016.
38. Davidson, P., *Turbulence: An Introduction for Scientists and Engineers*. Oxford University Press: 2015.
39. Secchi, E.; Rusconi, R.; Buzzaccaro, S.; Salek, M. M.; Smriga, S.; Piazza, R.; Stocker, R., Intermittent turbulence in flowing bacterial suspensions. *Journal of Royal Society Interface* **2016**, *13* (119), 20160175.
40. Simha, R. A.; Ramaswarmy, S., Hydrodynamic fluctuations and instabilities in ordered suspensions of self-propelled particles. *Physical Review Letters* **2002**, *89* (5), 058101.
41. Goodwin, J. W.; Hughes, R. W., *Rheology for Chemists: An Introduction*. Royal Society of Chemistry: 2008.
42. Lauga, E., *The Fluid Dynamics of Cell Moility*. Cambridge University Press: 2020.

43. Martinez, V. A.; Schwarz-Linek, J.; Reufer, M.; Wilson, L. G.; Morozov, A. N.; Poon, W. C. K., Flagellated bacterial motility in polymer solutions. *PNAS* **2014**, *111* (50), 17771–17776.
44. Zottl, A.; Yeomans, J. M., Enhanced bacterial swimming speeds in macromolecular polymer solutions. *Nature Physics* **2019**, *15* (6), 554–558.
45. Kamdar, S.; Shin, S.; Leishangthem, P.; Francis, L. F.; Xu, X.; Cheng, X., The colloidal nature of complex fluids enhances bacterial motility. *Nature* **2022**, *603* (7903), 819–823.
46. Constantino, M. A.; Jabbarzadeh, M.; Fo, H. C.; Bansil, R., Helical and rod-shaped bacteria swim in helical trajectories with little additional population from helical shape. *Science Advances* **2016**, *2* (11), e1601661.
47. Linz, B.; et al., An African origin for the intimate association between humans and *Helicobacter pylori*. *Nature* **2007**, *445* (7130), 915–918.
48. Georgiades, P.; Pudney, P. D. A.; Thornton, D. J.; Waigh, T. A., Particle tracking microrheology of purified gastrointestinal mucins. *Biopolymers* **2014**, *101* (4), 366–377.
49. Hart, J. W.; Waigh, T. A.; Lu, J. R.; Roberts, I. S., Microrheology and spatial heterogeneity of *Staphylococcus aureus* biofilms modulated by hydrodynamic shear and biofilm-degrading enzymes. *Langmuir* **2019**, *35* (9), 3553–3561.
50. Rogers, S. S.; van der Walle, C.; Waigh, T. A., Microrheology of bacterial biofilms in vitro: *Staphylococcus aureus* and *Pseudomonas aeruginosa*. *Langmuir* **2008**, *24* (23), 13549–13555.
51. Wucher, B. R.; Elsayed, M.; Adelman, J. S.; Kadouri, D. E.; Nadell, C. D., Bacterial predation transforms the landscape and community assembly of biofilms. *Current Biology* **2021**, *31* (12), 2643–2651.
52. Duffy, K. J.; Cummings, P. T.; Ford, R. M., Random walk calculations for bacterial migration in porous media. *Biophysical Journal* **1995**, *68* (3), 800–806.
53. de Anna, P.; Pahlavau, A. A.; Yawata, Y.; Stocker, R.; Juones, R., Chemotaxis under flow disorder shapes microbial dispersion in porous media. *Nature Physics* **2021**, *17*, 68–73
54. Gomand, F.; Mitchell, W. H.; Burgain, J.; Petit, J.; Borges, F.; Spagnolie, S. E.; Gaiani, C., Shaving and breaking bacterial chains with a viscous flow. *Soft Matter* **2020**, *16* (40), 9273.
55. Mushenheim, P. C.; Trivedi, R. R.; Tuson, H. H.; Weibel, D. B.; Abbott, N. L., Dynamic self-assembly of motile bacteria in liquid crystals. *Soft Matter* **2014**, *10* (1), 88–95.
56. Albersdorfer, A.; Sackmann, E., Swelling behaviour and viscoelasticity of ultrathin grafted hyaluronic acid films. *European Physical Journal B* **1999**, *10* (4), 663–672.
57. Morris, E. R.; Nishinari, K.; Rinaudo, M., Gelation of gellan – a review. *Food Hydrocolloids* **2012**, *28* (2), 373–411.
58. Ross Murphy, S. B., Structure-property relationships in food biopolymer gels and solutions. *Journal of Rheology* **1995**, *39* (6), 1451–1463.
59. Geisel, S.; Secchi, E.; Vermant, J., Experimental challenges in determining the rheological properties of bacterial biofilms. *Interface Focus* **2022**, *12* (6), 20220032.

60. Galy, O.; Latour-Lambert, P.; Zrelli, K.; Ghigo, J. M.; Beloin, C.; Henry, N., Mapping of bacterial biofilm local mechanics by magnetic microparticle actuation. *Biophysical Journal* **2012**, *103* (6), 1400–1408.
61. Picioreanu, C.; Blauert, F.; Horn, H.; Wagner, M., Determination of mechanical properties of biofilms by modelling the deformation measured using optical coherence tomography. *Water Research* **2018**, *145*, 588–598.
62. Hohne, D. N.; Younger, J. G.; Solomon, M. J., Flexible microfluidic device for mechanical property characterization of soft viscoelastic solids such as bacterial biofilms. *Langmuir* **2009**, *25* (13), 7743–7751.
63. Powell, L. C.; Abdulkarim, M.; Stokniene, J.; Yang, Q. E.; Walsh, T. R.; Hill, K. E.; Gumbleton, M.; Thomas, D. W., Quantifying the effects of antibiotic treatment on the extracellular polymer network of antimicrobial resistant and sensitive biofilms using multiple particle tracking. *npj Biofilms and Microbiomes* **2021**, *7* (1), 13.
64. Lieleg, O.; Caldara, M.; Baumgartel, R.; Ribbeck, K., Mechanical robustness of *Pseudomonas aeruginosa* biofilms. *Soft Matter* **2011**, *7* (7), 3307–3314.
65. Rusconi, R.; Lecuyer, S.; Autrusson, N.; Guglielmini, L.; Stone, H. A., Secondary flow as a mechanism for the formation of biofilm streamers. *Biophysical Journal* **2011**, *100* (6), P1392–1399.
66. Stoodley, P.; Cargo, R.; Rupp, C. J.; Wilson, S.; Klapper, I., Biofilm material properties as related to shear-induced deformation and detachment phenomena. *Journal of Industrial Microbiology and Biotechnology* **2002**, *29* (6), 361–367.
67. Aravas, N.; Laspidou, C. S., On the calculation of the elastic modulus of a biofilm streamer. *Biotechnology and Bioengineering* **2008**, *101* (1), 196.
68. Jana, S.; Charlton, S. G. V.; Eland, L. E.; Burgess, J. G.; Wipat, A.; Curtis, T. P.; Chen, J., Nonlinear rheological characteristics of single species bacterial biofilms. *npj Biofilms and Microbiomes* **2020**, *6* (10), 19.

22 Interacting Populations

Models for *population dynamics* are of clear importance in epidemiology to understand how infections travel through populations of host organisms (whether they are bacteria, eukaryotes or archaea) or become established in a particular multicellular individual. The infections modelled can be viral (e.g. bacteriophage), prokaryotic (e.g. tuberculosis) or eukaryotic (e.g. malaria) in origin. Emergent robust statistical phenomena are often observed in large communities of infectious organisms that are relatively insensitive to the exact details of the molecular biology going on at the nanoscale. The field of population dynamics is a key one in mathematical biology.[1,2] Here, some of the phenomena are introduced from a physical perspective.

The rate of spontaneous mutation is thought to be around 0.0033 base pairs per DNA replication event and it is found to be relatively invariant over many microbes[3] (recombination events occur at rates of 10^{-4}–10^{-6} nucleotides per generation, whereas single nucleotide replacements or insertions occur at rates of 10^{-7}–10^{-10} nucleotides per generation). This can provide a relatively constant source of genetic variation in communities of interacting bacteria to guide their evolution, although the repair mechanisms of each organism will tend to counteract this effect and they will be species dependent. Horizontal gene transfer is another important contributor to the genetic diversity of interacting bacteria providing up to 1% of the protein-coding sequences in a wide range of species.[4] Dormancy is thought to play a key role in nutrient-deficient environments for the maintenance of microbial diversity[5] i.e. dormant cells can maintain the species diversity over long time scales. Infections with bacteriophage can also drive genetic changes in the host populations of bacteria and bacteriophage DNA can become permanently integrated into bacterial genomes.

22.1 Modelling the Growth of Interacting Planktonic Bacteria

The growth of independent bacterial cells was covered in Chapter 16. Interactions between populations of cells are a rich source of new phenomena that modulate growth rates, drive evolution and give rise to a rich variety of ecological networks. Organisms often compete with one another for resources, but symbiotic interactions also commonly occur. Bacterial growth is a stochastic process and it is important to decide whether intrinsic or extrinsic factors in the environment lead to the observed variability (or both).

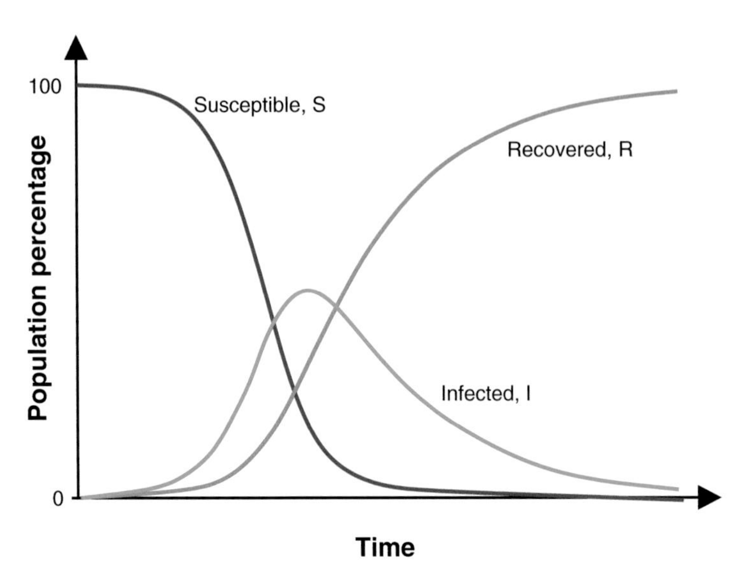

Figure 22.1 A *SIR* (susceptible/infectious/removed) model predicts the percentages of a population in each of the three states as a function of time during infections.

SIR (susceptible/infectious/removed) models are standard tools for understanding the progression of infections in a community of organisms (Figure 22.1) e.g. epidemics of bacteriophages that cross lawns of bacteria, bacterial infections of mammals or COVID-19 pandemics in humans. A standard SIR model consists of three coupled differential equations[6]:

$$\frac{dS}{dt} = \mu N - \beta SI - \mu S, \tag{22.1}$$

$$\frac{dI}{dt} = \beta SI - (\gamma + \mu) I, \tag{22.2}$$

$$\frac{dR}{dt} = \gamma I - \mu R, \tag{22.3}$$

where S is the number of susceptible organisms, I is the number of infectious organisms, R is the number of removed (dead/immune) organisms, N is the total population size, βSI is the number of new cases per unit time, γ is the mean duration of an infection and μ is the death rate.

Such SIR models can lead to stable or oscillatory solutions for the number of organisms, dependent on the exact choice of the parameters. The Lotka–Volterra model uses a simpler theoretical framework that is also often used for the study of mixed species population dynamics and also has oscillatory solutions (Section 22.3).

Spatial variations in populations of organisms can be described by the use of partial differential equations (Section 22.5 i.e. considering spatial derivatives of organism populations), but these equations can be challenging to solve analytically in realistic biofilm geometries and agent-based modelling tends to be more effective (Chapter 27).

22.2 Structured Populations in Biofilms

Agent-based models (ABMs) tend to be the standard tool to understand interacting bacterial populations since they can be used in a variety of realistic geometries (Chapter 27) and predict the development of structured populations. Agent-based models have been used to understand aging phenomena in populations of *E. coli* and their predictions were tested with measurements of protein damage.[7] Bacteria thus appear to have life spans that are determined by key molecules in their metabolisms (telomeres do not play a role in bacteria, since they have circular DNA chains).

A classic example of a structured bacterial biofilm is dental plaque where anaerobic bacteria are protected from the external aerobic environment around the teeth by other symbiotic bacterial species[8] (Section 25.1). The vast number of bacterial species (~500) and strains that occur in an individual's oral biofilm will require high throughput techniques and machine learning for rigorous wide-ranging studies. A hierarchical coarse-grained framework could be used to explain the synergy of the communities of bacteria. Predicting the long-term stability of dental bacterial communities is an interesting challenge e.g. a loss of species diversity is associated with oral disease in alcoholics.[9]

22.3 Mixed Species Populations

Microhabitat patches have been used to study the competition of mixtures of bacterial species.[10] The standard continuum model to describe populations in competition is due to Lotka and Volterra[11] (the LV model, Figure 22.2, where the bacteria

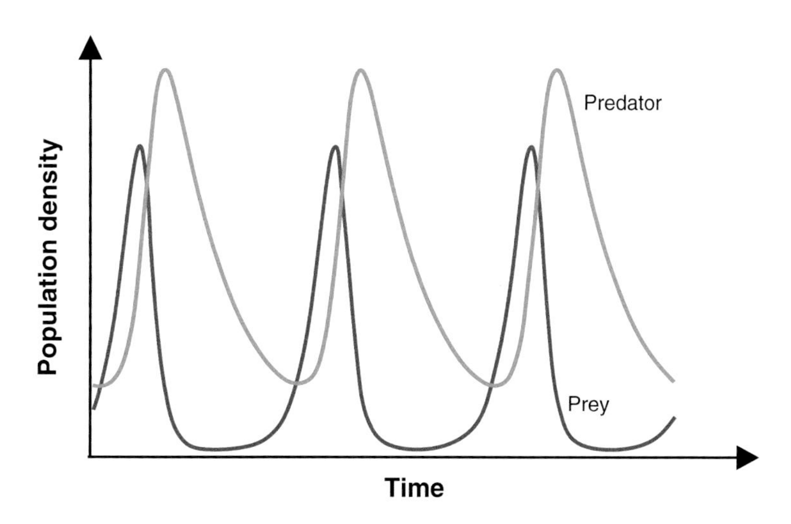

Figure 22.2 *Lotka–Volterra* model for the oscillatory population densities of predator and prey as a function of time.

experience logistic growth in the absence of their competitors). The LV model is provided by two symmetric ordinary differential equations that describe the populations of predator and prey. The *predator* equation is

$$\frac{dN_1}{dt} = r_1 N_1 \frac{(K_1 - N_1 - \beta_{12} N_2)}{K_1}, \tag{22.4}$$

and the *prey* equation is

$$\frac{dN_2}{dt} = r_2 N_2 \frac{(K_2 - N_2 - \beta_{21} N_1)}{K_2}, \tag{22.5}$$

where N_1 is number of predator organisms, N_2 is the number of prey organisms, β_{12} and β_{21} are competition coefficients, r_1 and r_2 are growth factors and K_1 and K_2 are the carrying capacities. Due to the symmetry of the equations, the labels predator and prey can be used interchangeably. Furthermore, N_1 and N_2 can be used for the populations of two competitive varieties of bacteria. Equations (22.4) and (22.5) are non-linear (due to N_1^2 and N_2^2 terms) and can lead to both stable and unstable oscillatory solutions over time (Chapter 12). The equations originally found success in modelling the oscillatory populations of snowshoe hares and lynxes in North America and fish catches in the Adriatic Sea. A weakness of the model is the prediction of unrealistically small populations of organisms, which would lead to extinction events. Furthermore, there is an implicit assumption that the microhabitat is constantly mixed (e.g. the bacterial suspensions are stirred) and spatial heterogeneity is neglected. To consider spatial variations of populations, extensions are required using partial differential equations that simultaneously consider space and time derivatives or numerical simulations, such as the agent-based models described in Section 22.2 (or combinations of the two).

Detailed developments of the LV model are part of classical mathematical biology.[1] It is possible to extend such models to k interacting species, although they still have the drawback that they neglect spatial heterogeneity,

$$\dot{X}_i = a_i X_i + \sum_{j=1}^{n} b_{ij} X_i X_j, \tag{22.6}$$

where X_i is the number of species i, a_i is a product of the growth factor and carrying capacity, b_{ij} are competition coefficients and n is the number of interacting species.

Contact killing of bacteria by different species in mixed populations can counter-intuitively lead to the stable coexistence of regions enriched in single species. A thin layer of dead cells limits the amount of bacteria killed.[12] The interactions between bacterial species can also be modulated by their response to bacteriophage.[13]

22.4 Evolution and Game Theory

Game theory has its origins in economics[14] and games were considered in which participants attempt to optimise their earnings. In biology, similar scenarios exist, but earnings are often substituted with population number to provide a parameter that

Table 22.1 Different *game theory strategies* proposed to understand phenomena with bacteria.[15]

Game theory strategy	Example bacterial systems
Rock, paper, scissors	Toxin-producing *E. coli.*[16]
Bet hedging	*E. coli* – lactose use and capsule expression, *Bacillus subtilis* competence development, persister cells[17] in *E. coli, S. aureus, M. tuberculosis,* and *Pseudomonas aeruginosa.*
Altruism	Siderophore production in *P. aeruginosa.*[18]
Red queen	Marine viruses in microorganisms,[19] bacteriophage coevolution with biofilms.[20]
Common goods process – with defectors and collaborators	Formation of biofilm EPS.[21]
Symbiosis	Creosote-eating pack rats, nitrogen fixing plants.

can be optimised. Game theory tends to be a good source of strategies that provide qualitative hypotheses to test with real organisms. Making the theories in quantitative agreement with experiments is tricky due to the huge range of stochastic influences on the behaviour that is modelled. Game theory can be thought of as a mathematical subfield of evolutionary biology.

Table 22.1 gives some examples of different strategies from game theory and examples of their occurrence with microorganisms. Many more sophisticated strategies are expected to be discovered and there are many other zoological examples already known that could be used for inspiration.[14]

Bet hedging and *bistability* have been observed in numerous bacterial species. Bet hedging involves the simultaneous use of two phenotypes to optimise the overall success of a population. Bistability involves abrupt switches between phenotypes and is a non-linear phenomenon (Chapter 12). Another common non-linear feature in populations of bacteria is that of *hysteresis* i.e. the transition from one state to another requires induction times greater than those for the reverse transition e.g. lactose use with *E. coli*, competence development in *B. subtilis* and persister cells with *Mycobacterium tuberculosis.*[22]

Altruism is superficially hard to understand in light of Darwin's theory of evolution i.e. how organisms begin to collaborate and help one another if they are programmed for competition. Biofilms promote altruism[23] under some conditions due to the trade-off between growth rate and growth yield (a common goods process). Agent-based models indicate biofilms can enrich populations for altruists, whereas enrichment culture within planktonic suspensions only selects for the highest growth rate. Altruism was studied in *Pseudomonas aeruginosa* in terms of siderophore production (iron-scavenging agents).[18] From an alternative perspective, it is also interesting to ask whether antagonistic interactions can also lead to stable populations rather than altruism.[24]

Rock, paper, scissors interactions are observed in toxin-producing (colicin) *E. coli* versus resistant cells versus sensitive cells.[16] The three populations are found to coexist in localised ecological niches and oscillations occur between the different strains of bacteria.[25] The populations can be quantified using light scattering with flow cytometry (Section 13.1.1).

Diauxic growth occurs when bacteria are cultured in combinations of two sugars (Section 16.3). The preferred sugar is exhausted first and then the second is used, leading to two-step sigmoidal growth. With *Lactococcus lactis*, diauxic growth is found to be due to the activity of bet hedging processes[26] i.e. two sub phenotypes of the bacteria occur rather than individual bacteria switching between the two sugars during growth.

Symbiotic bacteria can lead to the rapid evolution of their hosts e.g. creosote can suddenly become edible to pack rats if they are given the correct bacteria in their intestines.[27] This may be a driving factor observed in coprophagia (consumption of faeces) found in many organisms.

Kin selection is a mechanism to reduce the input of cheaters in interacting populations of organisms. However, experiments with *E. coli* have established cooperativity in populations of different strains in contradiction to simple game theory models.[28,29]

The *red queen hypothesis* (that organisms must constantly adapt, evolve and proliferate in competition with other organisms) has been used to explain bacteriophage coevolution with biofilms, although it is not thought to be the dominant driving strategy, since there is evidence that the bacteriophage evolve faster than the bacteria.[20] Instead, agent-based models suggest that spatial heterogeneity provides the populations with the necessary stability (phage populations can quickly explode, since 100 bacteriophages can be produced in the time for a single bacterial division event).

Cooperation and conflict occur in microbial biofilms with respect to the amount of extracellular polymeric substance (EPS) that is created by different interacting strains of bacteria (another *common goods process*).[21] Cheaters do not produce extracellular proteases.

Self-generated biodiversity (e.g. fast-growing biofilms) in *P. aeruginosa* is found in experiments that act as an insurance policy for the bacteria in the face of environmental stresses.[30] Extensive genetic diversification occurs during the short-term growth of the biofilms.

22.5 Pattern Formation in High Concentration Bacterial Solutions

A major motivation for understanding *pattern formation* with high concentration of bacteria is the potential use in diagnostic medicine with the analysis of microscopy images to classify different morphologies. A fundamental question is whether the patterns formed by the bacteria are intrinsic to the species or are determined by the

Figure 22.3 Concentric ring patterns from a *Proteus mirabilis* colony grown on an agar gel.[31] The shades of grey indicate the density of the cells growing on the surface. Reprinted figure with permission from Czirok, A.; Matsushita, M.; Vicsek, T., Theory of periodic swarming of bacteria: applications to Proteus mirabilis. *Physical Review E* **2001**, *63*, 031915. Copyright (2001) by the American Physical Society.

physical conditions e.g. whether the patterns are caused by microhydrodynamics. It is hoped that quantitative modelling of pattern formation in high-concentration bacterial solutions will help answer this question.

Concentric ring patterns in swarming suspensions of *Proteus mirabilis* have been studied[31] (Figure 22.3). Furthermore, rippling patterns of bacterial concentration occur in high-concentration suspensions with a characteristic wavelength when *Myxobacteria* grow on surfaces.[32] The ripples were successfully modelled as due to a refractory time caused by cell–cell collisions.

A delicate balance of chemoreception and motility occurs during pattern formation in high-concentration suspensions of bacteria.[33] The simplest description of a chemical receptor follows a sigmoidal form for the binding kinetics of the chemical on the receptor[34] (a Michaelis–Menten form, Equation (16.1)),

$$\frac{N_0}{N_t + N_0} = \frac{C}{K + C}, \tag{22.7}$$

where N_0 is the number of occupied receptors, N_t is the number of free receptors, C is the local concentration of the chemical sensed and K is a constant. The spatial gradient of Equation (22.7) describes the quantity that an organism can detect (called the receptor law),

$$\frac{\partial}{\partial x}\left(\frac{N_0}{N_t + N_0}\right) = \frac{K}{(K + C)^2}\frac{\partial C}{\partial x}. \tag{22.8}$$

This predicts that at very high concentrations of bacteria, the chemotaxic response vanishes due to saturation of the receptor (a sigmoidal-type behaviour). This is not observed experimentally and instead there is a process of adaptation that retunes the response of the sensory circuits in bacteria at higher concentrations.[35] This requires a more complex description and detailed models are available in the systems biology literature.[35]

A standard analytical framework for pattern formation is provided by *reaction–diffusion equations* (Chapter 5). A simple reaction–diffusion model to describe bacterial patterns is due to Fisher and Kolmogorov (FK, Equation (5.1)),

$$\frac{\partial u}{\partial t} = D\nabla^2 u + f(u), \tag{22.9}$$

where $f(u) = \alpha u(1-u)$ and u is the concentration of bacteria and α is a positive constant. This is just Fick's second law for diffusion with an added reaction term that has a specific form, $f(u)$. The FK model has classically been used to describe bacterial growth on agar.[34] More recent analytical and simulation studies have included a mechanical adhesive interaction between the bacteria and the substrate.[36] The adhesive force helped to explain the transitions between branched and circular morphologies.

A continuous family of travelling wave solutions is found for the Fisher–Kolmogorov model (Equation (22.9)). A question is how the front velocity is selected in the model. A stability analysis is needed to motivate the velocity and the FK model leads to compact (i.e. non-rough) growth patterns.[34] Often in experiments, bacterial culture can lead to rough growth fronts, so alternative models are needed to describe this phenomenon, such as diffusion-limited aggregation (DLA).[37] Models also need to consider the non-constant effective diffusion coefficients that are observed in experiments e.g. anomalous transport of bacteria can occur in which the motion is neither purely diffusive nor ballistic (Equation (1.8)). Anomalous reaction–diffusion equations have been developed by mathematicians, but their application to real-world problems is still relatively limited.[38]

More modern motile *agent-based models* (beyond the non-motile DLA models) can naturally lead to the rough growth patterns often observed experimentally and more accurate models for anomalous transport can be included (Chapter 2). A simple phenomenological method to simulate high concentration bacterial fluids is based on the Vicsek model (Chapter 27, a type of agent-based model),[39] where a simple stochastic alignment rule can lead to the flocking of motile agents i.e. directional coherence of the population. Many sophisticated elaborations of the Vicsek model have been proposed e.g. those that include chirality to describe the growth of bacterial colonies.[34,40] A fastest-growing morphology principle was suggested to choose between possible chiral vortex patterns. Knotted branched structures were observed with the growing structures of *Bacillus circulans* and were modelled with both reaction–diffusion and agent-based models.[41] The chiral patterns that form in biofilm structures have been modelled as due to the chirality of the flagella.[40] Since flagellar genes tend to be switched off in biofilms, the chirality of bacterial flagella would affect the formation of aggregates in the dense liquid state that precedes pattern formation.

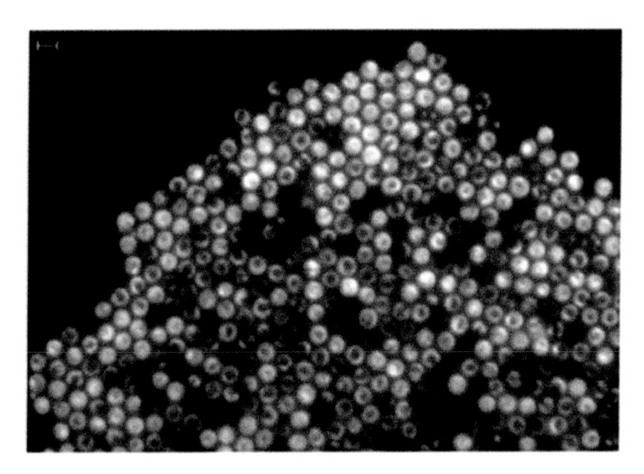

Figure 22.4 Hexagonal pattern formation in the collective dynamics of fast-moving *Thiovulum majus*.[43] The scale bar is 10 μm (top left). The brightness of the cells is modulated by the intracellular concentration of sulphur globules. Reprinted figure with permission from Petroff, A. P.; Wu, X. L.; Libchaber, A., Fast moving bacteria can self-organise into active 2D crystals of rotating cells. *Physical Review Letters* **2015**, *114*, 158102. Copyright (2015) by the American Physical Society.

The *dynamics of fusion* of two bacterial colonies of *Neisseria gonorrhoeae* were investigated experimentally using a combination of optical microscopy and optical tweezers.[42] Liquid-like behaviour was observed for the phases of bacterial colonies and motor activity strongly affected the surface tension and viscosity of the suspensions. Pili can also govern the liquid-like ordering and fusion dynamics of separate bacterial colonies.

Fast-moving bacteria (*Thiovulum majus*) in concentrated solutions can form a hexagonal pattern of collectively rotating cells[43] (Figure 22.4). The patterns are superficially reminiscent of convective roll patterns in Hele Shaw cells used to observe the flow of simple fluids,[44] but the self-organisation is driven by the internal motor activity of the bacteria, rather than external heating.

Surface waves were found to control bacterial attachment and growth in experiments with *E. coli*.[45] Patterned biofilms could be grown, where the bacteria avoided regions with turbulent motion and wave nodes in the bulk motility, whereas they preferred wave antinodes.

Pattern formation of non-motile *E. coli* has been explored as a process of *phase separation*.[46] On lattice agent-based simulations were performed with diffusing agents that experienced an attractive interaction and they were successful in motivating the rich variety of emergent patterns observed experimentally.

A clear demarcation line can form between two separate colonies of similar species of bacteria when they overlap during growth.[47] It is thought to be driven by the death of the bacteria at the point of overlap. Surprisingly, two colonies of *B. subtilis* were found to either merge or be clearly demarcated, depending on the substrate agar concentration and the initial distance of separation of the colonies.

Suggested Reading

McNamara, J. M.; Leimar, O. *Game Theory in Biology: Concepts and Frontiers*. Oxford University Press: 2020. A wide ranging book on biological game theory, although there is a niche in the market for something with more emphasis on microorganisms e.g. a textbook on 'Game theory with microorganisms' needs to be written.

Novak, M. A. *Evolutionary Dynamics: Exploring the Equations of Life*. Harvard University Press: 2006. Excellent intuitive introduction to game theory and evolutionary dynamics.

Otto, S. P., Day, T. *A Biologist's Guide to Mathematical Modelling in Ecology and Evolution*. Princeton University Press: 2007.

References

1. Otto, S.; Day, T., *A Biologist's Guide to Mathematical Modeling in Ecology and Evolution*. Princeton University Press: 2007.
2. Vynnycky, E.; White, R. G., *An Introduction to Infectious Disease Modelling*. Oxford University Press: 2010.
3. Drake, J. W., A constant rate of spontaneous mutation in DNA-based microbes. *PNAS* **1991**, *88* (16), 7160–7164.
4. Ochman, H.; Lawrence, J. G.; Groisman, E. A., Lateral gene transfer and the nature of bacterial innovation. *Nature* **2000**, *405* (6784), 299–304.
5. Jones, C. J.; Lennon, J. T., Dormancy contributes to the maintenance of microbial diversity. *PNAS* **2010**, *107* (13), 5881–5886.
6. Ellner, S. P.; Guckenheimer, J., *Dynamic Models in Biology*. Princeton University Press: 2006.
7. Koleva, K. Z.; Hellweger, F. L., From protein damage to cell aging to population fitness in *E. coli*: Insights from a multi-level agent-based model. *Ecological Modelling* **2015**, *301*, 62–71.
8. Wilson, M., *Bacteriology of Humans: An Ecological Perspective*. Blackwell: 2008.
9. Barb, J. J.; et al., The oral microbiome in alcohol use disorder. *Journal of Oral Microbiology* **2022**, *14* (1), 2004790.
10. Lambert, G.; Bergman, A.; Zhang, Q.; Bortz, D.; Austin, R., Physics of biofilms: The initial stages of biofilm formation and dynamics. *New Journal of Physics* **2014**, *16* (4), 045005.
11. Edelstein-Keshet, L., *Mathematical Models in Biology*. SIAM: 2005.
12. Steinbach, G.; Crison, C.; Ng, S. L.; Hammer, B. K.; Yunker, P. J., Accumulation of dead cells from contact killing facilitates coexistence in bacterial biofilms. *Journal of the Royal Society – Interface* **2020**, *17* (173), 20200486.

13. Li, Y. Y.; Lachnit, T.; Fraune, S.; Bosch, T. C. G.; Traulsen, A.; Sieber, M., Temperate phages as self-replicating weapons in bacterial competition. *Journal of the Royal Society – Interface* **2017**, *14* (137), 20170563.
14. McNamara, J. M.; Leimar, O., *Game Theory in Biology: Concepts and Frontiers*. Oxford University Press: 2020.
15. Li, X. Y.; Pietschke, C.; Fraune, S.; Altrock, P. M.; Bosch, T. C. G.; Traulsen, A., Which games are growing bacterial populations playing? *Journal of Royal Society Interface* **2015**, *12* (108), 20150121.
16. Kerr, B.; Riley, M. A.; Feldman, M. W.; Bohannan, J. M., Local dispersal promotes biodiversity in a real-life game of rock-paper-scissors. *Nature* **2002**, *418* (6894), 171–174.
17. Maisonneuve, E.; Gerdes, K., Molecular mechanisms underlying bacterial persisters. *Cell* **2014**, *157* (3), 539–548.
18. Griffin, A. S.; West, S. A.; Buckling, A., Cooperation and competition in pathogenic bacteria. *Nature* **2004**, *430* (7003), 1024–1027.
19. Ignazio-Espinoza, J. C.; Ahlgren, N. A.; Fuhrman, J. A., Long-term stability and red queen-like strain dynamics in marine viruses. *Nature Microbiology* **2020**, *5* (2), 265–271.
20. Heilmann, S.; Sneppen, K.; Krishnae, S., Coexistence of phage and bacteria on the boundary of self-organized refuges. *PNAS* **2012**, *109* (31), 12828–12833.
21. Xavier, J. B.; Foaster, K. R., Cooperation and conflict in microbial biofilms. *PNAS* **2007**, *104* (3), 876–881.
22. Veening, J. W.; Smits, W. K.; Kuipers, O. P., Bistability, epigenetics and bet-hedging in bacteria. *Annual Review of Microbiology* **2008**, *62*, 193–210.
23. Kreft, J. U., Biofilms promote altruism. *Microbiology* **2004**, *150* (8), 2751–2760.
24. Zapien-Campos, R.; Olmedo-Alvarez, G.; Santillan, M., Antagonistic interactions are sufficient to explain self-assemblage of bacterial communities in a homogeneous environment. *Frontiers in Microbiology* **2015**, *6*, 489.
25. Reichenbach, T.; Mobilia, M.; Frey, E., Mobility promotes and jeopardizes biodiversity in rock-paper-scissors games. *Nature* **2007**, *448* (7157), 1046–1049.
26. Solopova, A.; van Gestel, J.; Weissing, F. J.; Bachmann, H.; Teusink, B.; Kok, J.; Kuipers, O. P., Bet-hedging during bacterial diauxic shift. *PNAS* **2014**, *111* (20), 7427–7432.
27. Yong, E., *I Contain Multitudes: The Microbes Within Us and A Grander View of Life*. Vintage: 2017.
28. Lambert, G.; Vyawahare, S.; Austin, R. H., Bacteria and game theory: The rise and fall of cooperation in spatially heterogeneous environments. *Interface Focus* **2014**, *4* (4), 0029.
29. Whiteley, M.; Diggle, S. P.; Greenberg, E. P., Progress in and promise of bacterial quorum sensing research. *Nature* **2017**, *551* (7680), 313–320.
30. Bolen, B. R.; Thoendel, M.; Singh, P. K., Self-generated diversity produces an insurance effect in biofilm communities. *PNAS* **2004**, *101* (47), 16630–16635.
31. Czirok, A.; Matsushita, M.; Vicsek, T., Theory of periodic swarming of bacteria: Application to *Proteus mirabilis*. *Physical Review E* **2001**, *63* (3 Pt 1), 031915.

32. Borner, U.; Deutsch, A.; Reichenbach, H.; Bar, M., Rippling patterns in aggregates of myxobacteria arise from cell-cell collisions. *Physical Review Letters* **2002**, *89* (7), 078101.
33. Ben-Jacob, E.; Levine, H., Self-engineering capabilities of bacteria. *Journal of Royal Society Interface* **2006**, *3* (6), 197–214.
34. Ben-Jacob, E.; Cohen, I.; Levine, H., Cooperative self-organization of microorganisms. *Advances in Physics* **2010**, *49* (4), 395–554.
35. Alon, U., *An Introduction to Systems Biology: Design Principles of Biological Circuits*. 2nd ed. CRC Press: 2020.
36. Farrell, F.; Hallatschek, O.; Marenduzzo, D.; Waclaw, B., Mechanically driven growth of quasi-two-dimensional microbial colonies. *Physical Review Letters* **2013**, *111* (16), 168101.
37. Tronnolone, H.; et al., Diffusion-limited growth of microbial colonies. *Scientific Reports* **2018**, *8* (1), 5992.
38. Mendez, V.; Fedotov, S.; Horsthemke, W., *Reaction-transport Systems: Mesoscopic Foundations, Fronts and Spatial Instabilities*. Springer: 2012.
39. Vicsek, T.; Czirok, A.; Ben-Jacob, E.; Cohen, I.; Shochet, O., Novel type of phase transition in a system of self-driven particles. *Physical Review Letters* **1995**, *75* (6), 1226–1229.
40. Ben-Jacob, E.; Cohen, I.; Shochet, O.; Tenenbaum, A.; Czirok, A.; Vicsek, T., Cooperative formation of chiral patterns during growth of bacterial colonies. *Physical Review Letters* **1995**, *75* (15), 2899.
41. Wakano, J. Y.; Maenosono, S.; Komoto, A.; Eiha, N.; Yamaguchi, Y., Self-organized pattern formation of a bacteria colony modeled by a reaction diffusion system and nucleation theory. *Physical Review Letters* **2003**, *90* (25), 2581021.
42. Cronenberg, T.; Welker, A.; Zollner, R.; Meel, C., Molecular motors govern liquid-like ordering and fusion dynamics of bacterial colonies. *Physical Review Letters* **2018**, *121* (11), 118102.
43. Petroff, A. P.; Wu, X. L.; Libchaber, A., Fast moving bacteria can self-organize into active two-dimensional crystals of rotating cells. *Physical Review Letters* **2015**, *114* (15), 158102.
44. Guyon, E.; Hulin, J. P.; Petit, L.; Mitescu, C. D., *Physical Hydrodynamics*, 2nd ed. Oxford University Press: 2015.
45. Hong, S. H.; Gorce, J. B.; Punzmann, H.; Francois, N.; Shats, M.; Xia, H., Surface waves control bacterial attachment and formation of biofilms in thin layers. *Science Advances* **2020**, *6* (22), eaaz9386.
46. Thomen, P.; Valentin, J. D. P.; Bitbol, A. F.; Henry, N., Spatiotemporal pattern formation in *E. coli* biofilms explained by a simple physical energy balance. *Soft Matter* **2020**, *16* (2), 494–504.
47. Paul, R.; Ghosh, T.; Tang, T.; Kamar, A., Rivalry in *Bacillus subtilis*: Enemy or family? *Soft Matter* **2019**, *15* (27), 5400–5411.

23 Biofilm Formation

Biofilms can be simply defined as communities of microorganisms that are attached to a surface,[1] although fundamental disambiguation of states of bacterial physiology (e.g. biofilm versus planktonic forms of bacteria) requires the study of programmes of genetic expression. Many bacteria create biofilms when they experience nutrient-rich media and may return to the planktonic state when they become nutrient-deprived. Unfavourable conditions will often disrupt biofilms, since the bacteria will relocate themselves to avoid external stresses, e.g. exposure to ultraviolet (UV) light,[2] but the opposite phenomenon can also occur e.g. bacteria can upregulate extracellular polymeric substance (EPS) production to reinforce their biofilms in response to antibiotic stresses.[3] In general, most bacteria spend most of their time in communities encased in structured polymeric gels, which protect them from attack by other organisms and stabilise the environmental conditions the bacteria experience.

23.1 Structure of Biofilms

There are a large range of advantages for bacteria that commit themselves to the biofilm form of communal living and a relatively small number of disadvantages (Table 23.1).

Agent-based approaches (Chapter 27) can test hypotheses on the physical origin of biofilms (e.g. what set of rules does each bacterium need to follow to create the optimal biofilm?) and biofilms can be considered from an evolutionary perspective. The size and shape of biofilms can vary widely for a particular species of bacterium and they are affected by age, strain, chemical environment (e.g. nutrients and effluents), physical environment (temperature, pressure, geometry) and non-linear stochastic effects.

Once genetic evidence was available for the distinct biofilm state of bacterial cells, the next challenge was to understand the stages involved in the morphogenesis of biofilms driven by their genetic programmes. A prototypical example for biofilm formation was developed with *Pseudomonas aeruginosa*. The detailed research on *P. aeruginosa* biofilms was motivated by people with cystic fibrosis in which such infections can be a major factor in morbidity and the biofilms also commonly occur in infected wounds. *P. aeruginosa* biofilms are typically described using five stages of growth (Figure 23.1) that are associated with five separate phenotypes (free floating, reversible attachment, irreversible attachment, maturation and dispersion) and

Table 23.1 Advantages and disadvantages for bacteria living in the biofilm state of matter.[4]

Advantages of biofilms	Disadvantages of biofilms
An increased rate of mutation in biofilms occurs via horizontal gene transfer[5] i.e. biofilms facilitate genetic information transfer.	The time lag to relocate bacterial cells if conditions become unfavourable.
Biofilms can act as a sunshade to modulate the transmission of harmful UV light.	High-density populations are more susceptible to viral invasion in the same way multicellular organisms are vulnerable, such as humans in cities.
EPS can act as an electron/donor acceptor.	The metabolic cost for making the biofilm.
EPS is a sink for excess energy of the bacteria.	Once the cells are committed to a biofilm, they are slower to react to external responses e.g. exposure to UV light.[2]
Biofilms can absorb both polar and apolar organic compounds.	Absorption of the light required by photosynthetic species by biofilm EPS.
Intracellular communication can improve.	Osmotic gradients due to EPS could increase the uptake of unfavourable molecules through gel swelling e.g. antibiotics.
Improves adhesion to surfaces.[6]	More limited access to nutrients e.g. food or oxygen.
Biofilms acts as a reservoir of enzymes.	
Maintains the hydration levels of bacteria.	
Provides nutrient storage e.g. can store excess carbon.	
Can sequester toxic ions.	
Ion exchange enzymes housed in the biofilm can filter and collect essential nutrients.[7]	
Communal external digestion system.[8]	
Encourage tolerance to antibiotics.	
Help social communication.	
Provides protection from viruses (bacteriophages).[9]	
Mucoid-producing bacteria are harder to diagnose in human disease.[10]	
Substantial mechanical yield stresses of the biofilm stop bacteria from being washed away at low shear stresses, so they remain in a favourable niche.[11]	
Osmotic pressure gradients due to EPS can drive the expansion of bacterial communities.[12]	
Provides a fortress that offers protection from invader cells (bacteria, phagocytes, fungi, etc.).	

EPS denotes extracellular polymeric substance. Many similar considerations occur with slime layers and capsules (Chapter 20).

each stage corresponds to distinct patterns of genetic expression for the biofilm programme.[14] It must be borne in mind that this is a convenient fiction to simplify the behaviour, even for *P. aeruginosa,* and generalisation of this scheme to the formation

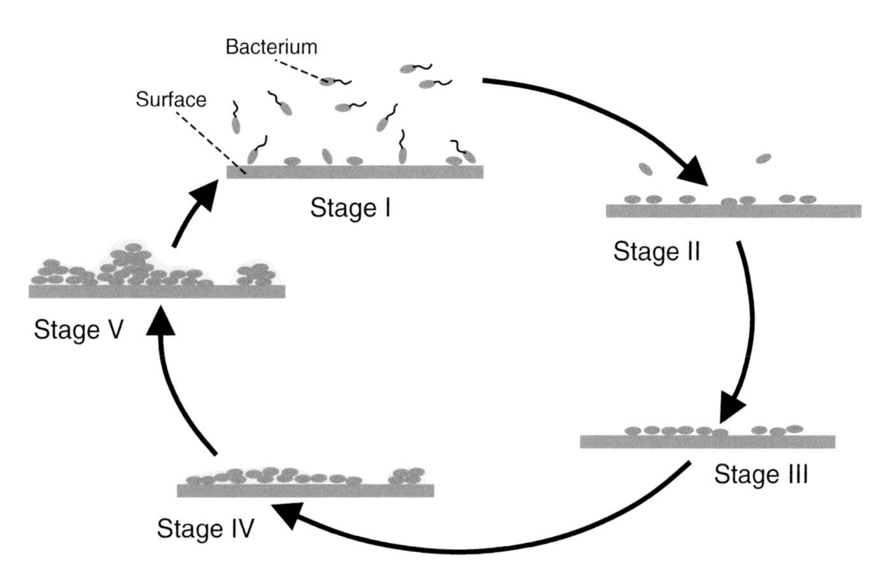

Figure 23.1 Generic stages in the life cycle of biofilm-forming bacteria: *I* (free floating), *II* (reversible attachment), *III* (irreversible attachment), *IV* (maturation) and *V* (dispersion). There are many exceptions to this general scheme dependent on the specific type of bacteria considered e.g. tuberculosis infections in lung tissue which have an intracellular phase. Furthermore, this scheme is more guided by the availability of convenient culture methods (e.g. agar gels) and it is not a rule for all biofilm-forming bacteria. *Pseudomonas aeruginosa* is considered the prototypical bacterium that conforms to this five-stage scheme and even here there is evidence for sub-stages e.g. planktonic bacteria shed from a stage V biofilm can be distinct from those in stage I.[13]

of all bacterial biofilms would be wrong regarding significant details in the majority of cases. For example, with *P. aeruginosa*, there is evidence that dispersed cells from stage V can be physiologically distinct from stage I cells.[13]

Figures 23.2 and 23.3 show schematic diagrams of examples of the interfaces on which biofilms grow e.g. inside organisms, inside cells (e.g. tuberculosis in lung tissue), on the outer surface of organisms, at solid/liquid interfaces, at solid/air interfaces and at air/water interfaces. Different interfaces have presented the bacteria with different evolutionary challenges to optimise the biofilms formed. Often, bacteria will only form biofilms on a specific type of interface. Diverse homes for bacteria are created in symbiosis with multi-cellular organisms (Figure 23.3) e.g. bobtail squid make specialised internal homes for *Vibrio fischeri* bacteria and use them for luminescent signalling. Symbiotic biofilms also occur in lichens (cyanobacteria), molluscs (spirochetes) and termites (spirochetes).[15]

One of the best-explored bacterial biofilms by biophysicists is created by *Bacillus subtilis*, since the bacteria are non-pathogenic (relatively safe for use by non-specialists), have applications in agriculture and the bacterium's biochemistry is understood with molecular detail, including its full genetics.[16] However, *B. subtilis* is unusual in that it prefers to grow pellicles at air/liquid interfaces, rather than films at solid/liquid interfaces that are common with other biofilm-forming, more medically relevant, bacteria.

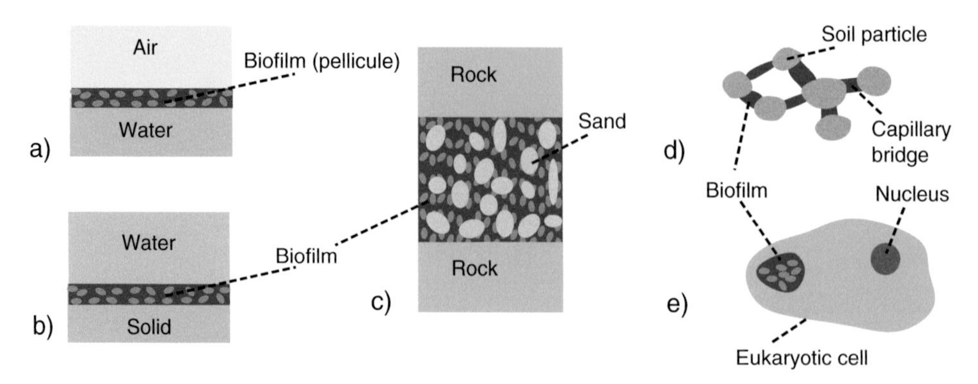

Figure 23.2 Some examples of environments in which biofilms grow. (a) Air/water interface, (b) solid/water interface, (c) granular bed, (d) soil and (e) inside eukaryotic cells.

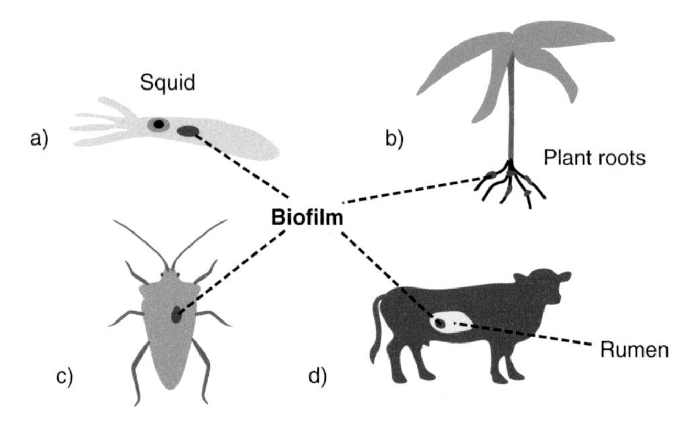

Figure 23.3 Formation of biofilms *in vivo*. (a) A bobtail squid, (b) plant roots, (c) a stink bug insect and (d) a cow (rumen). Not to scale.

The dispersal mechanism of *Staphylococcus aureus* biofilms involves clumped cells, whereas non-clumped dispersal occurs with *P. aeruginosa*[17] (stage V, Figure 23.1). Whole biofilms can also move actively and this form of surface dispersal is called *repelling migration* or *rafting*.

Bacteria are constantly monitored by the *host's immune system*, so if they are biochemically too loud, they risk a severe immunological response, which will lead to death. Bacteria have evolved to produce extremely potent exotoxins that can rapidly kill the host e.g. botulinum toxin. However, the immediate death of the host may cause the bacteria to lose both a source of long-term food and a vector to enable infection of other organisms. Therefore, more mild perturbations of the hosts' physiology are often superior in evolutionary terms e.g. a lung infection that very slowly debilitates the host over periods of 10–20 years can be better for the bacteria in evolutionary terms than one that kills within an hour since much larger numbers of bacteria are produced as a result. An example of this phenomenon is found in cystic fibrosis patients where *P. aeruginosa* infections can evolve to become less virulent.

There is some confusion over whether biofilms promote plasmid-mediated *horizontal gene transfer* (HGT) between bacteria or not.[18] The suggested resolution to the question is that HGT is encouraged during early events of biofilm formation when the bacteria first come into close contact with one another. However, this leads to small hot spots of activity and the development of biofilms' EPS then limits the global motion of HGT through the community i.e. a wave of HGT does not cross the mature biofilm.

A wide range of parameters can be used to *quantify the structure* of a biofilm, such as mean thickness, roughness, fractal exponent, density, porosity, surface area and mean pore radius.[19] There is a strong connection between the parameters that can be measured and the physics of the measuring instruments (Chapter 13). To quantify the structure of biofilms, mathematical tools are available developed in soft matter physics and biophysics, such as the pair correlation function, liquid structure factor and Ripley K function[20] (Chapter 1).

Bistable oscillations (a non-linear effect) have been observed in the growth of *B. subtilis* biofilms. It is hypothesised that the oscillations are due to electrical signalling in the biofilms (Chapter 29) due to nutrient starvation.[21] This demonstrates an alternative mechanism for biofilm communication in addition to quorum sensing.

Some bacteria are found to be magnetic, including marine and fresh water (swamp) varieties, and they use it to detect external magnetic fields[22] (Chapter 18). The relationship between magnetic sensitivity and biofilm growth (or biomat growth, the marine equivalent of a biofilm) is not well established.

Similar to the fundamental definition of a bacterial species, unambiguous identification of biofilm states depends on genetic studies and it is the programmes of gene expression that define the biofilm state. For example, simple aggregation of bacteria does not necessitate biofilm formation, since it can be due to separate genetic programmes. Some predatory bacteria (which actively kill and eat other bacteria) form multi-cellular fruiting bodies, e.g. *Myxobacteria*, and the signalling processes that coordinate the creation of fruiting bodies are similar to those used with biofilms (e.g. similar molecules are used) implying strong evolutionary links.

Even single bacterial species that form biofilms can demonstrate *genetic* and *phenotypic heterogeneity*, so biofilms are always physiologically and physically heterogeneous.[23] For example, the oxygen concentration can be measured through a biofilm using microelectrodes (Figure 23.4) and it is found to vary across the films

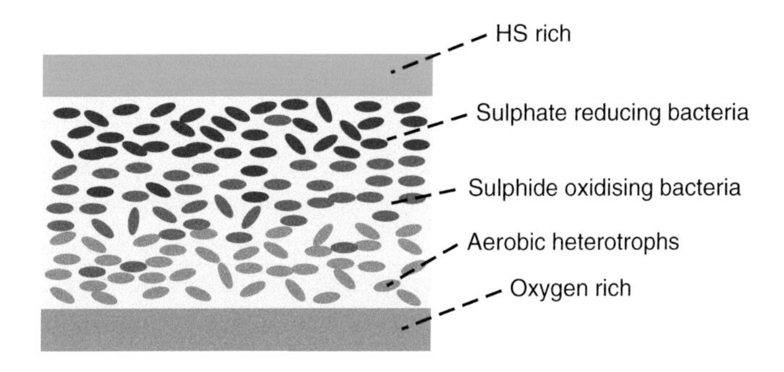

Figure 23.4 *Physiological heterogeneity* in a mixed-species bacterial biofilm. The bottom of the film is rich in oxygen, whereas the top of the film is rich in hydrogen sulphide (HS).[23]

Table 23.2 Advantages and disadvantages of standard experimental techniques for mapping the heterogeneous physiological activity of biofilms.[23]

Technique	Advantages	Disadvantages
Synthetic Nernstian dye (e.g. ThT or PI).	Can observe wavefronts during electrical signalling, a standard assay to determine whether cells are alive.	Can affect cellular division rates.
SIMS (secondary ion mass spectrometry).	Provides detailed molecular information using spatially resolved mass spectrometry.	Highly invasive, so the development of properties and dynamics cannot be probed in real time.
CTC (tetrazolium salt) staining to understand respiration.	Respiratory activity can be localised via fluorescence.	Only probes respiration.
Membrane permeability staining.	Can localise cell membrane integrity at the single-cell level. Can characterise the efficacy of antimicrobials. Commercial kits are available.	Artefacts possible such as incomplete staining, EPS staining and staining of live cells as dead.
Synthetic activity of DNA.	Via fluorescence. Little sample preparation is required.	Destructive.
Optogenetics via reporter genes.	Non-destructive, allows continuous monitoring of gene expression over time. Allows the expression of specific genes to be monitored. Several genes can be monitored simultaneously with multiple colours.	Genetically engineered strains need to be created. May perturb the cell physiology. Time delays for gene expression.
Fluorescence *in situ* hybridisation (FISH).	Can localise species diversity. A wide range of biofilms can be explored.	Good for calculating cell numbers. Not good for gene expression.
Fluorescent lectin staining.	Multiple polysaccharides can be measured at the same time.	Lectins may not properly enter the biofilms.
Laser microdissection.	Small groups of cells can be localised and analysed.	Destructive.

More modern nanotechnological techniques, such as quantum sensing with nanodiamonds, need to be established with biofilms.

e.g. obligate aerobic bacteria can become asphyxiated in the centres of large biofilms. The situation becomes even more complicated in mixed-species biofilms, which are the most common situation in nature. Some examples of techniques used to measure chemical heterogeneity in biofilms are described in Table 23.2.[23]

Injection of fluorescence dyes into biofilms followed by imaging with fluorescence microscopy indicates there can be substantial mass transfer within them.[24] Similarly, microsensors have measured the substantial motility of small molecules inside biofilms e.g. dissolved oxygen. The relative amount of RNA in a biofilm can be used to measure the physiological activity of the bacterial cells due to the relatively fast times for RNA degradation.[25]

Simple one- and two-dimensional *reaction–diffusion* (RD) models have successfully modelled the interplay of gradients of metabolic components and the physical heterogeneity of microorganisms[26] (Chapter 5). For example, RD models could be used to describe the growth rates of *P. aeruginosa* in cystic fibrosis sputum, the growth rate of *streptococcal* biofilms in endocarditis, the penetration of oxygen in *P. aeruginosa* biofilms, anabolic activity, iron use in the periphery of *P. aeruginosa* biofilms and the low growth rates in the interior of *Klebsiella pneumoniae* biofilms. Continuum RD models for biofilms do have a number of weaknesses e.g. failure to describe rough growth fronts and the overestimation of antibiotic permeation (Chapters 5 and 24).

Bacterial biofilms can have morphologies that are characteristic of the species producing them.[27] Biofilm morphology could thus be used for diagnostic purposes and machine learning could allow the automation of the process of diagnosis, improving its speed and fidelity (Chapter 14). Machine learning could also allow some of the challenges presented in understanding the extreme complexities of the intermediate structures of mixed-species biofilms to be tackled, which is the majority situation in naturally occurring biofilms.

23.1.1 Extracellular Polymeric Substance Components

Some of the standard polysaccharide components of the EPS in biofilms[28] are shown in Table 23.3. A very wide range of EPS molecules are made by bacteria and they are species dependent. Medically important bacteria are predominantly highlighted in the table. Bacterial exopolysaccharides from them can act as virulence factors and are the targets for drugs. The full range of EPS molecules is huge due to the vast number of possible biofilm-producing species and the extreme chemical heterogeneity of the molecules they produce (polysaccharides are particularly diverse due to the large number of ways the monomers can be interconnected).

Biopolymers produced by bacteria in the form of capsules, slimes or biofilms can be important in pathogenic bacteria.[29] However, in non-pathogenic bacteria, they can be an important source of materials for medicine and industry e.g. hyaluronate can be created, which is used as a replacement biomaterial with humans (e.g. synovial joints) and as additives in cosmetics, food or packaging.

In general, biofilms are complicated mixtures of molecules and hundreds of different molecules are often observed in a single biofilm. Seven categories of EPS have been defined: *structural*, *sorptive*, *surface-active*, *active*, *informative*, *redox active* and *nutritive*.[30]

Studies indicate that bacteria can alter the production of carbohydrates in their biofilms to more effectively absorb antibiotics in their environments.[31] This provides the bacteria with a sensitive control variable to adapt to antibiotic-induced stresses. Examples of the increased production of biofilm EPS due to stress are observed in

Table 23.3 Polysaccharides found in the extracellular polymeric substance (EPS) of some standard biofilms of pathogenic bacteria.[28]

Exopolysaccharide	Structure	Pathogen	Pathogenic mechanism	Potential drugs and targets	Targeted vaccines
Alginate	Heteropolymer of partially acetylated mannuronic acid and guluronic acid.	*Pseudomonas spp. e.g. P. aeruginosa*	Biofilm matrix. Antiphagocytic factor. Protection from free radicals and antibiotics.	Alginate oligomers, Alginate lyase.	Yes
Pel	Partially acetylated glycosidic linkages N-acetylgalactosamine and N-acetylglucosamine.	*P. aeruginosa.*	Biofilm matrix. Aminoglycoside antibiotic absorbing.	Glycoside hydrolase.	NA
Vibrio polysaccharide	Monomers of glucose, galactose, N-acetylglucosamine and mannose.	*Vibrio cholerae.*	Biofilm component.	NA	
PsI	Repeating pentasaccharide of D mannose, D glucose and L rhamnose.	*P. aeruginosa.*	Adhesion of cells to surfaces in biofilms.	Glycoside hydrolase.	Yes
Cellulose	Homopolymer of D glucose.	Many enterobacterial species e.g. *Escherichia coli.*	Biofilm component, which acts like mortar.	NA	NA
Hyaluronate	Heteropolymer of glucuronate and N-acetylglucosamine.	Group A and B *streptococcus.*	Biofilm component.	NA	Yes
K-antigen capsular polysaccharide	>80 serotypes.	*E. coli.*	Capsule component.	Synthesis inhibition.	NA
Colanic acid	Heteropolymer of fucose, glucose, glucuronate and galactose.	*E. coli, Shigella, Salmonella.*	Biofilm component.	NA	NA
GBS polysaccharides	9 serotypes.	Group B *streptococcus.*	Capsule component.	NA	Yes
Polysialic acid polysaccharides	Polymer of n-acetylneuraminic acid with sialic acid linkages.	*Neisseria meningitidis, E. coli K1.*	Capsule component.	Synthesis inhibition.	Yes
Pneumococcal polysaccharides	91 serotypes.	*S. pneumonia.*	Capsule component.	Glycoside hydrolase.	Yes
Staphylococcus polysaccharides	>10 serotypes.	*S. aureus.*	Capsule component.	NA	Yes
Levan	Fructose monomers.	*B. subtilis, S. mutans.*	Biofilm component.	NA	NA

Proteins and nucleic acids are also common components of the EPS.

colonic acid with *Escherichia coli* and alginate with *P. aeruginosa*.[32] This phenomenon is hard to differentiate from the formation of a slime layer without additional genetic information (slime layers and biofilms can be coexpressed).

Another example of the regulation of EPS is when *PIA* is created in *S. aureus* biofilms under standard culture conditions, but phenol-soluble modulins (PSMs, a type of peptide that forms amyloid), toxic to white blood cells, are created when the bacteria experience nutrient deficiencies. The detailed molecular biology of the PSMs is becoming better understood.[33] Seven different peptides are known to be involved in PSMs from *S. aureus* and *in vitro* experiments have studied their self-assembly into amyloid fibres. The different peptides experience varying mechanisms for self-assembly and modulate the resultant biofilm architectures, which has given evolution significant flexibility to create the optimal biofilm structures. Curli pili fibrils (Chapter 19) are well-established functional amyloids in other bacterial biofilms. Esp surface proteins also form amyloid-like fibres in biofilms of *Enterococcus faecalis*.[34]

PeI occurs in *Pseudomonas aeruginosa* biofilms. Biofilms that are created at the air/water interface are called pellicles and they need PeI for their formation in *P. aeruginosa*.[35] *Alginates* and *PsI* are also expressed in *P. aeruginosa* biofilms. Alginates can scavenge reactive oxygen species (e.g. ROS created by bleach) and inhibit cationic antimicrobials.[36] There is evidence from *P. aeruginosa* that carbohydrates in their biofilms provide sequestration of some standard antibiotics e.g. tobramycin, by physically interacting with them.[37] Molecular details of the exact architecture of most bacterial biofilms are still relatively badly defined, but some progress has been made in understanding the structuration of PsI in *P. aeruginosa* biofilms. PsI adopts a helical pattern attached to the bacterial cells, when observed using confocal microscopy with fluorescent lectin labels.[38]

Levan is produced by bacteria from sucrose in dental biofilms, e.g. *S. mutans*, and this biopolymer is a major component of plaque and provides an energy store for the bacteria. It is a major reason why eating lots of sugar rots the teeth i.e. it provides the bacteria with a secure fortress to encourage their growth by producing more levan.

Colonic acid occurs in *E. coli* biofilms and negative mutants have collapsed biofilms.[39] Colonic acid thus forms a type of scaffolding that is important for the three-dimensional structure of the biofilms.

Vibrio polysaccharide occurs in *Vibrio cholerae* biofilms. The biofilm morphology transforms from rough to smooth when the bacterium is unable to produce the polysaccharide.[40]

B. subtilis polysaccharides are essential for the creation of pellicles and smooth colonies in their biofilms.[28] *Cellulose* (a polysaccharide) can have a high tensile strength similar to that of steel and provides a scaffolding biopolymer in a wide range of bacterial biofilms, including *B. subtilis*, *Gluconacetobacter xylinus*, *E. coli*, *Salmonella enterica*, and *P. aeruginosa*.[41]

The attack of bacteriophage on bacteria in biofilms has been studied and under some circumstances, the biofilm matrix provides protection against the viruses because the phage adsorbs onto the matrix polymers. Specifically, the amyloid network of CsfA (curli) polymers can adsorb bacteriophage in *E. coli* biofilms.[9] Bacteriophages are often the predominant cause of death to bacteria in natural environments and thus bacteriophage adsorption is a major driver for biofilm evolution.

Amyloid diseases are a significant cause of morbidity in humans and include Alzheimer's, Parkinson's and Creutzfeldt–Jakob diseases. The name amyloid derives from the iodine stain used in early studies to diagnose the diseases in which protein aggregates in the brain looked similar to starch granules (which contain the polysaccharides amylose and amylopectin), which also stain purple with iodine. Beta sheets are the most thermodynamically stable secondary structural motif for proteins, so their preponderance in amyloids is perhaps not surprising. Less obvious is the fact that beta sheets can aggregate together to form giant one-dimensional assemblies (microns in length) and that a very wide range of peptides and proteins can demonstrate this mechanism of directed self-assembly. Amyloids also form in the biofilms of many bacteria, such as *S. aureus* and *B. subtilis.*[42] Amyloid aggregates are also found in fungal biofilms, so they represent a very general mechanism to reinforce biofilm gels.[43] As stated previously, amyloids in *E. coli* biofilms were found to provide protection against bacteriophage.[9] Amyloids also occur in the intracellular environment of healthy eukaryotic cells in humans, e.g. in P bodies, so they do not necessarily have to be associated with disease.[44]

The protein TsA can form amyloids in *B. subtilis* biofilms.[45] Purified TsA can self-assemble *in vitro* and mutant knockdowns lose the wild-type morphology of their biofilms, so the amyloid clearly plays an important role in biofilm formation.

Phenol-soluble modulins are found in some bacterial biofilms and they can form amyloids e.g. *S. aureus*. The modulins demonstrate antibacterial activity against foreign bacterial strains and cause lysis of host blood cells.[42] White blood cells play an important factor in combating infections in humans, so the modulins can thus act as a key virulence factor, since they stop the cells' activity. Synthetic amyloid analogues have also been created from short peptides that can act as potent antimicrobials causing bacterial cell lysis[46] (Chapter 24).

The causal link between bacterial amyloids and human amyloid diseases (Parkinson's, Alzheimer's, etc.) has not yet been rigorously proven, but there is lots of good circumstantial evidence that bacteria play a role, although they are possibly not the primary causative factor. If a bacterial origin can be proved for any disease, the prospects of treatment are often greatly improved, since the offending bacteria can be targeted with antimicrobials e.g. sleeping sickness is due to eukaryotic parasitic worms, but effective clinical targets were found when it was noticed the worms were obligate symbionts with bacteria; kill the bacteria with antibiotics, the worms also die and the patients can be cured.[47]

Marine bacteria have been found to concentrate metals into their biofilms in the form of precipitates e.g. Zn S granules are produced by sulphate-reducing bacteria and grey biofilms are observed as a result.[48] Conducting marine biofilms have been found to sequester copper from electrodes improving their conductivity (Chapter 29).[49]

Bacterial cells can produce a cocktail of enzymes and some of them are retained in biofilms and slime layers. For example, potent exotoxins and endotoxins are commonly created by bacteria.[50] They can form a protective shield when retained in a biofilm.

23.1.2 Quorum Sensing

Programmes of gene expression control the development of biofilms and the genes of individual bacteria in these programmes can interact via quorum sensing. Quorum molecules are released by bacteria in the vicinity of environments that are suitable for the growth of biofilms and a quorum is passed if the concentration of quorum signalling molecules exceeds a threshold value. A list of some quorum molecules is shown in Table 23.4. Many species of bacteria have multiple quorum-sensing molecules.

Table 23.4 The structures of some standard molecules that are detected during *quorum sensing*.[51]

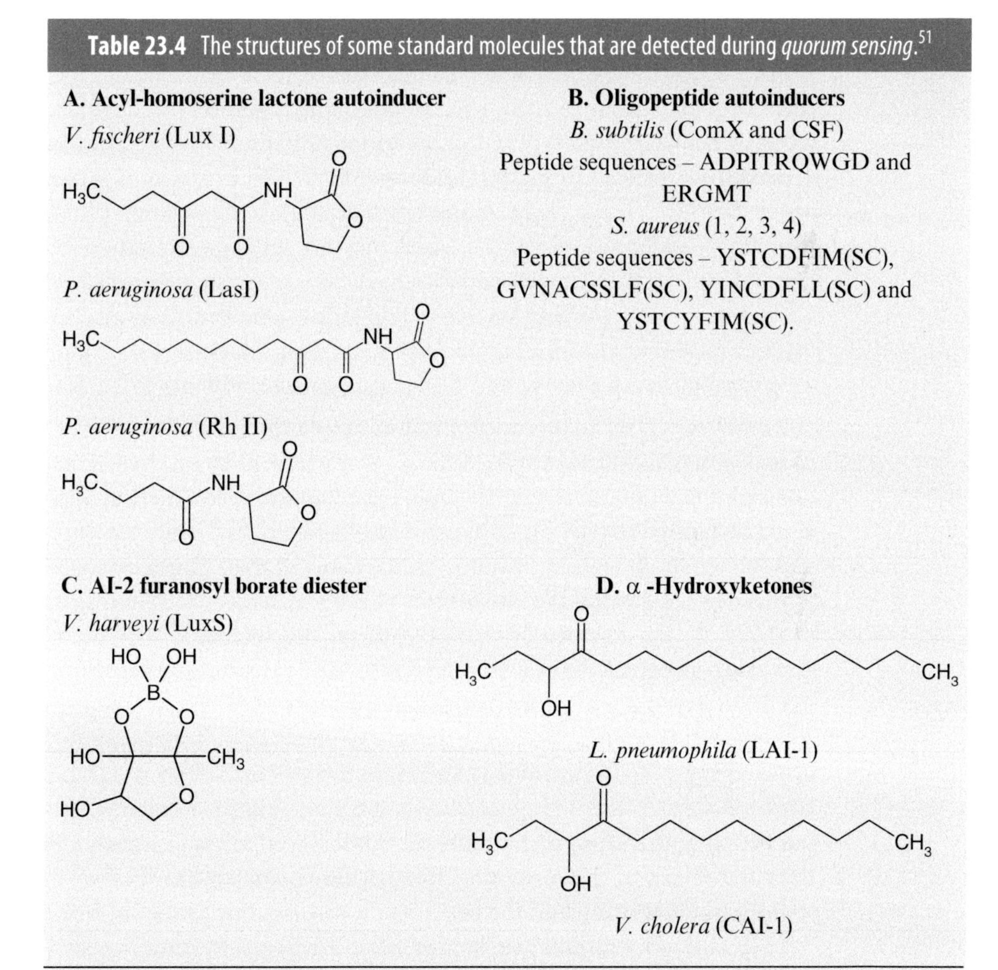

A. Acyl-homoserine lactone autoinducer
V. fischeri (Lux I)
P. aeruginosa (LasI)
P. aeruginosa (Rh II)
B. Oligopeptide autoinducers
B. subtilis (ComX and CSF)
Peptide sequences – ADPITRQWGD and ERGMT
S. aureus (1, 2, 3, 4)
Peptide sequences – YSTCDFIM(SC), GVNACSSLF(SC), YINCDFLL(SC) and YSTCYFIM(SC).
C. AI-2 furanosyl borate diester
V. harveyi (LuxS)
D. α -Hydroxyketones
L. pneumophila (LAI-1)
V. cholera (CAI-1)
H3C
NH
O
OH
HO
B
CH3

(A) Gram-negative bacteria mostly use AHL autoinducers. (B) Gram-positive bacteria mostly use modified peptides as autoinducers. (C) and (D) are some common exceptions.

Why there is such redundancy in many species is still not understood in detail, although quorum sensing is used to control gene expression of multiple virulence factors, not only the formation of biofilms e.g. sporation and loss of flagella. Small regulatory RNAs (srRNA) have also been shown to play a role in regulating the molecular processes involved in biofilm formation with *V. harveyi* and *V. cholerae.*[52] Similar mechanisms are expected to occur in many more bacteria, but the molecular biology of srRNAs is less well understood than with proteins. Thus, although it is an exciting emergent area of molecular biology, the programmes of gene expression are incompletely understood and quantitative studies are at an early stage of development (Section 18.3).

23.2 Surface Interactions of Biofilms

B. subtilis has three types of *lipopeptides* that it secretes: surfactin, iturins and fengycins. They play multiple roles, such as inhibiting digestion (they are toxic), facilitating motility, helping biofilm formation/development and in surface attachment. Surfactin is needed for pellicle formation in *B. subtilis* i.e. the creation of a biofilm at the air/water interface. These three lipopeptide surfactants show some promise as environmentally friendly antimicrobials, since they are anti-fungal, anti-mycoplasma bacterial and anti-viral.[53] The lipopeptides have very low critical micellar concentrations[54] which means they are very surface active at low concentrations and could be advantageous for many industrial applications e.g. as surfactants. For example, surfactin is exceptionally surface active and the surface tension of water $\left(72 \text{ mN m}^{-1}\right)$ is reduced to 27 mN m^{-1} at surfactin concentrations of as little as 20 μM.[55] Widespread adoption of surfactin in industrial applications is mostly limited by production costs, not performance. Lipopeptide surfactants would have numerous commercial applications if they could be created in bulk quantities at reasonable prices,[56] since they are biodegradable and thus environmentally friendly, unlike many of their synthetic equivalents.

Wetting is a general type of surface phase transition for a fluid at a surface, which is driven by the balance of free energies at the surface.[54,57] If the free energies of the components at the surface change so that no solution for the Young–Laplace equation exists, an abrupt (first-order) phase transition can occur called a *wetting transition*.[58] *B. subtilis* biofilms have an extremely large contact angle for the wetting of water e.g. 135–145°, which is higher than water on Teflon and they are thus highly hydrophobic surface[59] (Figure 23.5b). This wetting phenomenon provides a strong barrier to penetration of biofilms by water-based antimicrobials and is due to the hydrophobicity of the surfactins. Furthermore, some bacterial biofilms are resistant to both liquid wetting and the penetration of gases due to the surface-free energies.[59]

In *B. subtilis* biofilms, surfactins play a role as wetting agents (they promote wetting phenomena) and facilitate the motility of bacteria needed for pellicle formation. Surfactins are also important for the creation of liquid-filled channels in *B. subtilis* biofilms, which are used for nutrient transport (Figure 23.6a). Channels

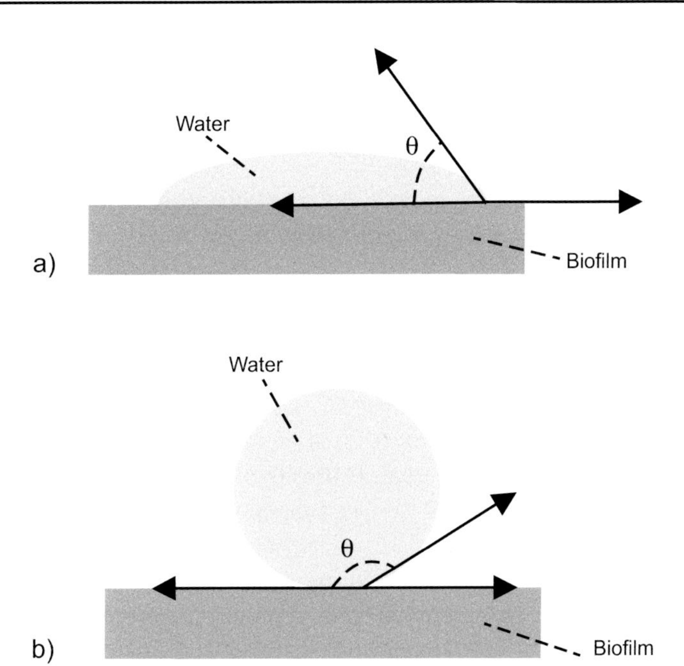

Figure 23.5 *Contact angle* (θ) of a water droplet on a biofilm. (a) A hydrophilic biofilm surface and (b) a hydrophobic biofilm surface (e.g. a *B. subtilis* biofilm). The contact angle is very large in the case of hydrophobic surfaces.

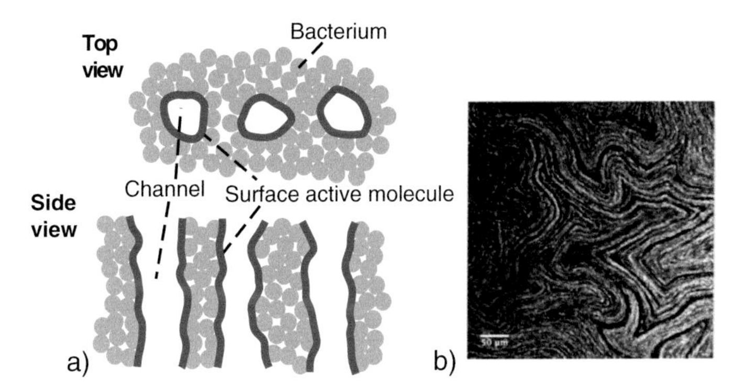

Figure 23.6 (a) Schematic diagram of vertical channels in a biofilm expressing surface active molecules. (b) Optical image of an *E. coli* biofilm showing channels.[60]

in *B. subtilis* biofilms are an evolved phenotype. The primary driver of the evolutionary process is still open for debate. It could be the ingress of food, the transfer of oxygen, the removal of waste products or communication.[61] The channels are ~ 90 μm in diameter and they seem to facilitate the transport of fluid through the biofilm. The exact molecular mechanism of channel formation is also currently unknown, but a process of liquid-liquid phase transition combined with tubular bilayer growth of the surfactins has been suggested.

Surfactant peptides (PSMs) are also found to be crucial for *S. aureus* biofilms.[62] The peptides are found to be necessary for the development of biofilm channels and biofilm expansion. The process of quorum sensing directs the creation of these surfactant peptides.

10 μm diameter intracolony microchannels were imaged in *E. coli* biofilms using a novel optical microscopy technique (based on a mesolens[60]) that was sensitive to a relatively large area of biofilm ~ 36 mm^2 (Figure 23.6b). The bacteria reform the microchannels if they are disrupted in a biofilm,[60] so these structures are actively maintained. The detailed mechanism of growth of the microchannels is currently unknown.

Substantial mechanical forces are built up in growing bacterial biofilms e.g. ~80 Pa in *B. subtilis.*[63] If cut, this causes the biofilms to spread by up to 20% of their initial sizes and thus encourages invasion of new environments and self-repair. The process of biofilm morphogenesis is thus dynamic and it is not just driven by the apposition of new particles. Cell death is found to precede wrinkle formation in *B. subtilis.*[64] Cell death is thus another driver for bacterial biofilm morphogenesis and similar mechanisms are observed in the morphogenesis of multi-cellular eukaryotic organisms e.g. it determines the number of digits in a human hand. The build-up of osmotic pressure in *B. subtilis* biofilms can drive spreading.[65] Bacterial flagella were not found to play a significant role. The self-healing properties of *B. subtilis* films have been found to facilitate applications e.g. three-dimensional printing of the bacteria.[66]

Buckling instabilities have been observed for *B. subtilis* pellicles at the air/water interface[67] (Figure 23.7). The buckling wavelength was calculated using a continuum model by considering the stress in an elastic biofilm. EPS is required for biofilm elasticity and thus for it to buckle. The buckling of *B. subtilis* biofilms leads to wrinkles that in turn cause them to have anisotropic wetting properties.[68] Wrinkles in *B. subtilis* biofilms are found to facilitate nutrient transfer. Capillary flows due to evaporation in *P. aeruginosa* biofilms grown on agar have also been shown to play a role in buckling.[69]

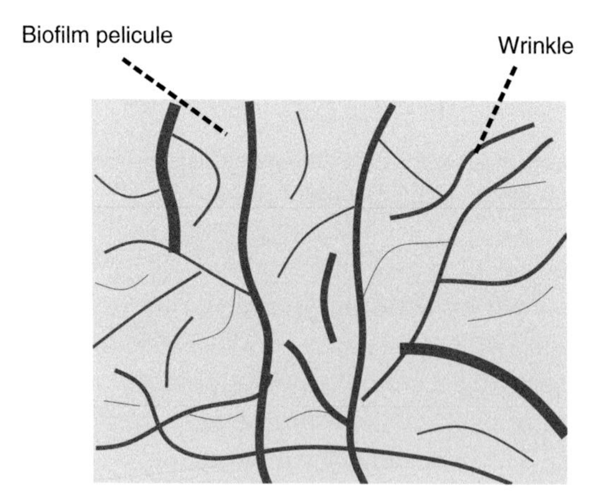

Figure 23.7 Schematic diagram showing a buckling instability (wrinkles) in *B. subtills* biofilms at the air/liquid interface (pellicles).[67]

23.3 Mixed-Species Biofilms

Completely monoculture biofilms (i.e. biofilms containing a single species) are often rare in practical cases e.g. chronic infections of humans are predominantly caused by polymicrobial biofilms.[70] Polymicrobial infections tend to lead to worse prognoses for disease (Section 31.6). Antimicrobial sensitivities can be reduced in polymicrobial biofilms e.g. β lactamase-producing strains of bacteria can help protect the whole community in the biofilm from antimicrobials. Quorum sensing signals and secondary metabolites are thought to mediate interspecies interactions in polymicrobial biofilms. A classic example of a mixed biofilm is seen with tooth decay due to dental caries in which aerobic bacterial species protect anaerobic species from oxygen exposure in a close process of symbiosis (Chapter 25). An average of 500 species of bacteria can take part in biofilm growth in the mouth of a single individual.

It has been found that invasive bacterial species can drill holes into the biofilms of other species, apply biocides and occupy spaces in the biofilms previously used by the other species.[71] Thus, radical changes in biofilm structures can occur in mixed populations of bacteria.

Suggested Reading

Ben Jacob, E. et al., Cooperative self-organisation of microorganisms. *Advances in Physics* 2000, *49* (4), 395–554.

Lewandowski, Z.; Beyenal, H., *Fundamentals of Biofilm Research*, 2nd ed. CRC Press: 2013.

Li, H. et al., Data driven quantitative modelling of bacterial active nematics. *PNAS* **2019** *116* (3), 777.

Mazza, M. G. The physics of biofilms – an introduction. *Journal of Physics D* **2016**, *49* (20), 203001.

References

1. O'Toole, G. O.; Kaplan, H. B.; Kolter, R., Biofilm formation as microbial development. *Annual Review of Microbiology* **2000**, *54*, 49–79.
2. Blee, J. A.; Roberts, I. S.; Waigh, T. A., Membrane potentials, oxidative stress and the dispersal response of bacterial biofilms to 405 nm light. *Physical Biology* **2020**, *17* (3), 036001.
3. Yin, W.; Wang, Y.; Liu, L.; He, J., Biofilms: The microbial 'protective clothing' in extreme environments. *International Journal of Molecular Sciences* **2019**, *20* (14), 3423.

4. Flemming, H. C., EPS – then and now. *Microorganisms* **2016**, *4* (4), 41.
5. Hoiby, N.; Bjarnsholt, T.; Givskov, M.; Molin, S.; Ciofu, O., Antibiotic resistance of bacterial biofilms. *International Journal of Antimicrobial Agents* **2010**, *35* (4), 322–332.
6. Mazza, M., The physics of biofilms – an introduction. *Journal of Physics D: Applied Physics* **2016**, *49* (20), 203001.
7. Pepper, I.; Gerba, C. P.; Gentry, T. J., *Environmental Microbiology*, 3rd ed. Academic Press: 2015.
8. Flemming, H. C.; Wingender, J.; Szewzyk, U.; Steinberg, P.; Rice, S. A.; Kjelleberg, S., Biofilms: An emergent form of bacterial life. *Nature Reviews Microbiology* **2016**, *14* (9), 563–575.
9. Vidakovic, L.; Singh, P. K.; Hartmann, R.; Nadell, C. D.; Drescher, K., Dynamic biofilm architecture confers individual and collective mechanisms of viral protection. *Nature Microbiology* **2018**, *3* (1), 26–31.
10. Torok, E.; Moran, E.; Cooke, F., *Oxford Handbook of Infectious Diseases and Microbiology*. Oxford University Press: 2016.
11. Jefferson, K. K., What drives bacteria to produce a biofilm? *FEMS Microbiology Letters* **2004**, *236* (2), 163–173.
12. Yan, J.; Nadell, C. D.; Stone, H. A.; Wingreen, N. S.; Bassler, B. L., Extracellular-matrix mediated osmotic pressure drives *Vibrio cholerae* biofilm expansion and cheater exclusion. *Nature Communications* **2017**, *8* (1), 327.
13. Chua, S. L.; et al., Dispersed cells represent a distinct stage in the transition from bacterial biofilm to planktonic lifestyles. *Nature Communications* **2014**, *5*, 4462.
14. Sauer, K.; Camper, A. K.; Ehrlich, G. D.; Costerton, J. W.; Davies, D. G., *Pseudomonas aeruginosa* displays multiple phenotypes during development as a biofilm. *Journal of Bacteriology* **2002**, *184* (4), 1140–1154.
15. Dyer, B. D., *A Field Guide to Bacteria*. Comstock Publishing Associates: 2003.
16. Vlamakis, H.; Chai, Y.; Beauregard, P.; Losick, R.; Kolter, R., Sticking together: Building a biofilm the *Bacillus subtilis* way. *Nature Reviews Microbiology* **2013**, *11* (3), 157–168.
17. Hall-Stoodley, L.; Costerton, J. W.; Stoodley, P., Bacterial biofilms: From the natural environment to infectious diseases. *Nature Reviews Microbiology* **2004**, *2* (2), 95–108.
18. Stalder, T.; Top, E., Plasmid transfer in biofilms: A perspective on limitations and opportunities. *npj Biofilms and Microbiomes* **2016**, *2*, 16022.
19. Lewandowki, Z.; Beyenal, H., *Fundamentals of Biofilm Research*. CRC Press: 2013.
20. Holmes, S.; Huber, W., *Modern Statistics for Modern Biology*. Cambridge University Press: 2019.
21. Martinez, R.; Liu, J.; Suel, G. M.; Garcia-Ojalvo, J., Bistable emergence of oscillations in growing *Bacillus subtilis* biofilms. *PNAS* **2018**, *115* (36), E8333–E8340.
22. Frankel, R. B.; Blakemore, R. P.; Wolfe, R. S., Magnetite in freshwater magnetotactic bacteria. *Science* **1979**, *203* (4387), 1355–1356.

23. Stewart, P. S.; Franklin, M. J., Physiological heterogeneity in biofilms. *Nature Reviews Microbiology* **2008**, *6* (3), 199–210.
24. Billings, N.; Birjiniuk, A.; Samad, T. S.; Doyle, P. S.; Ribbeck, K., Material properties of biofilms – key methods for understanding permeability and mechanics. *Reports on Progress in Physics* **2015**, *78* (3), 036601.
25. Magalhaes, A. P.; Franca, A.; Pereira, M. O.; Cerca, N., RNA-based qPCR as a tool to quantify and to characterize dual-species biofilms. *Scientific Reports* **2019**, *9* (1), 13639.
26. Stewart, P. S.; Zhang, T.; Xu, R.; Pitts, B.; Walters, M. C.; Roe, F.; Kikhney, J.; Moter, A., Reaction-diffusion theory explains hypoxia and heterogeneous growth within microbial biofilms associated with chronic infections. *npj Biofilms and Microbiomes* **2016**, *2*, 16012.
27. Heydorn, A.; Nielsen, A. T.; Hentzer, M.; Sternberg, C.; Givskov, M.; Ersboll, B. K.; Molin, S., Quantification of biofilm structures by the novel computer program COMSTAT. *Microbiology* **2000**, *146* (10), 2395–2407.
28. Limoli, D. H.; Jones, C. J.; Wozniak, D. J., Bacterial extracellular polysaccharides in biofilm formation and function. *Microbiology spectrum* **2015**, *3* (3), 10.1128.
29. Moradali, M. F.; Rehm, B. H. A., Bacterial biopolymers: From pathogenesis to advanced materials. *Nature Reviews Microbiology* **2020**, *18* (4), 195–210.
30. Flemming, H. C.; Neu, T. R.; Wozniak, D. J., The EPS Matrix: The house of biofilm cells. *Journal of Bacteriology* **2007**, *189* (22), 7945–7947.
31. Mah, T. F.; O'Toole, G. A., Mechanisms of biofilm resistance to antimicrobial agents. *Trends in Microbiology* **2001**, *9* (1), 34–39.
32. Hentzer, M.; Teitzel, G. M.; Balzer, G. I.; Heydorn, A.; Molin, S.; Givskov, M.; Parsek, M. R., Alginate overproduction affects *Pseudomonas aeruginosa* biofilm structure and function. *Journal of Bacteriology* **2001**, *183* (18), 5395–5401.
33. Zaman, M.; Andreason, M., Cross-talk between individual phenol-soluble modulins in *Staphylococcus aureus* biofilm enables rapid and efficient amyloid formation. *eLife* **2020**, *9*, e59776.
34. Taglialegna, A.; Matilla-Cuenca, L.; Dorado-Morales, P.; Navarro, S.; Ventura, S.; Garnett, J. A.; Lasa, I.; Valle, J., The biofilm-associated surface protein Esp of *Enterococcus faecalis* forms amyloid-like fibers. *npj Biofilms and Microbiomes* **2020**, *6* (1), 15.
35. Calvin, K. M., The Pel and Psl polysaccharides provide *Pseudomonas aeruginosa* structural redundancy within the biofilm matrix. *Environmental Microbiology* **2012**, *14* (8), 1913–1928.
36. Nizer, W. S.; Inkovskiy, V.; Versey, Z.; Strempel, N.; Cassol, E.; Overhage, J., Oxidative stress response in *Pseudomonas aeruginosa*. *Pathogens* **2021**, *10* (9), 1187.
37. Mah, T. F.; Pitts, B.; Pellock, B.; Walker, G. C.; Stewart, P. S.; O'Toole, G. A., A genetic basis for *Pseudomonas aeruginosa* biofilm antibiotic resistance. *Nature* **2003**, *426* (6964), 306–310.
38. Ma, L.; Conover, M.; Lu, H.; Parsek, M. R.; Bayles, K.; Wozniak, D. J., Assembly and development of the *Pseudomonas aeruginosa* biofilm matrix. *PLOS Pathogens* **2009**, *5* (3), e1000354.

39. Horvat, M.; Pannuri, A.; Romero, T.; Dogsa, I.; Stopar, D., Viscoelastic response of *Escherichia coli* biofilms to genetically altered expression of extracellular matrix components. *Soft Matter* **2019**, *15* (25), 5042.
40. Teschler, J. K.; Zamorano-Sanchez, D.; Utada, A. S.; Warner, C. J. A.; Wong, G. C. L.; Linington, R. G.; Yildiz, F. H., Living in the matrix: Assembly and control of *Vibrio cholerae* biofilms. *Nature Reviews Microbiology* **2015**, *13* (5), 255–268.
41. Thongsomboon, W.; Serra, D. O.; Possling, A.; Hadjineophytou, C.; Hengge, R.; Cegelski, L., Phosphoethanolamine cellulose: A naturally produced chemically modified cellulose. *Science* **2018**, *359* (6373), 334–338.
42. Peschel, A.; Otto, M., Phenol-soluble modulins and staphylococcal infection. *Nature Reviews Microbiology* **2013**, *11* (10), 667–673.
43. Mourer, T.; Ghalid, M. E.; d'Enfert, C.; Bachellier-Bassi, S., Involvement of amyloid proteins in the formation of biofilms in the pathogenic yeast *Candida albicans*. *Research in Microbiology* **2021**, *172* (3), 103813.
44. Alberts, B., *Molecular Biology of the Cell*, 6th ed. Garland Science: 2015.
45. Romero, D.; Aguilar, C.; Losick, R.; Kolter, R., Amyloid fibers provide structural integrity of *Bacillus subtilis* biofilms. *PNAS* **2010**, *107* (5), 2230–2234.
46. Gong, H.; et al., Aggregated amphiphilic antimicrobial peptides embedded in bacterial membranes. *ACS Applied Materials and Interfaces* **2020**, *12* (40), 44420–44432.
47. Geiger, A.; Fardeau, M. L.; Falsen, E.; Ollivier, B.; Cuny, G., *Serratia glossinae sp. nov.*, isolated from the midgut of the tsetse fly *Glossina palpalis gambiensis*. *International Journal of Systematic and Evolutionary Microbiology* **2009**, *60* (Pt 6), 1261–1265.
48. Labrenz, M.; et al., Formation of spalerite (ZnS) deposits in natural biofilms of sulfate-reducing bacteria. *Science* **2000**, *290* (5497), 1744–1747.
49. Beuth, L.; Pfeiffer, C. P.; Schroder, U., Copper-bottomed: Electrochemically active bacteria exploit conductive sulphide networks for enhanced electrogeneity. *Energy and Environmental Science* **2020**, *13* (9), 3102–3109.
50. Henkel, J. S.; Baldwin, M. R.; Barbieri, J. T., Toxins from bacteria. *EXS* **2010**, *100*, 1–29.
51. Wilson, M., *Bacteriology of Humans: An Ecological Perspective*. Blackwell: 2008.
52. Lenz, D. H.; Mok, K. C.; Lilley, B. N.; Kulkarni, R. V.; Wingreen, N. S.; Bassler, B. L., The small RNA chaperone Hfq and multiple small RNAs control quorum sensing in *Vibrio harveyi* and *Vibrio cholerae*. *Cell* **2004**, *118* (1), 69–82.
53. Ongena, M.; Jacques, P., Bacillus lipopeptides: Versatile weapons for plant disease biocontrol. *Trends in Microbiology* **2008**, *16* (3), 115–125.
54. Israelachvili, J. N., *Intermolecular and Surface Forces*. Academic Press: 2011.
55. Peyoux, F.; Bomatis, J. M.; Wallach, J., Recent trends in the biochemistry of surfactin. *Applied Microbiology Biotechnology* **1999**, *51* (5), 553–563.
56. Raaijmakers, J. M.; de Bruijn, I.; Nybroe, O.; Ongena, M., Natural functions of lipopeptides from *Bacillus* and *Pseudomonas*: More than surfactants and antibiotics. *FEMS Microbiology Reviews* **2010**, *34* (6), 1037–1062.

57. de Gennes, P. G.; Brochard-Wyart, F.; Quere, D., *Capillarity and Wetting Phenomena: Drops, Bubbles, Pearls and Waves*. Springer: 2003.
58. Waigh, T. A., *The Physics of Living Processes*. Wiley: 2014.
59. Epstein, A. K.; Pokroy, B.; Seminara, A.; Aizenberg, J., Bacterial biofilm shows persistent resistance to liquid wetting and gas penetration. *PNAS* **2011**, *108* (3), 995–1000.
60. Rooney, L. M.; Amos, W. B.; Hoskisson, P. A.; McConnell, G., Intra-colony channels in *E. coli* function as a nutrient uptake system. *The ISME Journal* **2020**, *14* (10), 2461–2473.
61. Wilking, J. N.; Zaburdaev, V.; De Volder, M.; Losick, R.; Brenner, M. P.; Weitz, D. A., Liquid transport facilitated by channels in *Bacillus subtilis* biofilms. *PNAS* **2013**, *110* (3), 848–852.
62. Periasamy, S.; Joo, H. S.; Duong, A. C.; Bach, T. H. L.; Tan, V. Y.; Chatterjee, S. S.; Cheung, G. Y.; Otto, M., How *Staphylococcus aureus* biofilms develop their characteristic structure. *PNAS* **2012**, *109* (4), 1281–1286.
63. Douarche, C.; Allain, J. M.; Raspaud, E., *Bacillus subtilis* bacteria generate an internal mechanical force within a biofilm. *Biophysical Journal* **2015**, *109* (10), 2195–2202.
64. Asally, M.; et al., Localized cell death focuses mechanical forces during 3D patterning in a biofilm. *PNAS* **2012**, *109* (46), 18891–18896.
65. Seminara, A.; Angelini, T. E.; Wilking, J. N.; Vlamakis, H.; Ebrahim, S.; Kolter, R.; Weitz, D. A.; Brenner, M. P., Osmotic spreading of *Bacillus subtilis* biofilms driven by an extracellular matrix. *PNAS* **2012**, *109* (4), 1116–1121.
66. Huang, J. D.; et al., Programmable and printable *Bacillus subtilis* biofilms as engineered living materials. *Nature Chemical Biology* **2019**, *15* (1), 34–41.
67. Trejo, M.; Douarche, C.; Bailleux, V.; Poulard, C.; Mariot, S.; Regeard, C.; Raspaud, E., Elasticity and wrinkled morphology of *Bacillus subtilis* pellicles. *PNAS* **2013**, *110* (6), 2011–2016.
68. Zhang, C.; Li, B.; Tang, J. Y.; Qin, Z.; Feng, X. Q., Experimental and theoretical studies on the morphogenesis of bacterial biofilms. *Soft Matter* **2017**, *13* (40), 7389–7397.
69. Si, T.; Ma, Z.; Tang, J. X., Capillary flows and mechanical buckling in a growing annular bacterial colony. *Soft Matter* **2018**, *14* (2), 301–311.
70. Orazi, G.; O'Toole, G. A., 'It takes a village': Mechanisms underlying antimicrobial recalcitrance of polymicrobial biofilms. *Journal of Bacteriology* **2019**, *202* (1), e00530-19.
71. Houry, A.; Gohar, M.; Deschamps, J.; Tischenko, E.; Aymerich, S.; Gruss, A.; Briandet, R., Bacterial swimmers that infiltrate and take over the biofilm matrix. *PNAS* **2012**, *109* (32), 13088–13093.

24 Action of Antibiotics and Antiseptics, a Physical Perspective

Methods to kill bacteria are important to control disease and to rebalance microbial communities towards more favourable compositions. Pharmacists tend to differentiate between drug molecules that act as *antibiotics* and those that are *antiseptics*. The difference is based on whether the materials can be used inside an organism or not. Antiseptics often have toxic side effects, which completely obstruct their *in vivo* applications i.e. there is not a therapeutic range of concentrations over which they can be safely used inside the organism. However, in the physical literature, the difference between antibiotics and antiseptics is often lost, since both treatments kill bacteria and this precedent is followed in the present discussion where many of the complex physiological issues are ignored (this is an important job for pharmacologists, so do not use any drugs solely developed by physicists).

The discussion will concentrate on the physical aspects of antibiotic and antiseptic activity. There are already extensive accounts of antimicrobials from an organic chemistry perspective, which are better covered in the specialist literature.[1] Traditional antiseptics and disinfectants have also been satisfactorily reviewed in the literature from an applied biochemistry perspective.[2] Due to the complexity of bacteria, a complete quantitative understanding of the action of even simple antiseptics (e.g. alcohols) is still lacking, although a qualitative picture is available and some quantitative progress has been made.

Standard antiseptics include detergents (e.g. soaps in shower gels, shampoos and washing up liquids), bleaches, alcohols and phenolics (e.g. TCP). Detergents have a physical mode of action and disrupt bacterial function by causing the disintegration of their membranes. Detergents are surface active (surfactants). In general, surfactants (e.g. soap) cannot be used internally in medical applications, since they would cause the human cell membranes to disintegrate. Standard cationic varieties of soaps are particularly potent against bacteria because bacterial membranes are negatively charged. These effective antiseptics are used widely as disinfectants in clinical settings e.g. quaternary ammonium compounds (QACs) used to clean surfaces in hospitals. Unfortunately, traditional cationic surfactants also tend to be toxic to human cells. Some amphiphilic surface-active proteins, such as defensins, function naturally *in vivo* as detergents e.g. in the stomach, but they have been carefully optimised by evolution to be selective for bacterial cells and leave human cells unharmed. Synthetic peptide surfactants can be more biocompatible than commonly available synthetic cationic surfactants and are being developed to help bridge the gap towards *in vivo* antibiotic applications as highly potent biocompatible detergents.[3,4]

There is a distinction between *bacteriocidal* (kills bacteria) and *bacteriostatic* (stops division) modes of action for antibiotics and antiseptics. Clearly, bacteriocidal activity is preferable, although bacteriostatic activity can also be effective to control bacterial populations. Furthermore, indiscriminate targeting of all the bacteria in a community can have adverse health effects, since many bacteria (commensals) fulfil valuable functions e.g. they produce vitamins and help guide processes involved in morphogenesis. Overuse of broadband antibiotics can unbalance bacterial populations and lead to their domination by problematic species e.g. *Clostridioides difficile* infections in the intestines that in extreme cases require faecal transplants to rebalance the populations. Healthy organisms tend to use naturally created antimicrobials to target problematic species and leave the rest of the bacterial flora unharmed, otherwise helpful bacteria will be lost and environmental niches created for recolonisation by problematic species.

The mechanisms of antibiotic resistance have been an important issue over the last 100 years and were encountered almost immediately after the introduction of the first widely used antibiotics. From the early 1960s to 2000, there was an innovation gap for the introduction of major new classes of antibiotics, which compounded some of the problems with antibiotic resistance. Most new antibiotics were chemically tailored derivatives of well-worn scaffolds.[5] Large profits were not available in the pharmaceutical industry from antibiotic research (a cheap single course of antibiotics was often sufficient to remove an infection) and more money could be made from the treatment of recurrent diseases which disincentivised innovations. Furthermore, newly approved antibiotics are often restricted as a last line of defence in serious hospital infections, where they are urgently needed, and there is less profit to be made from them due to smaller numbers of patients, again reducing incentives for their production.

Antibiotic resistance is to some extent a natural phenomenon due to the evolutionary pressure of competing organisms over 3.5 billion years. Microbes often create antimicrobials to kill their neighbours and have thus developed mechanisms to neutralise them. Random samples of soil bacteria found every strain was multidrug resistant on average to a third of 21 antibiotics.[6] Of the 480 bacterial isolates tested, 2 of them were resistant to 15 of the 21 antibiotics used. It is thus relatively easy for horizontal gene transfer to give genes for resistance to clinically relevant organisms when they are in contact with an environmental reservoir of other bacteria.

It is interesting to consider the timeline for resistant strains after antibiotics first become commercially available.[7] In 1940, penicillinase was first identified in bacteria i.e. an enzyme that bacteria use to neutralise the antibiotic penicillin. Five years later in 1945, 20% of *Staphylococcus aureus* in hospitals were found to have it.

Some of the least well understood and diverse classes of naturally occurring biomolecules are the polyphenols e.g. tanins.[8] Many of these have antimicrobial activity via detergency and this is a major factor driving their preponderance in plants. It is expected that many more useful antimicrobials could be discovered among the polyphenols, although challenges are posed by their complex chemistry.

Many biological surfaces profit from *in vivo* sloughing of mucus, which protects them from the buildup of adsorbed bacteria or their biofilms e.g. mucus adhered to the skin of dolphins, intestines and eyeballs.[9] Often dead cells are continually shed from the surfaces of organisms, which also deters the build-up of biofilms. A key issue with many synthetic materials used in regenerative medicine (e.g. replacement heart valves) is that their surfaces are static and they cannot renew themselves via a sloughing process. The surfaces of synthetic materials thus present ideal environments for prolonged bacterial colonisation, which has serious implications for surgical interventions.

Mucus and other components of the *extracellular matrix* (the ECM, e.g. collagen and aggrecan) of eukaryotic organisms are expected to adsorb and store substantial amounts of antibiotics and antiseptics[10] e.g. mucin sticks to a wide range of molecules. Thus, bacterial infections could profit from a close association with ECM components, since they act as a sink for antibiotics, although this will also regulate their ability to grow through ECM layers, such as mucus.

From a physical perspective, one of the most resistant bacteria discovered to date is *Deinococcus radiodurans*, which is able to survive starvation, oxidative stress and large amounts of DNA damage.[11] It can do this because multiple systems for DNA repair, DNA damage export, desiccation, starvation recovery and genetic redundancy are all present simultaneously in one cell.

24.1 Penetration of Molecules through Gels and Biofilms

Simple *reaction–diffusion* (RD) models (Chapter 5) can be used to calculate the motility of molecules through biofilms[12] and can be used to understand the penetration of antibiotics (Figure 24.1). A RD equation in one dimension used to describe the transport of molecules through a biofilm is

$$\frac{\partial c}{\partial t} = D\frac{\partial^2 c}{\partial x^2} - K^2 c, \tag{24.1}$$

where c is the concentration of the antibiotic, D is its diffusion coefficient of the antibiotic in the biofilm and K^2 is the reaction rate constant, which depends on sorption by the biofilm extracellular polymeric substance (EPS).

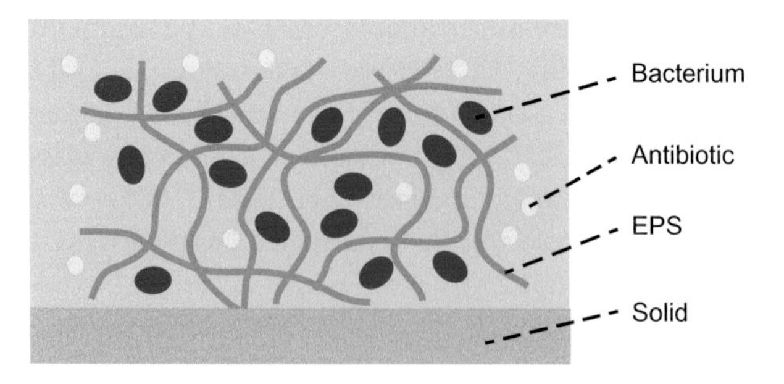

Figure 24.1 Schematic diagram of the penetration of antibiotics (yellow circles) through a bacterial biofilm on a solid/liquid interface.

A more detailed analysis of Equation (24.1) involves the *Thiele number* (ϕ), a characteristic dimensionless parameter, which compares the rates of sorption on to the biofilm components to the rate of diffusion through the biofilm,

$$\phi = \left(\frac{k_s X_s L_f^2}{D_e} \right)^{1/2}, \tag{24.2}$$

where k_s is the sorption rate constant, X_s is the concentration of binding sites, L_f is the thickness of the planar biofilm and D_e is the effective diffusion coefficient of the antibiotic in the biofilm. However, when these simple RD models are applied to antibiotics, rates of transport are often overestimated by an order of magnitude in comparison with experiment[13] i.e. diffusing antibiotics move much more slowly than expected. Clearly, an important factor is missing from the model. Alternative models have been put forward including anomalous diffusion of the antibiotics, adsorption onto both biofilm matrix polymers and dead bacterial contents and physiological changes in bacteria, such as the development of slow growing persister cells.[12] More detailed research is required to differentiate between the alternative models e.g. a much wider range of time scales and distances needs to be probed in experiments to identify the scaling regimes for anomalous diffusion of antibiotics.

A standard test of antibiotics is due to *Kirby-Bauer* (the disk method, Figure 24.2). It depends on the spatial diffusion of antibiotics across the surface of a gel and thus RD equations can be used to describe the circular regions in which bacterial growth is inhibited (agent-based model extensions are also possible, Chapter 27). However, in practice, the analysis is done semi-quantitatively using a comparison of the radii of the regions in which inhibition occurs after a certain time, to determine the relative antimicrobial efficacies. Inhibition extends to the point where the antibiotic concentration decreases below the MIC (minimum inhibitory concentration). The challenge is to properly characterise sub-diffusion,[14] bacterial motility,

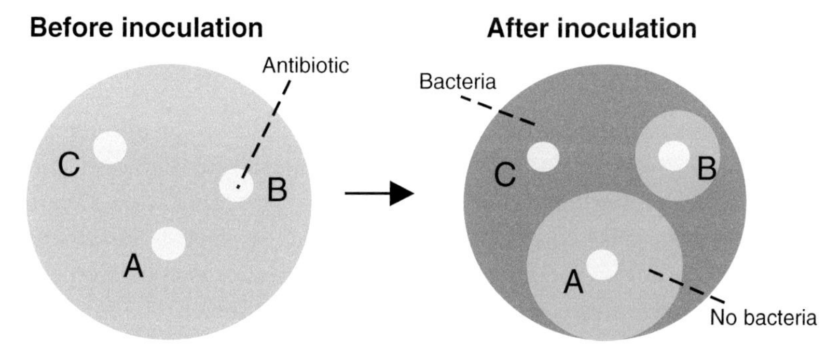

Figure. 24.2 The *Kirkby-Bauer test* (the disk method) is used to measure the efficacy of antibiotics to inhibit the growth of bacteria. Antibiotic A is more effective than B, which is more effective than C. Both transport of the antibiotic and its minimum inhibitory concentration (MIC) determine the radii of the disks of inhibition (the green circles, after inoculation).

convective flow due to evaporation and absorption effects on the antibiotic action in such tests.[15]

The resistance of biofilms to antibiotics is found to scale with their thickness as qualitatively expected from RD treatments.[16] Mature well-established biofilms are thus much harder to disrupt with antibiotics.

Diffusion through biofilms is still not understood in quantitative detail. An analogous eukaryotic system to bacterial biofilms is provided by mucin gels,[17] which have been extensively studied to understand the transport of pharmaceuticals. Detailed work on mucins demonstrates a size filter effect (only small molecules are transmitted) and transport processes are very sensitive to the relative charges of the matrix and the nanoparticles that are diffusing. Subdiffusion is commonly observed for nanoparticles in all types of hydrogel including those made from mucins, following Equation (1.8). Similar subdiffusive effects will present a strong kinetic barrier in bacterial biofilms to food capture, the excretion of toxic molecules and antibiotic penetration.

More sophisticated RD modelling was performed to describe experiments on ciprofloxacin antibiotic diffusion in *Pseudomonas aeruginosa* biofilms formed in an artificial sputum medium.[18,19] Diffusion inside gels can be anomalous due to a variety of effects (Chapter 1) and anomalous transport requires more sophisticated models to describe the fractional kinetics observed e.g. the mean square displacement of permeants follow $\langle \Delta r^2 \rangle = D_\alpha \tau^\alpha$, where α is a non-integer and D_α is a generalised diffusion coefficient (Equation(1.8)). A time fractional diffusion equation was used to describe the experimental data of ciprofloxacin in sputum,

$$\frac{\partial c}{\partial t} = D \frac{\partial^{1-\alpha}}{\partial t^{1-\alpha}} \left[\frac{\partial^2 c}{\partial x^2} - K^2 c \right], \tag{24.3}$$

where the symbols are as before. This is an extension of Equation (24.1) using fractional derivatives with respect to the time. RD models with fractional derivatives for the displacements are also possible.

The *advection-diffusion* equation has been widely used to describe antibiotic diffusion through bacteria grown on agarose gels that are used to test antibiotics,[20]

$$\frac{\partial c}{\partial t} = D \frac{\partial^2 c}{\partial x^2} + v \frac{\partial c}{\partial x}, \tag{24.4}$$

where v defines the dissipation rate slowing down the diffusion. However, the origin of the adjective term (v) is vague in this model and requires some justification. It could be describing convective motion due to the evaporation of water from the surface of the agar gels and could be considered an artefact in the experiment. In this case, the effect should be controlled by the reduction of evaporation, e.g. sealing the samples, to remove this complicating factor, although bacterial motility across the gels could also contribute to v. The fits to the data in the literature (Figure 24.3) are not overwhelmingly persuasive. The experiments need to be revisited over a wider range of concentrations and radii to differentiate between the alternative models e.g. Equation (24.4) versus Equation (24.3).

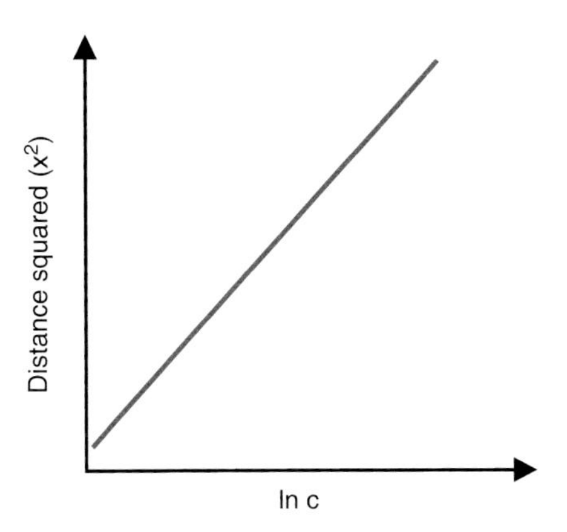

Figure 24.3 Schematic dependence of the disk radius $\left(x^2\right)$ as a function of the logarithm of the ampicillin concentration $\left(c\right)$ in a disk diffusion test[20] (Figure 24.2). The fit is slightly better with x^2 versus $\ln c$ than x versus $\ln c$, which is used as evidence for the *advection-diffusion model*, Equation (24.4).

24.2 Action of Antibiotics on Specific Bacteria and Biofilms

The main targets for common antibacterial drugs are cell wall synthesis, integrity of the cell wall, folic acid synthesis, RNA synthesis, protein synthesis and DNA replication and repair.[21] Minimum inhibitory concentrations (MICs) and kill rates are practically used to compare the potency of different antibiotics. The MIC is the lowest concentration of an antibiotic that prevents the growth of the bacteria. There is some evidence that very low concentrations of antibiotics (100× below the MICs) can select for resistant bacteria in a community, enriching their numbers e.g. tetracycline or streptomycin with *Salmonella typhimurium*.[22] However, accurate control experiments are tricky with such studies and similar behaviour has been found with no antibiotic[23] and the results may just be due to naturally occurring phenotypic variation e.g. the dynamic adaptation of the bacteria to better use the nutrients that are available in the culture media.

The specific nanoscale interactions of antibiotics with biofilms are not yet well defined. Mesoscopic interactions that inhibit antibiotic penetration into biofilms could be steric (excluded volume), charged (Derjaguin–Landau–Verwey–Overbeek [DLVO]), aromatic or hydrophobic (Chapter 4). For example, positively charged antibiotics will adsorb to negatively charged EPS and hydrophobic antibiotics could be adsorbed to hydrophobic regions of the EPS.

Analysis of diffusion profiles indicated the antibiotic tolerance of *P. aeruginosa* was due to its low metabolic activity and the low mobility of oxygen in the centres of the biofilms.[24] Starvation of *P. aeruginosa* directly affects antibiotic resistance i.e. the bacteria become more resistant as they are starved, since their metabolic activity reduces.

The process of starvation is controlled by a specific genetic programme in the bacteria called the *starvation signalling stringent response*.[25]

Mycoidal strains (slime-producing) of *P. aeruginosa* are associated with antibiotic resistance to tobramycin.[26] Mycoidal strains are characterised by the overproduction of the exopolysaccharide alginate. Such mycoidal strains of *P. aeruginosa* are thought to be important in cystic fibrosis pathogenesis.

Klebsiella pneumoniae biofilms were treated with ampicillin and ciprofloxacin[27] (small-molecule antibiotics). The ampicillin did not penetrate the biofilm, but the ciprofloxacin did. β-lactam-deficient mutants of the bacteria allowed the ampicillin to penetrate the biofilm, but antibiotic resistance of the bacteria was still observed. Thus, penetration into the biofilm is not the primary driver of resistance in this case.

Two standard synthetic fluorophores (without antibacterial activity), fluorescein and rhodamine B, diffused into *Staphylococcus epidermis* biofilms with 32% and 11% of the values of the diffusion coefficients in pure water, respectively, when observed with confocal microscopy.[28] These molecules are similar in size to standard antibiotics. There is thus a substantial slowdown of the diffusive kinetics, but these molecules still get through biofilms. Whether the kinetics of penetration were anomalous was not quantified. Anomalous transport is expected for particles that strongly interact with the biofilm EPS and/or particles above a size threshold (the mesh size of the polymers in the EPS).

There is clear evidence with *Bacillus subtilis* that the bacteria transit to the biofilm state in response to antibiotic-induced stress e.g. caused by treatment with rapamycin.[29] Evidence was also provided that the antibiotic induces a process of motility-driven phase separation, which directs the transition from swarms into biofilms.

Bacteriophages can act as extremely effective and selective antibiotics, since they can target specific strains of bacterial species, and have been lifesaving in recent applications e.g. with cystic fibrosis patients.[30] A problem is that the viruses themselves can elicit an immune response from patients and thus tend to be only single use when administered intravenously i.e. if they are used a second time, they are ineffective, since they are cleared by the host's immune system. There are perhaps slightly better prospects for routine bacteriophage treatments for intestinal bacterial infections or infected topical wounds, since the intestines and the skin provide environments that are controlled less stringently by the immune system.[30]

The growth of plaques of T7 bacteriophage on surfaces covered with *Escherichia coli* has been modelled using a RD formalism[31] (Figure 24.4). Wavefronts of bacteriophage infection develop across the lawns of bacteria.

The Ames test is used to understand whether a chemical causes mutation in the DNA of a bacterium.[32] It is used as a preliminary test to categorise molecules that are carcinogenic to eukaryotic organisms. As single-celled organisms, bacteria cannot experience cancer themselves, but there is some good evidence that bacteria play a role in a small number of cancers, e.g. *Helicobacter pylori* is involved in the development of some stomach cancers.

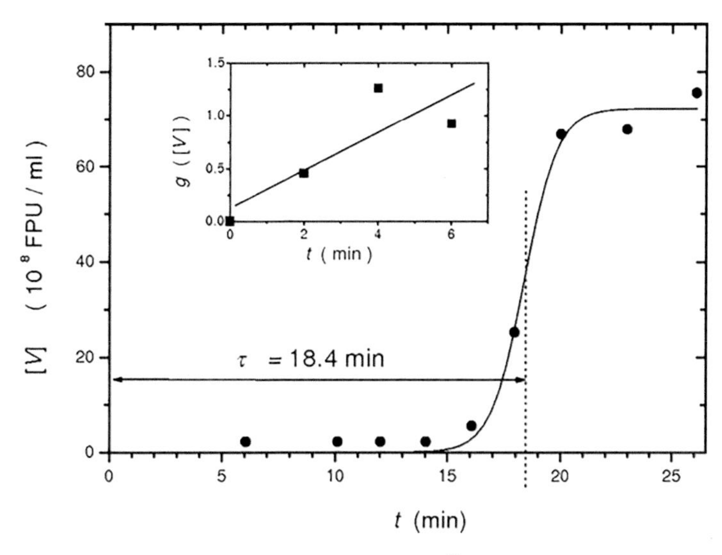

Figure 24.4 Wavefronts of *bacteriophage infections* move across lawns of bacteria.[31] The growth of T7 virus on *Escherichia coli* lawns is considered. The virus concentration (*[V]*) is plotted as a function of time. A time delay of 18.4 min is required for the propagation of the wavefronts. Reprinted figure with permission from Fort, J.; Mendez, V., Time-delayed spread of viruses in growing plaques. *Physical Review Letters* **2002,** *89* (17), 178101. Copyright (2002) by the American Physical Society.

24.3 Persister Cells

Separate from the phenomenon of sporulation (Chapter 15), there is the question of *persister cells*, which is still the subject of some controversy. The persister state is ascribed to bacteria both in the planktonic and biofilm phases.[33,34] Persister cells are linked with states of low metabolism, although there are no dramatic changes in cell morphology, such as those observed with sporulation.

Persister cells are hotly debated in the literature and a case can be made for their non-existence under some circumstances i.e. some experiments just demonstrate generic mechanisms of antibiotic adsorption by EPS and dead bacteria, not multi-drug resistance of persister cells. Persister cells are expected to be killed by antiseptics, but not by antibiotics, since antibiotics predominantly depend on active physiological processes for their functioning[33] and persister cells are associated with low bacterial metabolisms.

It is also important to understand the difference between *tolerance* (where there is very strong genetic evidence) and *persistence* following antibiotic treatment (weaker fundamental evidence), since it is still the subject of some controversy. Tolerance is a genetically determined ability for a population of bacteria to tolerate a particular antibiotic (e.g. the bacteria survive at elevated antibiotic concentrations) and it can be genetically inherited. The phenomenon of persistence is thought to have a different origin. Cloned bacteria (genetically identical bacteria) can grow at a range of different rates due to a wide range of factors, including internal stochastic effects (e.g. due to large fluctuations in the number of intracellular molecules), phenotypical variability (different genetic circuits are activated due to stochastic effects) and different external triggers (e.g. different phases of growth occur due to interactions with their environment, such as exponential growth versus plateau regimes or commitment to a biofilm state). This leads to the phenomenon of persistence. Many antibiotics target metabolically active bacteria, so bacteria that are growing slower are more likely to survive treatments. Such surviving bacteria are called *persisters*.

Practically, persisters are differentiated from tolerant cells in that new generations of cultured persister cells are still sensitive to antibiotics at similar levels to the original cultures. Evolutionary experiments with persisters have sometimes eventually achieved heritable tolerance,[35] so there appears to be a continuum of states between the extremes of pure persisters and pure antibiotic-tolerant bacteria. Clearly, mutants that produce antibiotic-tolerant phenotypes have a strong evolutionary pressure guiding their creation.

Slow bacterial growth is inversely correlated with the rate of respiration. Low rates of respiration are thus found to encourage persister formation in *E. coli*,[36] as expected from the hypothesis linking persisters with slow bacterial metabolisms.

Poor modelling of the chemical kinetics is one contributor to the ongoing puzzle of persister cells. Many plateau regimes on kill curves should be modelled using the adsorption of antibiotics on to exposed surfaces (e.g. adsorption onto EPS, slime, capsules and dead bacteria) and are not indicative that a population of persister cells exists[37] (Figure 24.5). Antibiotic concentrations experienced by bacteria have thus been

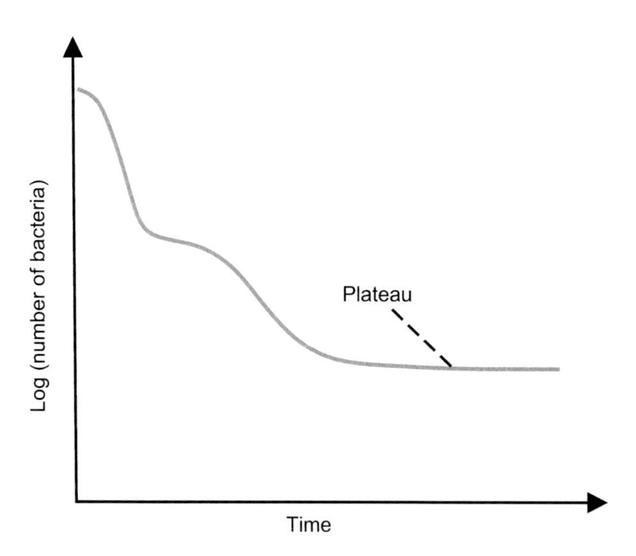

Figure 24.5 Schematic *kill curve* for bacteria exposed to different concentrations of antibiotic.[37] A plateau phase is observed for the resistance of bacteria to antibiotics at long times and is sometimes ascribed to persister cells. An alternative explanation is that the plateau is due to absorption by the EPS and/or the contents of dead bacteria.

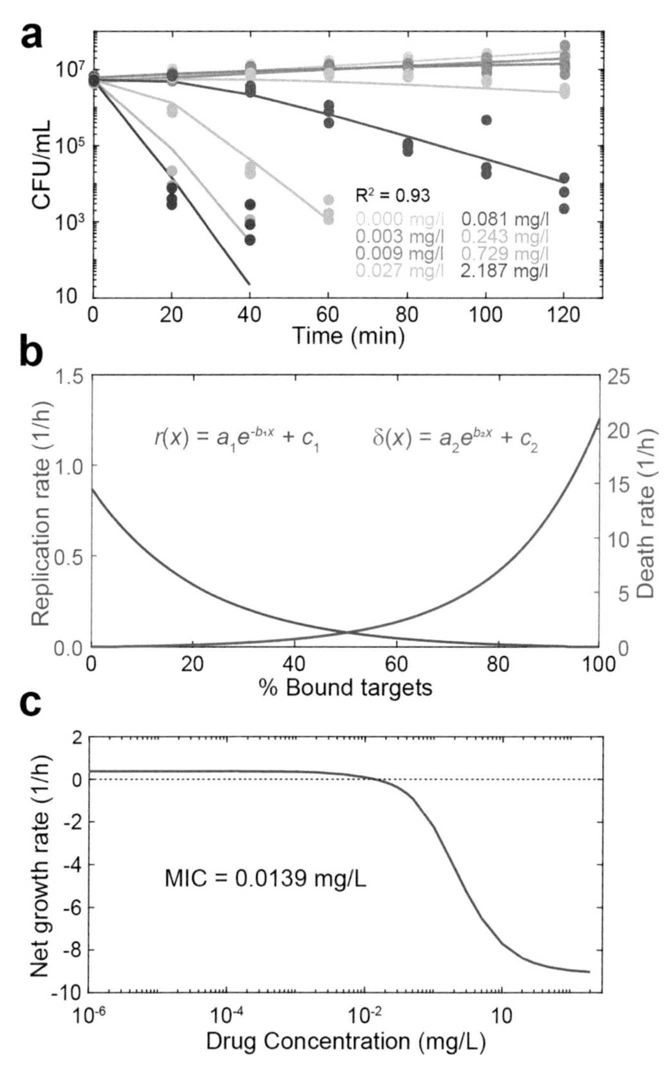

Figure 24.6 Fits to the *COMBAT model* for the activity of ciprofloxacin on *E. coli*.[38] (a) Kill curves as a function of time for different ciprofloxacin concentrations. Symbols show the experimental results. (b) The replication rate and death rate of the bacteria as a function of the binding of the ciprofloxacin. (c) Net growth rate of the bacteria as a function of the drug concentration.

overestimated due to these effects. Specifically, dead bacteria often explode due to the osmotic pressure differences during lysis, when the integrity of their membranes is compromised, and the large surface area of their contents exposed to the solutions is efficient in adsorbing antibiotics, which reduces the effective concentration of antibiotics.

Issues with assays for the antibiotic resistance of bacteria and biofilms also relate to the question of persister cells.[39] Many instances of claimed persister cells are better explained using more detailed kinetic frameworks that invoke adsorption by the biofilm EPS or dead bacteria and help explain plateaus on kill curves. For example, the COMBAT numerical model can describe these effects (Figure 24.6),[37,38]

$$\frac{\partial B}{\partial t}+\frac{\partial}{\partial x}(v_B B)=-rBF_{\lim}+S_B F_{\lim}-\delta B, \tag{24.5}$$

$$\frac{dA}{dt}=-k_f A\int_0^{\theta}(\theta-x)Bdx+k_r\int_0^{\theta}xBdx, \tag{24.6}$$

where A is the concentration of antibiotics inside the bacteria, B is the number of bacteria, $v_B=v_f-v_r$, $v_f=k_f A(t)(\theta-x)$, $v_r=k_r x$, r is the replication rate of the bacteria, $F_{\lim}$ is related to the carrying capacity, S_B is related to the binding targets on the bacteria, δ is the death rate, θ is the targets per bacterium, k_f is the rescaled binding rate, k_r is the unbinding rate, x is the number of bound targets and t is the time.

Biofilm-forming fungi (*Candida albicans*) also have plateaus in their kill curves that can be interpreted as persister cells, although again there is some controversy in this interpretation.[40] Biofilms are considered an important virulence factor and, in contrast to bacteria, the plateaus were not found to be present in stationary planktonic culture with fungi. It is concluded that planktonic fungi do not need this extra level of bet hedging.

Persistence is also observed with bacteria inside eukaryotic cells e.g. salmonella in human cells. Furthermore, persistence is found in human cancer cells in response to chemotherapy.[41] These observations are all consistent with a generic mechanism due to variable growth rates. In terms of growth curves, the persistence of bacteria is associated with an increased proportion of the bacteria in the stationary phase of growth (plateau phase III, Figure 16.1).

A practical solution to the problem of cells persisting after exposure to antibiotics is to mechanically clean away the dead cells, slimes and biofilms from the system, since all these molecules can act as a sink for antibiotics reducing their effective concentrations. Another solution is to increase the antibiotic concentrations locally in a process of targeted delivery e.g. using magnetic particles to carry antibiotics that are steered to the targets with a magnetic field.[42] Regular mechanical cleaning followed by reapplication of antibiotics is often used in wound treatment in a clinical setting (Chapter 31) and is a standard method to treat persistent infections in topical (e.g. skin) applications.

An alternative approach has considered bacterial persistence as a *phenotypic switch* i.e. the persistence corresponds to discrete changes in the phenotypes of the bacteria.[43] Two varieties of persister (I and II) were classified with slightly different switching kinetics (two coupled ordinary differential equations) i.e. normal and persister cells. However, the experimental data presented could be equally well motivated by a continuum of metabolic rates and genetic evidence is lacking for the distinct phenotypes assumed in this model. The simplest and most general interpretation for persistence is thus that the dead or live states of the bacteria are the only dichotomy in the problem and there are no distinct phenotypic differences involved with persistence and persister cells.

24.4 Surface Active Antiseptics

Surface active molecules (surfactants), such as soaps and polyphenols, are extremely useful in a range of technological applications e.g. in personal care products. Owing to the reduction of dimensionality at surfaces, exceedingly small amounts of surfactant can have huge effects on surface behaviour if the molecules are preferentially attracted to the surfaces and they can kill many bacteria in proportion to their mass as a result. Surface active antimicrobial peptides are widespread throughout the animal and plant kingdoms. Such peptides are used as defensive weapons against bacteria, viruses and fungi.[44] Bacteria have evolved some mechanisms to tolerate surface active peptides, such as releasing proteases that break down the molecules. *De novo* small cyclic cationic peptides (CAPs) have been created and have the advantage that they are proteolytically stable i.e. it is much harder for bacteria to break them down with enzymes.[45]

Some *cationic peptide surfactants* can bind to lipopolysaccharides and thus can reduce the occurrence of sepsis.[46] Consideration of the evolutionary connections between antimicrobial peptides and their activities can inform the creation of new materials.[47]

Antimicrobial peptides are key components of the innate immune systems of a wide range of organisms. Bacteria have evolved other mechanisms to evade antimicrobial peptides in addition to the release of proteases. Sensory proteins have been found to detect antimicrobial peptides and can induce the production of modified membrane lipids (e.g. lipopolysaccharides [LPS]) that have increased resistance to attack.[48] *S. typhimurium* regulates its LPS to provide resistance against cationic antimicrobials.[49] In general, Gram-negative bacteria have been found to modify LPS against CAPs. However, extensive CAP usage in clinical settings, such as hospitals, does not inevitably lead to resistance and no resistance after 10 years was found with the CAP tyrothricin. Although such evolutionary pressures might lead to a higher preponderance of biofilm-expressing phenotypes (with upregulated slime and EPS production helping to absorb the CAPs), this was not specifically explored.

There are currently 3,180 *antimicrobial peptides* in the antimicrobial peptide database and they are created naturally by prokaryotes, protists, fungi, plants and animals.[50,51] Some examples of antimicrobial peptides are shown in Figure 24.7. Peptides can also be used to target biofilm formation, although to date only a small number have been identified: protegrin 1, pleurocidin, LL-37, indolicidin, SMAP-29 and human β-defensin-3. These antibiofilm peptides have a variety of modes of action e.g. degradation of EPS, interference with stress signalling to downregulate the production of EPS and binding proteins and reduction of polysaccharide adhesins. Similar to MICs for bacterial growth, it is possible to define a *minimum biofilm inhibitory concentration* (MBIC) to help quantify the activity of the antibiofilm molecules.

Dissipative particle dynamics (DPD) simulations have been successful in demonstrating how non-ionic surfactants punch holes in model bacterial membranes.[52] Ionic surfactants are more challenging systems to model due to the long-range nature of the

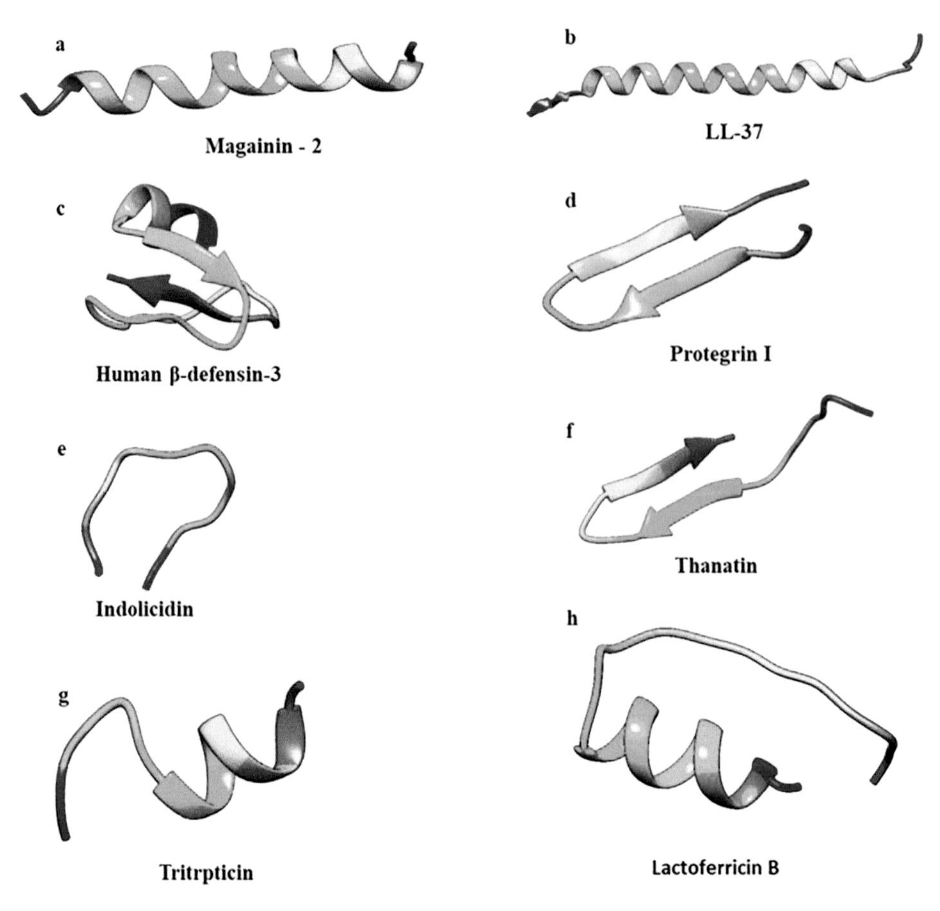

Figure 24.7 Crystalline structure of some *antimicrobial peptides*[50] including (a) magainin-2, (b) LL-37, (c) human β-defensin-3, (d) protegrin I, (e) indolicidin, (f) thanatin, (g) tritpticin and (h) lactoferricin B.

charge interactions (often it can be an intractable many-body problem). In addition to punching holes in membranes, ionic surfactants are thought to bind to DNA and proteins, which can also kill bacteria and make them more potent antibiotics.

Turgor pressure in bacteria is a crucial factor in the action of antimicrobial surfactants, but it is only just starting to be quantified in the context of detergency.[53] In particular, the dynamics of bacterial membrane failure is not well understood e.g. do the bacteria burst like balloons (or possibly fried sausages if the hoop stress is the key driver)[54]? Furthermore, how the failure of membranes relates to the hydrodynamics of cytoplasmic leakage is not well understood. Some initial work has been performed using MD simulations to explore the disruption of membranes with Gram-negative bacteria i.e. those that have two membranes.[55]

Bacterial turgor pressure can be measured using atomic force microscopy.[56] The release of osmotic pressure due to surfactant activity is not well understood. However, the macromolecules on the membrane surface (e.g. peptidoglycans) are known to be the primary origin of the shell strength that withstands the large osmotic pressure differences across bacterial cell walls.

The *lytic actions* of peptides on bacteria were followed with atomic force microscopy.[57] Antimicrobial peptides caused the turgor of *K. pneumoniae* to increase and then decrease. The Young's modulus of the bacterial membranes increased during this process, presumably due to the adsorption of the peptide, although dehydration may cause similar phenomena.

Peptides can *self-assemble* into aggregates due to their amphiphilic nature. This in turn can affect their interaction with bacteria and aggregates can be more effective in disrupting membranes[58] e.g. synthetic cationic surfactant peptides can self-assemble into giant beta sheet aggregates that cause membrane lysis.

Surface active *lipopeptides* are produced in three different varieties in *Bacillus*: surfactin, turin and fengycins. They are known to have a broad range of antimicrobial properties: bacterial, fungal and viral.[59]

Herniation of bacteria is caused by the antibiotic vancomycin that disrupts peptidoglycan synthesis[60] i.e. it causes large-scale remodelling of membrane structure. There are thus structural intermediates during bacterial membrane disruption by vancomycin before complete failure of the membranes and death of the cells.

Dead bacteria can adsorb antimicrobial peptides and lead to a collective tolerance mechanism.[61] Mixtures of the antimicrobial peptides with an adjuvant were found to decrease the protective effect of the dead bacteria.

Non-surfactant cationic molecules can also demonstrate good antimicrobial properties e.g. the polyelectrolyte chitosan from shellfish finds applications in wound dressings.[62] Surfaces with covalently attached cationic polymers (e.g. PVNP Br) are also able to kill substantial amounts of bacteria on contact.[63]

24.5 Physical Sterilisation

In common with other living organisms, bacteria can only survive under a limited range of physical conditions. Thus, sterilisation can be performed by adjusting the physical conditions outside their viable ranges e.g. autoclaving (using elevated temperature and pressure) is a standard procedure to sterilise equipment prior to microbial experiments. Most physical sterilisation techniques are also effective at killing human cells, so their *in vivo* use is forbidden. Bacterial spore cells tend to be the most resistant type of bacterial cell to physical sterilisation and often they require more extreme physical conditions than normal bacterial cells to kill them.

Convenient forms of physical sterilisation include ionising radiation (ultraviolet [UV]/visible, X-rays and gamma rays). These methods can damage bacterial molecules directly or indirectly via the creation of reactive oxygen species (ROS). Exposure to UV light causes growth to stop with *B. subtilis*[64] and a wide range of other bacteria. The morphological transition with *B. subtilis* was modelled using reaction–diffusion equations. Ozone created by plasmas is also effective for sterilisation i.e. directly creating ROS.

Sunlight is able to inactivate many microorganisms, including bacteria, leading to the parasol idea[65] i.e. pigmented molecules are created by bacteria to protect themselves. Photosynthetic bacteria have the opposite challenge, since they need to optimise the amount of light received to absorb their energy, but they also must protect themselves from damage.

The protective effects of bacterial biofilms under extreme physical conditions have been considered e.g. pressure, UV light, temperature, pH and salinity.[66] Clearly, this is an important evolutionary pressure for extremophiles.

Photothermal effects have been used with gold nanoparticles to kill bacteria.[67] Heating and cavitation were the main causes of bacterial death, and specific types of bacteria could be targeted using antibodies attached to the nanoparticles.

Magnetic nanoparticle techniques are promising to kill bacteria for *in vivo* applications[42,68] including those in biofilms (Chapter 30). The nanoparticle positions can be manipulated non-invasively using magnetic fields. They can carry antimicrobial payloads, but Curie heating (a magnetic phenomenon) can also be used for killing bacteria. Antibodies attached to the magnetic nanoparticle surfaces could make the particles strain or species selective. Silver- and gentamicin-coated magnetic nanoparticles have shown good anti-biofilm properties.[69]

24.6 Simple Chemical Antiseptics

A wide range of simple chemicals can act as antiseptics (Table 24.1). They are commonly used in medicine, the food industry, wound dressings and aseptic laboratory techniques.

Alcohols are widely used against bacteria. Ethanol and isopropyl alcohol can be very effective as antiseptics and are in standard use for sterilisation of laboratory equipment (60–90% volume fraction solutions with water are optimal). From the perspective of physical chemistry, these alcohols are polar solvents and have a reduced dielectric constant compared with water. Alcohols are thus readily miscible with water and can penetrate into bacterial cells but disrupt the electrostatic and hydrogen-bonding forces experienced by the molecules inside the cells. The alcohols thus disrupt protein folding and the activity of most of the other molecular machinery inside the bacterial cells. As a result, proteins will denature and coagulate, and similar dramatic restructuring will happen for other biomolecules. Disruption of the cells' membranes by alcohols can also lead to lysis. Alcohol is not very effective against spores or viruses that do not have lipid coats and some water is needed to allow penetration into bacteria. Alcoholics suffer from a reduced diversity of bacterial flora in their intestines as a result of the antimicrobial activity of ethanol, which is expected to negatively impact their health (similar to an overuse of general antibiotics[70]) in addition to other conditions (cirrhosis, etc.).

Silver particles are widely used in wound dressings to combat bacterial infections and have even found applications as odour suppressants in socks. The concentration of intracellular heavy metals is carefully controlled in many species of bacteria, and this

Table 24.1 A range of some commonly used *antiseptic molecules* that include phenolic compounds, alcohols, halogens, heavy metals, formaldehyde, ethylene oxide, β propiolactone and basic dyes.[15]

Antiseptic	How it functions	Application
Surfactants		
Wetting agents	Detergency.	Personal care products.
Cationic agents	Detergency.	QUATS are used clinically against bacteria and fungi in hospitals.
Anionic agents	Detergency.	Laundering, dishwashing and shampoo.
Phenolic compounds		
Phenol	Disrupts membranes and causes protein coagulation.	Active against bacteria, some fungi and viruses. Can be toxic to humans. A detergent. Found in carbolic soap.
Cresols	Disrupt membranes.	Can be toxic to humans.
Hexachlorophene	Inhibits respiration and disrupts membranes.	Gram-positive. Used in drops and creams.
Resorcinol	Interferes with the iodination of tyrosine and oxidation of iodine.	Antiseptic and disinfectant used for skin disorders e.g. acne.
Hexylresorcinol	Binds to tyrosinase.	Skin infections, antiaging creams.
Thymol	Inhibits bacterial growth and lactate production.	Food preservation.
Alcohols		
Ethanol Isopropanol	Denatures proteins and disrupts membranes.	Antibacterial, but not effective against spores or non-lipid enteroviruses.
Halogens Chlorine compounds	Oxidises proteins, lipids and carbohydrates.	Antibacterial, antifungal and sporicidal at high concentrations.
Iodine compounds	Disrupts proteins and nucleic acids.	Antiseptics on skin or tissue.
Heavy metals		
Mercury compounds Inorganic/organic	Protein denaturation, enzyme poisoning and so on.	Used by R. Koch as an antimicrobial. Poisonous to humans.
Silver compounds	Punctures cell walls, inhibits metabolism and disrupts DNA.	Burns and water treatment.
Acids and alkalis	Phase stability.	Food preparation.
Formaldehyde	Reacts with proteins and nucleic acids.	Very poisonous and carcinogenic. Bactericidal, fungicidal, virucidal and sporicidal.
Ethylene oxide	Alkylation of proteins, DNA and RNA.	Sterilisation.
Propiolactone	Damages DNA and proteins.	Disinfectant.
Basic dyes	Binds to DNA.	Use on skin. Toxic when ingested.

has found applications in bioremediation e.g. heavy metal pollution in uranium mines can be controlled using bacteria.[71] Bacterial resistance to silver is often due to heavy metal efflux pumping and it is observed in burn wards with *Salmonella* and *E. coli* infections that have been exposed to silver-containing dressings over long periods of time.

Chlorine ingress into biofilms (e.g. from bleach) has been measured using microelectrodes. Less than 20% of the bulk concentration penetrated into mixed *Pseudomonas* and *Klebsiella* biofilms.[72] Thus, biofilms can provide substantial protection from some simple chemical antiseptics.

24.7 Antibiotics – Chemicals for Internal Use

A wide range of *antibiotics* are used to treat human infections.[15] A few of them are shown in Table 24.2. Biofilms, slimes and capsules are created by bacteria to increase their tolerance to antibiotics, since they can adsorb antibiotics, reduce the permeation kinetics and are often associated with lower metabolic states that cannot be targeted by the antibiotics. Thus, antibiotic stresses can modify biofilm structure.[73]

Starvation makes bacteria much more antibiotic-resistant. Decreased metabolic activity through starvation often occurs in biofilms and the bacteria can activate a distinct genetic programme called the starvation signalling stringent response.[74] Starvation leads to the inactivity of many antibiotic targets as the bacteria decrease their metabolisms. A specific biochemical circuit to handle oxidative stress is also important, since it is in turn involved with the inactivity of many antimicrobials[25] e.g. bleach. In some species of bacteria, the extreme phenotypic limit is obtained called a *spore cell* (e.g. sporulation in *B. subtilis*) where the bacterial metabolism approaches zero. Spore cells (Section 15.1.1) are invulnerable to most antibiotics, although physical antiseptics can still be effective e.g. detergents can sometimes disrupt their membranes.

Antibacterial lectins are used to segregate microbiota from the intestine wall in mammals.[75] Thus, populations of bacteria in different parts of the intestine are carefully controlled *in vivo*.

Table 24.2 Some *prototypical antibiotics* with some modes of action.[15]

Antibiotic	Mode of action
Penicillin	Disrupts an enzyme needed for cell wall synthesis.
Streptomycin	Binds to ribosomes and inhibits protein synthesis.
Chloramphenicol	Binds to ribosomes and inhibits protein synthesis.
Tetracyclines	Binds to ribosomes and inhibits protein synthesis.
Bacitracin	Inhibits cell wall synthesis.
Polymyxin	Disrupts membranes.
Rifampicin	Disrupts RNA polymerase.
Quinolone	Disrupts DNA synthesis via topoisomerases.

All of these antibiotics can have side effects that range from mild to life-threatening, depending on dosages and the patients' specific physiology.

24.8 Antibiofilm Molecules

Lactoferrin is released by many human cells and it is commonly a part of mucosal surfaces where it is involved in iron sequestration. It has been shown to act as an anti-biofilm molecule in *P. aeruginosa* i.e. it stops them from producing biofilms, but does not significantly affect the division or morbidity of the bacterial cells.[76]

Depolymerisation of the macromolecular components of EPS can be used to disrupt biofilm components, such as proteins, polysaccharides and nucleic acids, via proteases, amylases (or other carbohydrate splitting enzymes) or DNaases, respectively. DNaases have been commercialised to treat infections in cystic fibrosis lungs using aerosol formulations.

D-amino acids trigger biofilm disassembly in a range of bacteria.[77] One contributing factor is that amyloid formation is disrupted by the D-amino acids, specifically the mechanism by which the amyloids are anchored onto bacterial cell walls. Removing the anchor points for the amyloids disrupts the mechanical integrity of the biofilms.

Some molecules that inhibit quorum sensing are produced by seaweeds that lack more sophisticated immune systems e.g. furanose is released.[78] These molecules tend to inhibit the colonisation of seaweed by bacteria and synthetic furanose has been successfully used to inhibit the virulence of *P. aeruginosa*.

Suggested Reading

McDonnell, G. E., *Antisepsis, Disinfection and Sterilization: Types, Action and Resistance*. ASM: 2017. Currently the definitive account of the action of antiseptics, including detergents, from an applied microbiology perspective.

Walsh, C., Wencewicz, T., *Antibiotics: Challenges, Mechanisms and Opportunities*. ASM: 2016. Detailed approach to current developments in antibiotic chemistry.

References

1. Walsh, C. T.; Wencewicz, T., *Antibiotics: Challenges, Mechanisms, Opportunities*. ASM Books: 2016.
2. McDonnell, G. E., *Antisepsis, Disinfection and Sterilization: Types, Action and Resistance*, 2nd ed. ASM Press: 2017.
3. Gong, H.; et al., Hydrophobic control of the bioactivity and cytotoxicity of de novo designed antimicrobial peptides. *ACS Applied Materials and Interfaces* **2019**, *11* (38), 34609–34620.

4. Gong, H.; et al., Aggregated amphiphilic antimicrobial peptides embedded in bacterial membranes. *ACS Applied Materials and Interfaces* **2020**, *12* (40), 44420–44432.
5. Fischbach, M. A.; Walsh, C. T., Antibiotics for emerging pathogens. *Science* **2009**, *325* (5944), 1089–1093.
6. D'Costa, V.; McGrann, K. M.; Hughes, D. W.; Wright, G. D., Sampling the antibiotic resistome. *Science* **2006**, *311* (5759), 374–377.
7. Taubes, G., The bacteria fight back. *Science* **2008**, *321* (5887), 356–361.
8. van Vranken, D.; Weiss, G. A., *Introduction to Bioorganic Chemistry and Chemical Biology*. Garland: 2012.
9. Costerton, J. W.; Lewandowki, Z.; Caldwell, D. E.; Korber, D. R.; Lappin-Scott, H. M., Microbial biofilms. *Annual Review of Microbiology* **1995**, *49*, 711–745.
10. Zhang, X. L.; Hansing, J.; Netz, R. R.; DeRouchey, J. E., Particle transport through hydrogels is charge asymmetric. *Biophysical Journal* **2015**, *108* (3), 530–539.
11. White, O.; et al., Genome sequence of the radioresistant bacterium *Deinococcus radiodurans* R1. *Science* **1999**, *284* (5444), 1571–1577.
12. Stewart, P. S., Theoretical aspects of antibiotic diffusion into microbial biofilm. *Antimicrobial Agents and Chemotherapy* **1996**, *40* (11), 2517–2521.
13. Stewart, P. S., Diffusion in biofilms. *Journal of Bacteriology* **2003**, *185* (5), 1485–1491.
14. Metzler, R.; Klafter, J., The restaurant at the end of the random walk. *Journal of Physics A: General Physics* **2004**, *37* (31), R161–R208.
15. Capuccino, J. G.; Welsh, C., *Microbiology: A Laboratory Manual*. Pearson: 2018.
16. Mah, T. F.; O'Toole, G. A., Mechanisms of biofilm resistance to antimicrobial agents. *Trends in Microbiology* **2001**, *9* (1), 34–39.
17. Bansil, R.; Turner, B. S., Mucin structure, aggregation, physiological functions and biomedical applications. *Current Opinion in Colloid and Interface Science* **2006**, *11* (2–3), 164–170.
18. Kosztolowicz, T.; Metzler, R.; Wasik, S.; Arabski, M., Modelling experimentally measured of ciprofloxacin antibiotic diffusion in *Pseudomonas aeruginosa* biofilm formed in artificial sputum medium. *PLOS One* **2020**, *15* (12), e0243003.
19. Kosztolowicz, T.; Metzler, R., Diffusion of antibiotics through a biofilm in the presence of diffusion and absorption barriers. *Physical Review E* **2020**, *102* (3-1), 032408.
20. Bonev, B.; Hooper, J.; Parisot, J., Principles of assessing bacterial susceptibility to antibiotics using the agar diffusion method. *Journal of Antimicrobial Chemotherapy* **2008**, *61* (6), 1295–1301.
21. Walsh, C., Molecular mechanisms that confer antibacterial drug resistance. *Nature* **2000**, *406* (6797), 775–781.
22. Gullbert, E.; Cao, S.; Berg, O. G.; Ilback, C.; Sandegren, L.; Hughes, D.; Andersson, D. I., Selection of resistant bacteria at very low antibiotic concentrations. *PLOS Pathogens* **2011**, *7* (7), e1002158.
23. Knoppel, A.; Nassall, J.; Andersson, D. I., Evolution of antibiotic resistance without antibiotic exposure. *Antimicrobial Agents and Chemotherapy* **2017**, *61* (11), e01495.

24. Walters III, C.; Roe, F.; Bugnicourt, A.; Franklin, M. J.; Stewart, P. S., Contributions of antibiotic penetration, oxygen limitation, and low metabolic activity to tolerance of *Pseudomonas aeruginosa* biofilms to ciprofloxacin and tobramycin. *Antimicrobial Agents and Chemotherapy* **2003**, *47* (1), 317–323.
25. Nguyen, D.; et al., Active starvation responses mediate antibiotic tolerance in biofilms and nutrient-limited bacteria. *Science* **2011**, *334* (6058), 982–986.
26. Hentzer, M.; Teitzel, G. M.; Balzer, G. I.; Heydorn, A.; Molin, S.; Givskov, M.; Parsek, M. R., Alginate overproduction affects *Pseudomonas aeruginosa* biofilm structure and function. *Journal of Bacteriology* **2001**, *183* (18), 5395–5401.
27. Andel, J. N.; Franklin, M. J.; Stewart, P. S., Role of antibiotic penetration limitation in *Klebsiella pneumoniae* biofilm resistance to ampicillin and ciprofloxacin. *Antimicrobial Agents and Chemotherapy* **2000**, *44* (7), 1818–1824.
28. Rani, S. A.; Pitts, B.; Stewart, P. S., Rapid diffusion of fluorescent tracers into *Staphylococcus epidermis* biofilms visualized by time lapse microscopy. *Antimicrobial Agents and Chemotherapy* **2005**, *49* (2), 728–732.
29. Grobas, I.; Polin, M.; Asally, M., Swarming bacteria undergo localized dynamic phase transitions to form stress induced biofilms. *eLife* **2021**, *10*, e62632.
30. Seifert, A.; Kashi, Y.; Livney, Y. D., Delivery to gut microbiota: A rapidly proliferating research field. *Advances in Colloid and Interface Science* **2019**, *274*, 102038.
31. Fort, J.; Mendez, V., Time-delayed spread of viruses in growing plaques. *Physical Review Letters* **2002**, *89* (17), 178101.
32. Vijay, U.; Gupta, S.; Mathur, P.; Suravajhala, P.; Bhatnagar, P., Microbial mutagenicity assay: Ames test. *Bio Protocols* **2018**, *8* (6), e2763.
33. Lewis, K., Persister cells, dormancy and infectious disease. *Nature Reviews Microbiology* **2007**, *5* (1), 48–56.
34. Lewis, K., Multidrug tolerance of biofilm and persister cells. In *Current Topics in Microbiology and Immunology*, Romeo, T., Ed.; Springer: 2008; Vol. 322; pp. 107–131.
35. Windels, E. M.; Michiels, J. E.; Fauvart, M.; Wenseleers, T.; Van den Bergh, B.; Michiels, J., Bacterial persistence promotes the evolution of antibiotic resistance by increasing survival and mutation rates. *The ISME Journal* **2019**, *13* (5), 1239–1251.
36. Orman, M. A.; Brynildsen, M. P., Inhibition of stationary phase respiration impairs persister formation in *E. coli*. *Nature Communications* **2015**, *6*, 7983.
37. Zur Wiesch, P. A.; et al., Classic reaction kinetics can explain complex patterns of antibiotic action. *Science Translational Medicine* **2015**, *7* (287), 287ra73.
38. Clarelli, F.; et al., Drug-target binding quantitatively predicts optimal antibiotic dose levels in quinolones. *PLOS Computational Biology* **2020**, *16* (8), e1008106.
39. Balaban, N. Q.; et al., Definitions and guidelines for research on antibiotic persistence. *Nature Reviews Microbiology* **2019**, *17* (7), 441–448.
40. Denega, I.; D'Enfert, C.; Backellier-Bassi, S., *Candida albicans* biofilms are generally devoid of persister cells. *Antimicrobial Agents and Chemotherapy* **2019**, *63* (5), e01979.

41. Drescher, K.; Dunkel, J.; Cisneros, L. H.; Ganguly, S.; Goldstein, R. E., Fluid dynamics and noise in bacterial cell-cell and cell-surface scattering. *PNAS* **2011**, *108* (27), 10940–10945.
42. Hwang, G.; et al., Catalytic antimicrobial robots for biofilm eradication. *Science Robotics* **2019**, *4* (29), eaaw2388.
43. Balaban, N. Q.; Merrin, J.; Chait, R.; Kowalik, L.; Leibler, S., Bacterial persistence as a phenotypic switch. *Science* **2004**, *305* (5690), 1622–1625.
44. Zasloff, M., Antimicrobial peptides of multicellular organisms. *Nature* **2002**, *415* (6870), 389–395.
45. Fernandez-Lopez, S.; et al., Antibacterial agents based on the cyclic D,L alpha peptide architecture. *Nature* **2001**, *412* (6845), 452–455.
46. Hancock, R. E.; Scott, M. G., The role of antimicrobial peptides in animal defenses. *PNAS* **2000**, *97* (16), 8856–8861.
47. Lazzaro, B. P.; Zasloff, M.; Rolff, J., Antimicrobial peptides: Application informed by evolution. *Science* **2020**, *368* (6490), 487–494.
48. Bader, M. W.; et al., Recognition of antimicrobial peptides by a bacterial sensor kinase. *Cell* **2005**, *122* (3), 461–472.
49. Guo, K.; Lim, K. B.; Gunn, J. S.; Bainbridge, B.; Darveau, R. P.; Hackett, M.; Miller, S. L., Regulation of lipid A modifications by *Salmonella typhimurium* virulence genes phoP-phoQ. *Science* **1997**, *276* (5310), 250–253.
50. di Somma, A.; Moretta, A.; Cane, C.; Cirillo, A.; Duilio, A., Antimicrobial and antibiofilm peptides. *Biomolecules* **2020**, *10* (4), 652.
51. Antimicrobial peptide database. http://aps.unmc.edu/AP.
52. Groot, R. D.; Rabone, K. L., Mesoscopic simulation of cell membrane damage, morphology change and rupture by nonionic surfactants. *Biophysical Journal* **2001**, *81* (2), 725–736.
53. Hwang, H.; Paracini, N.; Parks, J. M.; Lakey, J. H.; Gumbart, J. C., Distribution of mechanical stress in *Escherichia coli* cell envelope. *Biochimica et Biophysica Acta* **2018**, *1860* (12), 2566–2575.
54. Boal, D., *Mechanics of the Cell*. CUP: 2012.
55. Parkin, J.; Chavert, M.; Khalil, S., Molecular simulations of Gram-negative bacterial membranes: A vignette of some recent successes. *Biophysical Journal* **2015**, *109* (3), 461–468.
56. Arnoldi, M.; Fitz, M.; Bauerlein, E.; Fritz, M.; Radmacher, M.; Sackmann, E.; Boulbitch, A., Bacterial turgor pressure can be measured by atomic force microscopy. *Physical Review E* **2000**, *62* (1 Pt B), 1034–1044.
57. Mularski, A.; Wilksch, J. J.; Wang, H.; Hossain, M. A.; Wade, J. D.; Separovic, F.; Strugnell, R. A.; Gee, M. L., Atomic force microscopy reveals the mechanobiology of lytic peptide action on bacteria. *Langmuir* **2015**, *31* (22), 6164–6171.
58. Huffner, S. M.; Malmsten, M., Influence of self-assembly on the performance of antimicrobial peptides. *Current Opinion in Colloid and Interface Science* **2018**, *38*, 56–79.
59. Ongena, M.; Jacques, P., Bacillus lipopeptides: Versatile weapons for plant disease biocontrol. *Trends in Microbiology* **2008**, *16* (3), 115–125.

60. Huang, K. C.; Mukhopadhyay, R.; Wen, B.; Gitai, Z.; Wingreen, N. S., Cell shape and cell-wall organization in gram-negative bacteria. *PNAS* **2008**, *105* (49), 19282–19287.
61. Wu, F.; Tau, C., Dead bacterial adsorption of antimicrobial peptides underlies collective tolerance. *Journal of Royal Society Interface* **2019**, *16*, 20180701.
62. Rabea, E. I.; Badawy, E. T.; Stevens, C. V.; Smagghe, G.; Steurbaut, W., Chitosan as antimicrobial agent: Applications and mode of action. *Biomacromolecules 4* (6), 1457–1465.
63. Tiller, J. C.; Liao, C. J.; Lewis, K.; Klibanov, A. M., Designing surfaces that kill bacteria on contact. *PNAS* **2001**, *98* (11), 5981–5985.
64. Delprato, A. M.; Samadani, A.; Kudrolli, A.; Tsimring, L. S., Swarming ring patterns in bacterial colonies exposed to ultraviolet radiation. *Physical Review Letters* **2001**, *87* (15), 158102.
65. Nelson, K. L.; et al., Sunlight-mediated inactivation of health-relevant microorganisms in water: A review of mechanisms and modelling approaches. *Environmental Science Process Impacts* **2018**, *20* (8), 1089–1122.
66. Yin, W.; Wang, Y.; Liu, L.; He, J., Biofilms: The microbial 'protective clothing' in extreme environments. *International Journal of Molecular Sciences* **2019**, *20* (14), 3423.
67. Zharov, V. P.; Mercer, K. E.; Galitovskaya, E. N.; Smeltzer, M. S., Photothermal nanotherapeutics and nanodiagnostics for selective killing of bacteria targeted with gold nanoparticles. *Biophysical Journal* **2006**, *90* (2), 619–627.
68. Peyer, K. E.; Zhang, L.; Nelson, B. J., Bioinspired magnetic swimming microrobots for biomedical applications. *Nanoscale* **2013**, *5* (4), 1259.
69. Wang, X.; et al., Microenvironment-responsive magnetic nanocomposites based on silver nanoparticles/gentamicin for enhanced biofilm disruption by magnetic field *ACS Applied Materials and Interfaces* **2018**, *10* (41), 34905–34915.
70. Day, A. W.; Kumamoto, C. A., Gut microbiome dysbiosis in alcoholism. *Frontier in Cellular Infectious Microbiology* **2022**, *12*, 840164.
71. Silver, S., Bacterial silver resistance: Molecular biology and uses and misuses of silver compounds. *FEMS Microbiology Reviews* **2003**, *27* (2–3), 341–353.
72. de Beer, D.; Srinivasan, R.; Stewart, P. S., Direct measurement of chlorine penetration into biofilms during disinfection. *Applied and Environmental Microbiology* **1994**, *60* (12), 4339–4344.
73. Stewart, E. J.; Satorius, A. E.; Younger, J. G.; Solomon, M. J., Role of environmental and antibiotic stress on *Staphylococcus epidermidis* biofilm microstructure. *Langmuir* **2013**, *29* (23), 7017–7024.
74. Hoiby, N.; Bjarnsholt, T.; Givskov, M.; Molin, S.; Ciofu, O., Antibiotic resistance of bacterial biofilms. *International Journal of Antimicrobial Agents* **2010**, *35* (4), 322–332.
75. Vaishnava, S.; et al., The antibacterial lectin RegIII-Gamma promotes the spatial segregation of microbiota and host in the intestine. *Science* **2011**, *334* (6053), 255–258.

76. Singh, P. K.; Parsek, M. R.; Greenberg, E. P.; Welsh, M. J., A component of innate immunity presents bacterial biofilm development. *Nature* **2002**, *417*, 552–555.
77. Kolodkin-Gal, I.; Romero, D.; Cao, S.; Clardy, J.; Kolter, R.; Losick, R., D-Amino acids trigger biofilm disassembly. *Science* **2010**, *328* (5978), 627–629.
78. Hentzer, M.; et al., Attenuation of *Pseudomonas aeruginosa* virulence by quorum sensing inhibitors. *EMBO Journal* **2003**, *22* (15), 3803–3815.

25 Bacterial Diseases

Biofilms, *slime layers* (e.g. shed capsular materials that form thick viscoelastic fluids) and *capsules* play an important role in many bacterial diseases and are thus classified as important virulence factors. These protective mechanisms for bacteria are significant threats in modern medicine, since their treatment is very challenging and infections can become untreatable. Specifically, standard antibiotics are often much less effective if biofilms, slime layers and capsules are present (Chapter 24).

For detailed diagnostics of bacterial diseases, there are a number of good resources for medics.[1,2] Instead, here the relationship between physical principles and disease is emphasised.

Table 25.1 gives some examples of disease-causing bacteria. Around 100 species of bacteria are directly implicated in human disease, which is a very small subset of the total number of bacterial species ($\sim10^{12}$). However, a huge range of bacteria are opportunistic pathogens if they are not actively controlled by the host's immune system i.e. they will eat human cells if given the chance but are not specialised to do so. There are many more virulence factors for bacteria beyond biofilms, slimes and capsules e.g. flagella, secretion systems and DNA repair enzymes, but biofilms, slimes and capsules are emphasised since they play a crucial role in determining intercellular forces and thus the physical phenomena the bacteria experience.

In general, *immunology* is a fascinating, but extremely complicated area of biology.[3] The present coverage thus needs to restrict itself to relatively crude physical descriptions and by necessity ignores much of the biological detail. There is thus a wide range of processes that cannot be discussed in detail, such as Toll receptors on eukaryotic cells that grab onto bacterial DNA for oxidation, granulysins, the induction of antibodies in the innate immune system and neutrophil extracellular traps.[4,5] An extremely large area of immunology still needs to be considered from a biophysical perspective, but many of the mechanisms are only just starting to be named, let alone understood thoroughly with quantitative detail. Furthermore, the experiments are very exacting and hard to reproduce, leading to substantial controversy in the field.

Nosocomial infections are those that originate in a hospital and are a significant threat, since virulent infections can rapidly be transmitted in environments with high densities of immunocompromised individuals e.g. antibiotic-resistant bacteria that specialise in wound infections found in a ward of burns patients. Thus, people arrive in hospital and end up suffering from an infection they pick up there, which can be

Table 25.1 Examples of some bacteria implicated in human disease[1,2] (more extensive lists are available in the medical literature).

Species	Disease	Existence of strains with biofilms and capsules (B, C)
Streptococci	Group A – strep throat, skin infections, necrotising fasciitis, scarlet fever, streptococcal toxic shock syndrome and rheumatic fever.	B, C
	Group B – baby diseases e.g. meningitis.	B, C
	Viridans – dental infections, endocarditis and abscesses.	B, no capsule
	Group D – urinary tract infections, wound infections and endocarditis.	B, C
	Pneumoniae – pneumonia in adults and otitis media in children.	B, C
Staphylococci e.g. *Staphylococcus aureus*	Gastroenteritis, toxic shock syndrome, scalded skin syndrome, pneumonia, meningitis, osteomyelitis, endocarditis, septic arthritis, skin infection and blood infection.	B, C
Bacillus and **clostridium** (spore-forming rods)	*Bacillus anthracis* causes anthrax.	B, C, Spores
	Bacillus cereus causes gastroenteritis.	B, C, Spores
	Clostridium botulinum causes botulism.	B, C, Spores
	Clostridium tetani causes tetanus.	No biofilm, No capsules, Spores
	Clostridium perfringens causes gas gangrene.	B, C, Spores
	Clostridium difficile causes intestinal diseases.	B, C, Spores
Corynebacterium and *listeria* (non-spore-forming rods)	*Corynebacterium diphtheriae* – diphtheria. *Listeria monocytogenes* – meningitis.	B, no capsule, no spores
Neisseria	Meningococcal disease e.g. meningitis, sepsis and gonorrhoea.	B, C
Enterica	Diarrhoea, urinary tract infections, pneumonia, bacteraemia, sepsis, e.g. *Escherichia coli*, *Klebsiella pneumoniae*, *Shigella*, *Salmonella*, *Vibrio cholerae* and *Bacteroides fragilis*.	B, C
Hospital-acquired Gram negatives	*Pseudomonas aeruginosa* – pneumonia, osteomyelitis, wounds, sepsis, urinary tract infections and endocarditis.	B, C
Haemophilus, *Bordetella* and *Legionella*	Legionella disease.	B, C
Yersinia, *Francisella*, Brucellosis, *Pasteurella*	Bubonic plague.	B, C
Chlamydia, *Rickettsia*	Chlamydia.	B, C
Spirochetes	Syphilis and Lyme disease.	B, no capsule
Mycobacteria	Tuberculosis and leprosy.	B, C
Mycoplasma	Bronchitis and pneumonia.	B, C

worse than their original condition e.g. geriatric wards or neonatal wards can experience huge morbidity due to insufficient cleanliness.

Reasonably biocompatible antibiotics exist (Chapter 24), but antibiofilm drugs for internal use are much more limited. For example, if a biofilm is found to be coating a replacement heart valve, the patient will likely need to have another extremely invasive operation to have it replaced, since most antibiotics only function well with planktonic (free-swimming) bacteria and biofilm-based infections can be untreatable. Thus, in general, with regenerative medicine, replacement materials (stents, hip joints, heart valves, etc.) that can become home to a biofilm will resist antibiotic treatment and can threaten the life of an individual.

Replacement biomaterials placed inside organisms can hide bacterial infections away from the natural immune responses that constantly monitor potential threats. This is a key insight into many of the problems encountered in regenerative medicine e.g. post-operative infections. Tissue needs to be sufficiently well integrated with the biomaterial so that no immuno-hidden environments are presented to infections.[6] Implant-related infections are extremely common. For example, 1 million cases per year were recorded in the USA in 2006.[7] Lots of effort has been expended to create efficient antimicrobial coatings. Foreign body material is often found stuck to a wide range of regenerative materials causing significant complications on materials that include: contact lens, sutures, ventilation devices, heart valves, vascular grafts, arteriovenous shunts, endovascular catheters,[8] spinal fluid shunts, peritoneal dialysis, urinary tract infections (UTIs), intrauterine devices (IUD) and orthopaedic prostheses.[9] A fundamental challenge is that antimicrobial surface coatings can kill bacteria during initial colonisation events, but the dead bacteria screen the surface from future waves of colonisation and provide a convenient source of nutrition. Thus, without surface sloughing, bacteria can eventually grow on most surfaces. Reducing bacterial growth kinetics through surface modifications is however a key task, because it can give the immune system additional time to combat and conquer infections.

Many bacteria that live intracellularly can stealthily invade cells, so they do not alert the immune mechanism of the host organisms. They can then exploit the host's resources, kill the host's cells and break into neighbouring cells to spread the infections. An exceedingly complicated immunological story is involved with bacterial infections of human cells. Often, biofilms or capsular structures are switched on when the bacteria are inside the eukaryotic cells to provide additional protective mechanisms against intracellular defences e.g. capsules in *Escherichia coli* UTIs or biofilms in tuberculosis (TB) infections.

Zipper (*Yersinia pseudotuberculosis* and *Listeria monocytogenes*) and *trigger* (*L. monocytogenes*) mechanisms are used to describe how pathogenic bacteria enter cells by manipulating the self-assembly of motor proteins[10] (Figure 19.1). Bacteria also have a range of mechanisms to escape specialised phagocytotic cells in the immune systems to propagate the infections.[11] For example, *Shigella* escapes phagosomes with some sophisticated biochemistry.[11]

25.1 Bacterial Biofilms in Disease

Many pathogenic bacteria actively use biofilms in diseases and some examples are given in Table 25.2.

Plaque on teeth is a mixed-species bacterial biofilm and it is very important for the progress of tooth decay. People brush their teeth to remove plaque biofilms. Classification of all the different bacterial species in the mouth became possible with the combination of microfluidic techniques with 16S RNA sequencing.[13] More than 700 different bacterial species have been found in the oral cavity.[14] More than 500 different bacterial taxa are contained within a typical dental biofilm from an individual.[15] *Streptococcus mutans* produces a highly structured biofilm and is a key species involved in pathogenic dental bacteria.

Dental caries affects 2.3 billion people globally and is particularly prevalent in underprivileged children.[16] The primary cause of the disease is *S. mutans*, which

Table 25.2 Partial list of some human infections that involve biofilms.[12]

Infection	Biofilm forming bacterial species
Infections of muscles, bones and joints.	Gram-positive cocci.
Bone infections	A range of bacterial species.
Dental infections	Acidogenic Gram-positive cocci.
Gum disease	Gram-negative anaerobic oral bacteria.
Ear infections	Strains of *Haemophilus influenzae*.
Heart valve infections	Viridans group streptococci.
Flesh-eating infections	*Staphylococcus aureus* and Group A streptococci.
Bile duct infections	Enteric bacteria e.g. *E. coli.*
Lung disease in cystic fibrosis	*Pseudomonas aeruginosa*.
Prostate infections	*E. coli* and other Gram-negative bacteria.
Melioidosis	*Pseudomonas pseudomallei.*
Hospital-acquired infections	
Orthopaedic devices/replacement heart valves	*Staphylococcus epidermis* and *S. aureus.*
Dialysis	*A range of bacteria and fungi.*
Intensive care unit pneumonia	Gram-negative rods.
Sutures/shunts	*S. epidermis* and *S. aureus.*
Eye surgery	Gram-positive cocci.
Contact lens	*P. aeruginosa* and Gram-positive cocci.
Urinary catheter	*E. coli* and other Gram-negative rods.
Intrauterine devices	*Actinomyces israelii* and others.
Tracheal (breathing) tubes	A range of bacteria and fungi.
Catheters	*S. epidermidis*
Vascular grafts	Gram-positive cocci.
Stent blockage	A range of enteric bacteria.

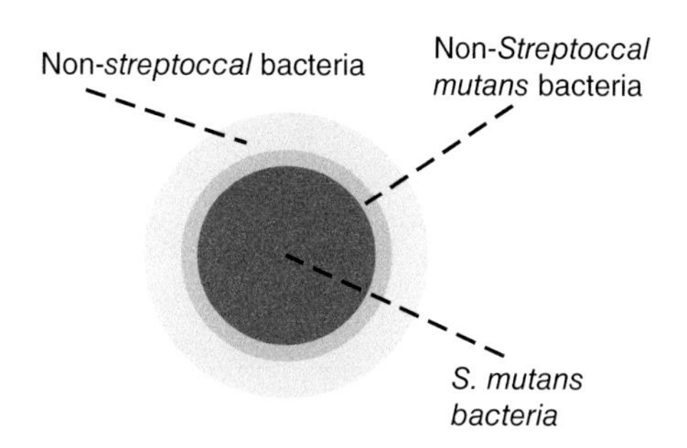

Figure 25.1 Schematic architecture of the rotund morphology found in mixed bacterial biofilms from *dental carries teeth.*[16] *S. mutans* is found in the centre of the mixed biofilm.

produces both biofilm (plaque) and acid that erodes the teeth. In an extensive study of biofilm architecture on teeth in dental caries using confocal microscopy and SEM, four principle types of mixed biofilm architecture were found: rotund, corncob, hedgehog and seaweed. The rotund architecture was the most common and consisted of a corona-like arrangement (Figure 25.1). *S. mutans* was in the centre, followed by non-mutans streptococci and then non-streptococcal bacteria. Some progress has been made in describing the specific binding proteins for biofilm and cell adhesion with dental caries e.g. mutanofactin proteins.[17]

The genetic evolution of biofilm-forming *P. aeruginosa* in long-term infections (over 8 years) in cystic fibrosis patients has been studied with molecular detail.[18] Surprisingly, the bacteria are found to experience loss of function mutations during the years that they infect the host patient. The virulence factors are selected against in these persistent infections after the initial stage of infection and the rationale is that they do this to avoid the host's immune responses i.e. the bacteria are most successful once they have established an infection if they lose virulence factors that will activate the host's immune system. A separate study, however, has demonstrated that *Pseudomonas aeruginosa* biofilms adjust their mechanical toughness by upregulating PsI expression as the infection progresses,[19] so clearly all the virulence factors are not selected against.

Quorum-sensing molecules can be used to diagnose *P. aeruginosa* infections in cystic fibrosis patients.[20] However, the specificity of the quorum-sensing molecules will likely be too limited to accurately determine the strains of bacteria involved and genetic studies are preferable in this case.

The dynamics of salmonella infections in macrophages have been explored at the single-cell level.[21] The probability of salmonella infections in individual cells when a bacterial cell encounters a macrophage cell is found to be surprisingly low (less than 5%).

Neutrophils have been found to form extracellular fibre taps to kill Gram-positive and Gram-negative bacteria.[5] Thus, phagocytosis (engulfing) by neutrophils is not obligatory for their bactericidal activity.

Adhesion of bacterial cells to human cells is a crucial step in many diseases (Chapter 4). Afimbrial adhesins are important i.e. membrane-associated proteins that

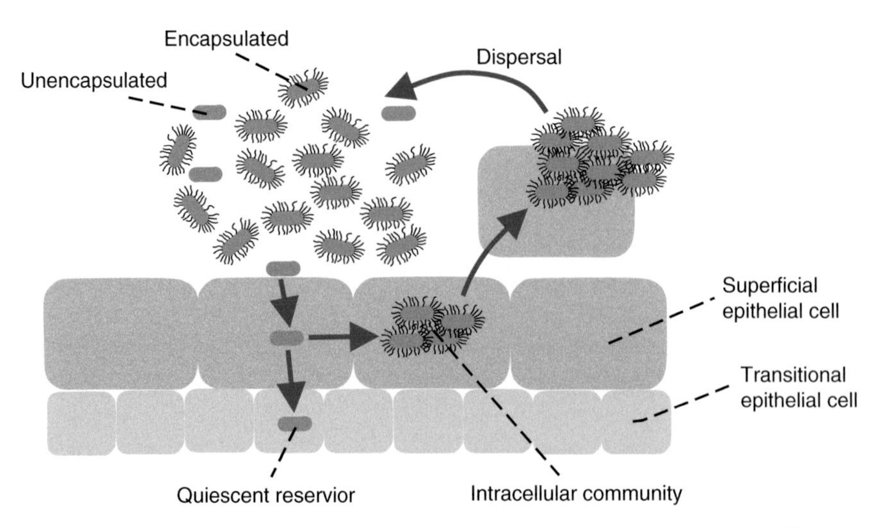

Figure 25.2 The cycle of infection in *urinary pathogenic* (UPEC) bacteria in epithelial cells on the wall of the bladder.[28] Encapsulated bacteria tend to switch off their capsules in order to gain entry into the epithelial cells.

extend over much shorter ranges than pili.[6] Furthermore, some bacteria induce vacuole formation and then directly enter inside cells. Some others dispense with these intermediate steps and enter the cytoplasm directly.

Virulent strains of *Anthrax bacillus* have a polyglutamic capsule and they can rapidly create a potent toxin *in vivo*.[22] The bacteria thus need to be treated with antibiotics before a critical concentration of the toxin is created that will kill the host. Anthrax is a reasonably common disease of livestock, since spores commonly exist in the soil and are often encountered by vets. Spores when dormant are very resistant to environmental conditions, such as UV, ionising radiation and heat. In contrast to spores, normal bacilli cells are relatively easy to kill. Removal of the virulence factors from *B. anthrax* (2 plasmids) allows the creation of a live vaccine that is used for animals, although there is still some innate virulence, stopping its use in humans.

Urinary tract infections are frequently antibiotic-resistant and can lead to morbidity in extreme cases[23,24] (Figure 25.2). They are commonly caused by *E. coli*, *Klebsiella pneumoniae*, *Proteus mirabilis*, *Enterococcus faecalis* and *Staphylococcus saprophyticus*. Biofilms can form in the bladder leading to antibiotic resistance and capsular strains are common, inhibiting phagocytosis (Chapter 20). With UTIs caused by *E. coli*, type I pili are important for them to stick to the luminal surface of the bladder.[25,26] Furthermore, *E. coli* can form biofilm pods inside epithelial bladder cells[27] i.e. intracellular biofilms.

Hydrophobic extracellular amyloid fibres (curli) are important for the mechanical properties of *E. coli* that cause UTIs when they grow at the air/water interface.[29] Interfacial rheometry experiments were performed in combination with Brewster angle microscopy.

Staphylococcus aureus is immotile and thus needs to depend on its host for its dissemination (e.g. via the pumping of blood or flaking of skin). The slow division processes that push subsequent generations of bacteria away from the initial site of infection need to be augmented by the host if the bacterium is to infect new environments. Perhaps surprisingly, *S. aureus* is the leading cause of infections in humans.[30] A series of virulence factors make clinical strains of *S. aureus* resistant to phagocytosis by neutrophils (e.g. they ultimately cause neutrophil lysis), which are upregulated when in contact with neutrophils i.e. global changes in gene expression occur. In *S. aureus* infections, surfactant peptides can also be produced by the bacteria and microcapsules are expressed.[31]

A seventh of the total number of deaths at the time of Robert Koch (~1870) were due to TB and infections with this bacterium have seen a recent resurgence in immunocompromised patients e.g. due to HIV (human immunodeficiency virus). Surprisingly, detailed biophysics research on TB biofilms is relatively recent, perhaps due to their complexity, health and safety issues, and challenges to form good model systems for human lungs. Granulomas occur in TB as the bacteria invade the lung cells of the host and form intracellular biofilm aggregates.[32] Currently, 2 billion people are thought to be infected with TB and 2 million die per year. It is a latent infection in most patients due to the granulomas. Granulysin in cytolytic T cells is important for killing extracellular *Mycobacterium tuberculosis*.[33]

Sepsis is a huge issue with bacterial infections since it is an extremely damaging immunological response of the host.[34] Multiple clinical trials are in progress for pharmaceuticals to reduce mortality from sepsis and septic shock.

Suggested Reading

Gladwin, M.; Trattler, B., *Clinical Microbiology Made Ridiculously Simple*. Medmaster: 2011.

Goering, R. et al., *MIMS Microbiology*, 6th ed. Elsevier: 2018.

Moran, E.; Cooke, F.; Torok, E., *Oxford Handbook of Infectious Diseases and Microbiology*. Oxford University Press: 2017.

Murphy, K.; Weaver, C., *Janeway's Immunobiology*, 9th ed. Garland Science: 2016.

Parham, P., *The Immune System*. Garland Science: 2014.

Smith, P. D.; Blumberg, R. S.; MacDonald, T.T., *Principles of Mucosal Immunology*. Garland Science: 2020.

Strelkauvkas, A.; Edwards, A.; Fahnert, B.; Pryor, G.; Strelkauskas, J., *Microbiology: A Clinical Approach*, 2nd ed. Garland Science: 2016.

Wilson, B. A.; Salyers, A. A.; Whitt, D. D.; Winkler, M. E., *Bacterial Pathogens: A Molecular Approach*, 4th ed. ASM Press: 2019.

Wilson, M., *Bacterial Disease Mechanisms: An Introduction to Cellular Microbiology*. Cambridge University Press: 2010.

References

1. Torok, E.; Moran, E.; Cooke, F., *Oxford Handbook of Infectious Diseases and Microbiology*. Oxford University Press: 2016.
2. Gladwin, M. T.; Trattler, W.; Mahan, C. S., *Clinical Microbiology Made Ridiculously Simple*. 7th ed. MedMaster: 2019.
3. Parham, P., *The Immune System*. Garland Science: 2014.
4. Hooper, L. V.; Littman, D. R.; MacPherson, A. J., Interactions between the microbiota and the immune system. *Science* **2012**, *336* (6086), 1268–1273.
5. Brinkmann, V.; Reichard, U.; Goosmann, C.; Fauler, B.; Uhlemann, Y.; Weiss, D. S.; Weinrauch, Y.; Zychlinky, A., Neutrophil extracellular traps kill bacteria. *Science* **2004**, *303* (5663), 1532–1535.
6. Finlay, B. B.; Cossart, P., Exploitation of mammalian host cell functions by bacterial pathogens. *Science* **1997**, *276* (5313), 718–725.
7. Hetrick, E. M.; Schoenfisch, M. H., Reducing implant-related infections: Active release strategies. *Chemical Society Reviews* **2006**, *35* (9), 780–789.
8. Faustino, C. M. C.; Lemor, S. M. C.; Monge, N.; Ribeiro, I. A. C., A scope at antifouling strategies to prevent catheter-associated infection. *Advances in Colloid and Interface Science* **2020**, *284*, 102230.
9. Filipovic, U.; Dahmane, R. G.; Ghannouchi, S.; Zore, A.; Bohinc, K., Bacterial adhesion on orthopedic implants. *Advances in Colloid and Interface Science* **2020**, *2020* (283), 10228.
10. Cossart, P.; Sansonetti, P. J., Bacterial invasion: The paradigms of enteroinvasive pathogens. *Science* **2004**, *304* (5668), 242–248.
11. Ogawa, M.; Yoshimori, T.; Suzuki, T.; Sagara, H.; Mizushima, N.; Sasakawa, C., Escape of intracellular *Shigella* from autophagy. *Science* **2005**, *307* (5710), 727–731.
12. Costerton, J. W.; Stewart, P. S.; Greenberg, E. P., Bacterial biofilms: A common cause of persistent infections. *Science* **1999**, *284* (5418), 1318–1322.
13. Marcy, Y.; et al., Dissecting biological 'dark matter' with single-cell genetic analysis of rare and uncultivated TM7 microbes from the human mouth. *PNAS* **2007**, *104* (29), 11889–11894.
14. Cugini, C.; Shanmugam, M.; Landge, N.; Ramasubbu, N., The role of exopolysaccharides in oral biofilms. *Journal of Dental Research* **2019**, *98* (7), 739.
15. Mah, T. F.; O'Toole, G. A., Mechanisms of biofilm resistance to antimicrobial agents. *Trends in Microbiology* **2001**, *9* (1), 34–39.
16. Kim, D.; et al., Spatial mapping of polymicrobial communities reveals a precise biogeography associated with human dental caries. *PNAS* **2020**, *117* (22), 12375–12386.
17. Li, Z. R.; et al., Mutanofactin promotes adhesion and biofilm formation of cariogenic Streptococcus mutans. *Nature Chemical Biology* **2021**, *17* (5), 576–584.
18. Smith, E. E.; et al., Genetic adaptation by *Pseudomonas aeruginosa* to the airways of cystic fibrosis patients. *PNAS* **2006**, *103* (22), 8487–8492.

19. Kovach, K.; et al., Evolutionary adaptations of biofilms infecting cystic fibrosis lungs promote mechanical toughness by adjusting polysaccharide production. *npj Biofilms and Microbiomes* **2017**, *3*, 1.
20. Singh, P. K.; Schaefer, A. L.; Parsek, M. R.; Moninger, T. O.; Welsh, M. J.; Greenberg, E. P., Quorum-sensing signals indicate that cystic fibrosis lungs are infected with bacterial biofilms. *Nature* **2000**, *407* (6805), 762–764.
21. Gog, J. R.; et al., Dynamics of Salmonella infection of macrophages at the single cell level. *Journal of Royal Society Interface* **2012**, *9* (75), 2696–2707.
22. Mock, M.; Fouet, A., Anthrax. *Annual Review of Microbiology* **2001**, *55*, 647–671.
23. Flores-Mireles, A. L.; Walker, J. N.; Caparon, M.; Hultgren, S. J., Urinary tract infections: Epidemiology, mechanism of infection and treatment options. *Nature Reviews Microbiology* **2015**, *13* (5), 269–284.
24. Johnson, J. R., Virulence factors in *Escherichia coli* urinary tract infection. *Clinical Microbiology Reviews* **1991**, *4* (1), 80–128.
25. Mulvey, M. A.; Lopez-Boado, Y. S.; Wilson, C. L.; Roth, R.; Parks, W. C.; Heuser, J.; Hultgren, S. J., Induction and evasion of host defenses by type 1-piliated uropathogenic *Escherichia coli*. *Science* **1998**, *282* (5393), 1494.
26. Connell, I.; Agace, W.; Klemm, P.; Schembri, M.; Marild, S.; Svanborg, C., Type 1 fimbrial expression enhances *Escherichia coli* virulence for the urinary tract. *PNAS* **1996**, *93* (18), 9827–9832.
27. Anderson, G. G.; Palermo, J. J.; Schilling, J. D.; Roth, R.; Heuser, J.; Hultgren, S. J., Intracellular bacterial biofilm-like pods in urinary tract infections. *Science* **2003**, *301* (5629), 105–107.
28. King, J. E.; Roberts, I. S., Bacterial surfaces: Front lines in host-pathogen cell-surface interactions. In *Biophysics of Infection*, Leake, M. C., Ed. Springer: 2016; pp. 129–156.
29. Wu, C.; Lim, J. Y.; Fuller, G. G.; Cegelski, L., Quantitative analysis of amyloid-integrated biofilms formed by uropathogenic *Escherichia coli* at the air-liquid interface. *Biophysical Journal* **2012**, *103* (3), 464–471.
30. Voyich, J. M.; et al., Insights into mechanisms used by *Staphylococcus aureus* to avoid destruction by human neutrophils. *Journal of Immunology* **2005**, *175* (6), 3907–3919.
31. Otto, M., Staphylococcal infections: Mechanisms of biofilm maturation and detachment as critical determinants of pathogenicity. *Annual Review of Medicine* **2013**, *64* (1), 175–188.
32. Segovia-Juarez, J. L.; Ganguli, S.; Kirschner, D., Identifying control mechanisms of granuloma formation during *M. tuberculosis* infection using an agent-based model. *Journal of Theoretical Biology* **2004**, *231* (3), 357–376.
33. Stenger, S.; et al., An antimicrobial activity of cytolytic T cells mediated by granulysin. *Science* **1998**, *282* (5386), 121.
34. Cohen, I., The immunogenesis of sepsis. *Nature* **2002**, *420* (6917), 885–891.

26 Systems Biology and Synthetic Biology with Populations of Bacteria

Systems biology on its own can be very useful to describe the physiology of cells (Chapter 11), but in practice it often lacks rigour and contains a fair amount of guesswork. Researchers tend to create indicative models with many floating variables, rather than exact quantitative solutions to physiological problems. One large challenge is the sensitivity of reaction kinetics to molecular environments and standard methodologies assume well-mixed kinetics i.e. spatial and geometrical effects are neglected.

The combination of systems biology and *synthetic biology* has been much more successful, since hypotheses can be tested with quantitative detail in newly engineered organisms where a small number of genes are altered i.e. a quantitative control experiment can be performed. Synthetic biology has been used with *Escherichia coli* cells to demonstrate that quorum sensing can lead to the synchronisation of genetic clocks in a community.[1] This was successfully modelled using a one-dimensional array of delay-differential equations coupled to a field of extracellular quorum molecules. Another notable example was the use of a synthetic biology approach to programme a lawn of *E. coli* cells to form a bullseye pattern based on the insertion of lactone signalling molecules and sensory units.[2]

Further, classic examples of *synthetic biology* include bistable switches[3] (e.g. the toggle switch), oscillators (e.g. the repressilator[4]) and logical operators (e.g. AND, NOR, etc.).[5] This simple functional programming has allowed quantitative models for gene networks to be rigorously tested and the underlying non-linear physics to be understood (Chapter 12).

Bacteria naturally excrete a range of industrially useful biopolymers e.g. cellulose, dextran, xanthan, polyesters, polyphosphates, alginate, hyaluronate, PGA, PHAs, CPS, PolyP and PsI.[6] Bacteria can also contain intracellular granules of polymers, which include polyamides, polyesters and polyphosphates. Biopolymers sourced from bacteria are thus an industrially relevant source of feedstock and provide synthetic biology with a well-defined industrial driver. Foreign molecules can be genetically expressed in *E. coli* and such methodologies are important in the pharmaceutical industry to create drugs. However, there are limits to the molecules that can be expressed in *E. coli* e.g. Chinese hamster cells are needed to create monoclonal antibodies.

Robust methods to understand the effect of heterogeneity on reactions in time and space in systems biology are still fairly limited e.g. to understand the subdiffusive transport of cargoes inside cells[7] (Chapter 1). Systems biology approaches that explicitly include the subdiffusive spatial dependence of kinetics still remain at a basic level of development i.e. the math is challenging and the experiments are tricky to perform *in vivo*, since accurate sampling of the ensemble behaviour with a sufficient number of measurements is required and most bacteria are very small in size.[8] Well-defined positioning and compartmentalisation occur for many molecules inside the environment of the bacterial cell and they occur at extremely high concentrations, so subdiffusion is an inevitable phenomenon.[9] The principles driving the organisation of intracellular regions inside bacteria are still badly understood due to the small length scales involved (a wide range of cytoskeletal proteins are implicated).[10] High-resolution microscopy techniques, such as super-resolution fluorescence microscopy, are only starting to provide sufficiently detailed data sets to solve some of the problems involved.

Biofilm gene expression can complicate the phenotypes observed in bacteria. Thus, gene expression needs to be considered in the context of biofilm formation in synthetic biology studies. Spatial and temporal patterns of expression of quorum-sensing genes have been observed in *Vibrio fischeri* cultures.[11] They are determined by a combination of diffusion, bacterial growth and regulatory feedback. A variety of useful systems biology models exist to describe quorum sensing,[12,13] synchronisation, communication, spatial patterning, pulse generation, logical gates and signal transduction.

Amyloid created in *E. coli* biofilm has been modified using synthetic biology techniques.[14] Gellan and xanthan are produced in bacterial biofilms and are frequently used in the food industry.[15]

Fluorescence microscopy experiments with *Bacillus subtilis* expressing a luciferase (luminescent) reporter indicated the existence of a *circadian clock*.[16] Circadian clocks are well established in photosynthetic bacteria, but their discovery in non-photosynthetic bacteria, such as *B. subtilis*, is more surprising. Such clocks will have important generic implications for bacterial physiology if they can be rigorously established.

The *stochastics of gene expression* were measured on the single-cell level with *E. coli* expressing β-galactosidase.[17] Dedicated simulation techniques based on stochastic expression in systems biology often use the Gillespie algorithm.[18] This provides an accurate numerical method to handle fluctuations in systems modelled with differential equations.

Suggested Reading

Alon, U., *An Introduction to Systems Biology*. Chapman and Hall: 2019.

Forger, D. B., *Biological Clocks, Rhythms and Oscillations, the Theory of Biological Time Keeping*. MIT Press: 2017.

Golding, I.; Cox, E.C., Physical nature of bacterial cytoplasm. *Physical Review Letters* **2006**, *96* (9), 098102. Evidence for sub-diffusion in bacteria. Many other studies indicate this is a common phenomenon.

Ingalls, B. P. *Mathematical Modelling in Systems Biology*. MIT Press: 2013.

References

1. Danino, T.; Mandragon-Palomino, O.; Tsimring, L.; Hasley, J., A synchronised quorum of genetic clocks. *Nature* **2010**, *463* (7279), 326–330.
2. Basu, S.; Gerchman, Y.; Collins, C. H.; Arnold, F. H.; Weiss, R., A synthetic multicellular system for programmed pattern formation. *Nature* **2005**, *434* (7037), 1130–1134.
3. Gardner, T. S.; Cantor, C. R.; Collins, J., Construction of a genetic toggle switch in *Escherichia coli*. *Nature* **2000**, *403* (6767), 339–342.
4. Elowitz, M. B.; Leibler, S., A synthetic oscillatory network of transcriptional regulators. *Nature* **2000**, *403* (6767), 335–338.
5. Alon, U., *An Introduction to Systems Biology: Design Principles of Biological Circuits*, 2nd ed. CRC Press: 2020.
6. Moradali, M. F.; Rehan, B. H. A., Bacterial biopolymers: From pathogenesis to advanced materials. *Nature Reviews Microbiology* **2020**, *18* (4), 195–210.
7. Hofling, F.; Franosch, T., Anomalous transport in the crowded world of biological cells. *Reports on Progress in Physics* **2013**, *76* (4), 046602.
8. Mendez, V.; Fedotov, S.; Horsthemke, W., *Reaction-transport Systems: Mesoscopic Foundations, Fronts and Spatial Instabilities*. Springer: 2012.
9. Golding, I.; Cox, E. C., Physical nature of bacterial cytoplasm. *Physical Review Letters* **2006**, *96* (9), 098102.
10. Ramos-Leon, F.; Ramamurthi, K. S., Cytoskeletal proteins: Lessons learned from bacteria. *Physical Biology* **2022**, *19* (2), 021005.
11. Patel, K.; Rodriguez, C.; Stabb, E. V.; Hagen, S. J., Spatially propagating activation of quorum sensing in *Vibrio fischeri* and the transition to low population density. *Physical Review E* **2020**, *101* (6-1), 062421.
12. James, S.; Nilsson, P.; James, G.; Kjelleberg, S.; Fagerstrom, T., Luminescence control in the marine bacterium *Vibrio fischeri*: An analysis of the dynamics of lux regulation. *Journal of Molecular Biology* **2000**, *296* (4), 1127–1137.
13. Ingalls, B. P., *Mathematical Modeling in Systems Biology: An Introduction*. MIT Press: 2013.
14. Chen, A. Y.; Deng, Z.; Billings, A. N.; Seker, U. O. S.; Lu, M. Y.; Citorik, R. J.; Zakeri, B.; Lu, T. K., Synthesis and patterning of tunable multiscale materials with engineered cells. *Nature Materials* **2014**, *13* (5), 515–523.
15. Morris, E. R.; Nishinari, K.; Rinaudo, M., Gelation of gellan – a review. *Food Hydrocolloids* **2012**, *28* (2), 373–411.
16. Eelderink-Chen, Z.; Bosman, J.; Sartor, F.; Dodd, A. N.; Kovacs, A. T.; Merrow, M., A circadian clock in a nonphotosynthetic prokaryote. *Science Advances* **2021**, *7* (2), eabe2086.
17. Cai, L.; Friedman, N.; Xie, S., Stochastic protein expression in individual cells at the single molecule level. *Nature* **2006**, *440* (7082), 358–362.
18. Gillespie, D. T., Exact stochastic simulation of coupled chemical reactions with delays. *The Journal of Physical Chemistry* **1977**, *25* (12), 2340–2361.

27 Simulations of Cells and Biofilms

Bacterial biofilms are extremely complex entities (similar in some respects to the complex tissues of multicellular eukaryotic organisms, although cell differentiation is less well defined), so it is clear that simple analytical theories will only be useful approximations in fairly restrictive scenarios.[1] Instead, a clear case can be made for simulations of biofilms to include more realistic details in models, particularly when non-linear phenomena are observed (where analytic solutions for even compact toy models are the exception not the rule).[2]

Simulations of bacteria that include all their molecular details are still far from being computationally tractable and it is doubtful that they ever will be due to a hierarchy of intractable many-body problems and non-linear effects. Molecular dynamics simulations are useful to understand small numbers of molecules e.g. the interaction of antimicrobial peptides with membranes or the docking of an antibody with an antigen. Molecular dynamics simulations have been reviewed in the literature satisfactorily.[4,5] For larger systems, different simulation tools are required. Recent successes have been seen in simulations of biofilms that focus on *agent based models* (ABMs) in which an array of agents (the bacteria) are prepared that interact with one another using an identical set of rules, which can include stochastic factors[2,6] (Figure 27.1).

The terminology, ABM, is used interchangeably with individual-based models (IBMs) in the literature. The first group to use agent-based modelling with biofilms was led by Jan Kreft[7,8] (BacSim, 1998). However, ABMs were first used in ecology in the early 1970s,[9] so they were reasonably well established in the biological modelling community before their adoption in biofilm research. Both the continuum and discrete approaches to the computational modelling of biofilms have been reviewed in detail.[10] In general, ABMs are successful for simulations of the emergent collective behaviour of large ensembles of agents where the agents act as discrete units.[2] Typically, the agents are assumed to all follow the same simple rules in the simulations for their interactions with neighbouring individuals. This simplicity allows many emergent phenomena to be motivated in a transparent manner that would otherwise be intractable due to the complex behaviours of the agents and thus the high computational costs.[2] ABMs can allow the modelling of the interactions of people in a society, animals in an ecosystem or bacteria in a biofilm.

A distinction is made between coarse-grained ABM simulations and more conventional techniques used to simulate colloids (Chapter 7), such as dissipative particle dynamics (DPD), Monte Carlo and molecular dynamics.[11] ABMs can provide many useful insights due to the simple rules chosen for the agents, but they risk

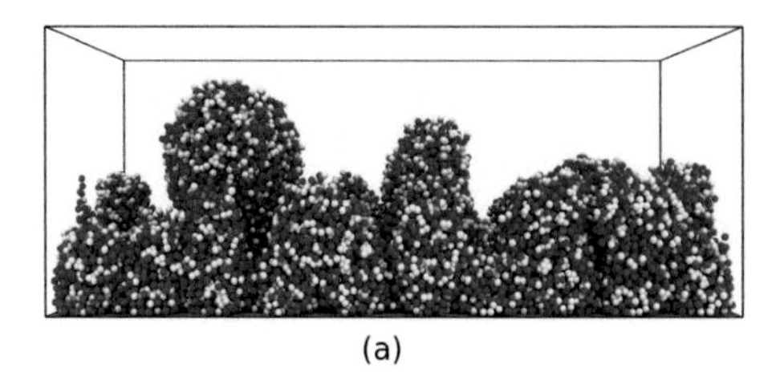

(a)

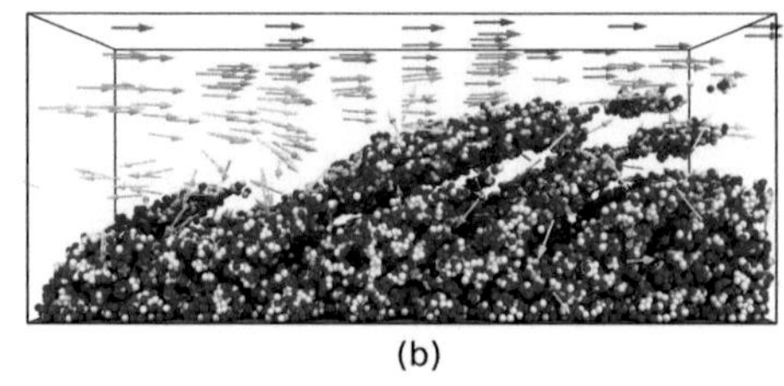

(b)

Figure 27.1 An *agent-based model* (ABM) for the deformation of biofilms under shear flow. (a) No shear flow. (b) Shear flow.[3]

missing many of the fundamental details of the physical interactions, since they are relatively naïve coarse-grained approaches. However, colloidal simulations can be too slow to provide sufficient detail to describe biofilms and fundamental problems still exist in simulating the glassy, non-ergodic phenomena that commonly occur in both colloidal and polymeric gels that are thought to be relevant to biofilm research (Chapters 6 and 7). Further challenges to theoretical modelling are provided by the active nature of the growth processes that occur in bacterial biofilms. Significant challenges are presented for simulations of active condensed matter[12,13] at the level of single cells, even without considering cellular interactions. Thus, it is hard to include all the relevant mesoscopic physics in simulations and often agent-based modelling only manages to be reasonably accurate in handling some components e.g. the excluded volume interactions of each individual bacterium due to the large computational requirements i.e. the bacteria are unable to overlap. Thus, the majority of the interactions included in ABM are empirical approximations. However, ABM often gives reasonably realistic descriptions of biofilms by only considering simple rules for excluded volume and the division of the bacterial cells during growth. They are thus useful minimal models.

The initial BacSim software[7] introduced a *shoving algorithm* to update bacterial positions when the cells are divided and similar algorithms are common in other bacterial ABMs. A challenge is to make this process less arbitrary and give it more robust physical foundations. Shoving algorithms are needed in the majority of ABMs to model bacterial growth (Figure 27.2). They need to be given more fundamental foundations via mechanistic descriptions of bacterial division and the response of neighbouring cells.

Multiscale ABMs are often integrated with systems biology packages, which can describe gene networks inside the bacteria.[14] Well-mixed kinetics are assumed for the intracellular components in the systems biology calculations (Chapter 11), whereas the spatial dependence of the kinetics is more accurately described at the scale of single cells using the ABM.

Cellular automata have a long history for the simulation of biofilms and can be considered a discrete lattice equivalent of ABMs.[6] They have advantages in terms of computational speed, but it is hard to make them particularly realistic due to their discrete nature, so they have become less common compared with continuous ABMs for bacterial simulations. Historically, Conway's game of life, a type of

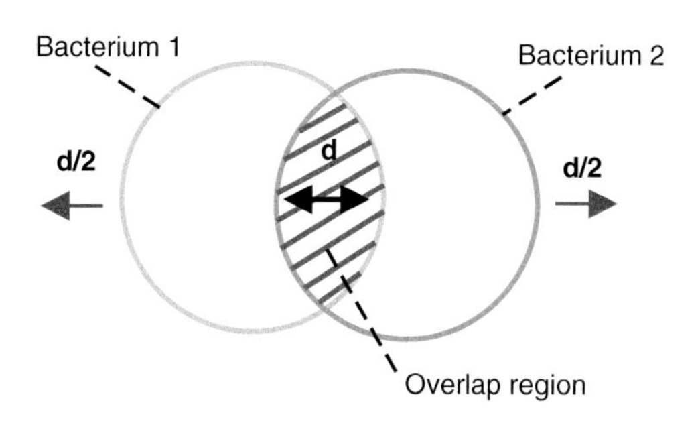

Figure 27.2 The *shoving algorithm* moves the cells in equal amounts to relax the overlap condition during a new time step in an ABM e.g. after the divisions of cells. For large numbers of overlapping cells, the cells are moved in the opposite direction of the vector sums of their overlap displacements δ is the size of the overlap region.

cellular automata, played an important role as an archetypal model in mathematical biology, since simple computational rules were seen to give rise to complex emergent phenomena.[15] Hybrid ABMs exist with both discrete and continuous components e.g. off-lattice positions of the bacteria with on-lattice diffusion calculations of the chemicals they release. Cellular automata models are able to describe the growth of channels and the mushroom structures of biofilms.[16]

The effects of extracellular polymeric substance (EPS) and capsules on biofilm structure were simulated with an extended version of the ABM BacSim.[17] Inclusion of realistic motility mechanisms and aerotaxis have also been successful in modelling *Bacillus subtilis* biofilms at the air/water interface.[18]

ABMs can be conveniently combined with decision trees from machine learning.[15] A much wider range of hybrid machine learning with ABMs is possible and this could be a rich area of future study (Chapter 14).

Diffusion-limited aggregation (DLA) was initially developed by physicists working with synthetic colloids and provides a simple mechanism that leads to characteristic fractal-like structures observed in some biofilms and bacterial aggregates (Chapter 7) i.e. sticky particles move via diffusion until they make contact with another particle at which point they stick together to form large aggregates.[19] Most bacterial motion is not purely diffusive, but at long time scales, the motion can approach Brownian scaling with time (Equation (2.11); although the effective diffusion coefficient is much larger than that due to thermal forces), so DLA might be expected to be a reasonable model in the long time limit for some bacterial aggregates. Thus, DLA has had some success in describing biofilm aggregation behaviour.[20] Simple rules for bacterial aggregation can thus give rise to complex self-similar fractal morphologies (Figure 7.3).

The *Vicsek model* is another archetypal ABM used to understand flocking transitions in assemblies of motile agents.[21] In the model, for each time step, an agent is driven with a constant speed and it adopts the average direction of motion of the agents in its neighbourhood (within a radius r) and some random perturbations are added to its direction $(\Delta\theta)$ at each time step,

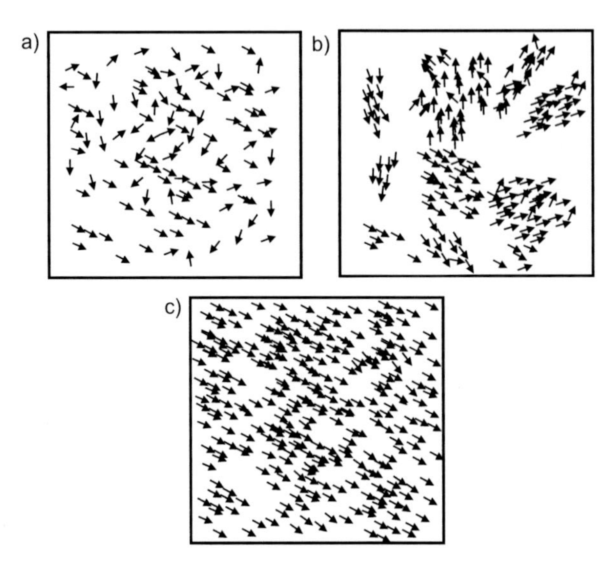

Figure 27.3 The *Vicsek model* can describe dynamic phase transitions of motile agents in the bulk.[21] (a) Initial arrangement of swimmer particles. (b) Small densities of particles and small angular noise terms $(\Delta\theta)$ create coherently swimming flocks. (c) Higher densities of particles and small angular noise terms creates ordered phases with a single direction of motion.

$$x_i(t+1) = x_i(t) + v_i(t)\Delta t, \quad (27.1)$$

$$\theta(t+1) = \langle\theta(t)\rangle_r + \Delta\theta, \quad (27.2)$$

where $x_i(t)$ is the position of agent i at time step t, θ is the direction of the agent at time step t and $v_i(t)$ is the velocity of the agent at time step t. Δt is the time increment. Some of the resultant flocking behaviour is shown in Figure 27.3 and dynamic phase transitions can occur for the swimmers.

Agent based models have been used to explore why at a high cell density, some bacteria activate polymer secretion, whereas others stop secreting polymers.[22] It is thought to be a key difference between chronic long-lived infections and acute short-lived infections caused by bacterial biofilms.

Another ABM was used to study the evolution of biofilm-producing strains in mixed-strain systems. They found polymer secretion into the EPS is altruistic (Section 22.4) to the upper cells closer to the interfaces, since it pushes higher generations into better oxygen conditions, but it suffocates non-polymer producers.[23]

ABMs have been used to investigate the inhibition of quorum sensing in biofilms (Section 18.3). Inhibition is found to depend on the form of positive feedback in the quorum sensing circuit and the time at which the inhibitory molecules are added is found to be crucial.[24]

Surface roughening with and without flow was modelled with ABMs that included intracellular adhesive linkages.[25] A sophisticated ABM based on LAMMPs (a molecular dynamics package) was used to simulate changes in three-dimensional (3D) biofilms under shear including flattening of the biofilm and detachment of the biofilm.[26]

The surfing of faster-growing strains of *Escherichia coli* in growth fronts during cell culture has been modelled.[27] The phenomenon is thought to be a general one in rod-shaped microorganisms growing in two dimensions.

ABMs were used to explain microcolony formation with *Pseudomonas aeruginosa* biofilms. Microcolonies are thought to be a motility-driven phenomenon, since their motility is reduced by the self-produced polymers creating the microcolonies.[28,29]

A two-dimensional ABM (called GRO) was developed that focuses on synthetic biology aspects of bacterial communication.[30] It was used to model electrical signalling in *B. subtilis* biofilms.[31] Simbiotics is a more recent development (2017) for the simulation of 3D biofilms.[32] It includes more extensive cell geometries, physical force dynamics, genetic circuits, metabolic pathways, chemical diffusion and cell interactions. BSim is another specialist platform for simulating bacteria and was used to describe the electrical signalling of *E. coli* biofilms in three dimensions.[33]

Agent based models were used to investigate the mechanisms of bacterial aggregation. Specifically, *Myxococcus xanthus* was explored, which has a complex life cycle including multi-cellular mounds, spores and fruiting bodies.[34]

The spreading behaviour of initially aggregated bacterial cells that nucleate biofilms was modelled.[35] However, there is lots more species-specific biology that probably needs to be included in such models to result in a very realistic representation.

Reaction–diffusion phenomena can be conveniently included in most ABMs (Chapter 5). Reactions are described using standard models of chemical kinetics (following the law of mass action or Hill-type enzyme kinetics) and diffusion is typically calculated from Fick's second law numerically solved on a lattice (Equation (2.4)). Extensions of ABMs to describe anomalous transport of whole cells are fairly limited in the literature e.g. multifractal FBM[36] is not yet well established (Chapter 2). Furthermore, it would be good to extend the modelling of fractional diffusion to generalise the Fickian models for intracellular transport.

IdynoMics is another popular 3D agent-based simulator for biofilms and is an update based on BacSim.[37] It has improved handling of the pressure field, allows continuous secretion of EPS, enables comparison of spatially heterogeneous systems and allows condition dependent switching of individual cell metabolisms.

BacArena is an R-based ABM that allows detailed modelling of bacterial metabolism. Spatial differentiation of *P. aeruginosa* in biofilms fed with different fermentation products could be modelled. A seven-species model of bacteria in the human gut was used to study *E. coli* overgrowth associated with disease.[38]

The spreading of immotile bacteria on agar surfaces was modelled using ABMs and compared with the Fisher equation ($\frac{\partial c}{\partial t} = D\frac{\partial^2 c}{\partial t^2} + \alpha c(1-c)$, Chapter 5).[39] Colony growth was found to be very sensitive to the initial concentration of nutrients. Transitions occurred from finger-like to branched structures during bacterial growth and the roughness decreased with initial nutrient concentration.

Suggested Reading

It is reasonably straightforward to start simulating a biofilm using an ABM. It only requires a couple of hours to download some freeware and to run some simple simulations from scratch on a standard pc.

Mounfield, C. C., *The Handbook of Agent Based Modelling*. Independent Publishing: 2020. Useful collection of models discussed from a physics perspective.

O'Sullivan, D.; Perry, G. L. W., *Spatial Simulation: Exploring Patterns and Processes*, 1st ed. Wiley: 2013. Useful introductory approach for the development of agent based models to solve physics problems.

Wilensky, U.; Rand, W., *An Introduction to Agent Based Modelling: Modelling Natural, Social and Engineered Complex Systems with NetLogo*. MIT Press: 2015. Focuses on NetLogo. Has the advantage that it is in three dimensions and is mathematically simple. It is accessible to non-scientists.

References

1. Dzianach, P. A.; Dykes, G. A.; Strachan, N. J. C.; Forbes, K. J.; Perez-Reche, F. J., Challenges of biofilm control and utilisation: Lessons from mathematical modelling. *Journal of the Royal Society – Interface* **2019**, *16* (155), 20190042.
2. Jensen, H. J., *Complexity Science: The Study of Emergence*. Cambridge University Press: 2023.
3. Li, B.; et al., NUFEB: A massively parallel simulator for individual-based modelling of microbial communities. *PLOS One* **2019**, *15* (12), e1007125.
4. Karplus, M.; McCammon, J. A., Molecular dynamics simulations of biomolecules. *Nature Structural Biology* **2002**, *9* (9), 646–652.
5. Frenkel, D.; Smit, B., *Understanding Molecular Simulation: From Algorithms to Applications*. Academic Press: 2001.
6. Mounfield, C. C., *The Handbook of Agent Based Modelling*. Independent Publishing: 2020.
7. Kreft, J. U.; Booth, G.; Wimpenny, J. W. T., BacSim, a simulator for individual-based modelling of bacterial colony growth. *Microbiology* **1998**, *144* (12), 3275–3287.
8. Kreft, J. U.; Picioreanu, C.; Wimpenny, J. W. T.; van Loosdrecht, M. C. M., Individual-based modelling of biofilms. *Microbiology* **2001**, *147* (11), 2897.
9. Grimm, V., *Individual-based Modeling and Ecology*. Princeton University Press: 2005.
10. Mattei, M. R.; Frunzo, L.; D'Acunto, B.; Pechaud, Y.; Pirozzi, F.; Esposito, G., Continuum and discrete approach in modeling biofilm development and structure: A review. *Journal of Mathematical Biology* **2018**, *76* (4), 945–1003.

11. Allen, M. P.; Tildesley, D. J., *Computer Simulation of Liquids*. Oxford University Press: 2017.
12. Marchetti, M. C.; Joanny, J. F.; Ramaswamy, S.; Liverpool, T. B.; Prost, J.; Rao, M.; Simha, R. A., Hydrodynamics of soft active matter. *Review of Modern Physics* **2013**, *85* (3), 1143.
13. Pismen, L., *Active Matter Within and Around Us: From Self-propelled Particles to Flocks and Living Forms*. Springer: 2021.
14. Latif Jr, M.; May, E. E., A multiscale agent-based model for the investigation of *E. coli* K12 metabolic response during biofilm formation. *Bulletin of Mathematical Biology* **2018**, *80* (11), 2917–2956.
15. Wilensky, U.; Rand, W., *An Introduction to Agent-based Modeling: Modeling Natural, Social and Engineered Complex Systems with NETLogo*. MIT Press: 2015.
16. Picioreanu, C.; Van Loosdrecht, M. C. M.; Heijnen, J. J., Mathematical modeling of biofilm structure with a hybrid differential-discrete cellular automaton approach. *Biotechnology and Bioengineering* **1998**, *58* (1), 101–116.
17. Kreft, J. U.; Wimpenny, J. W. T., Effect of EPS on biofilm structure and function as revealed by an individual-based model of biofilm growth. *Water Science and Technology* **2001**, *43* (6), 135–141.
18. Ardre, M.; Henry, H.; Douarche, C.; Plapp, M., An individual-based model for biofilm formation at liquid surfaces. *Physical Biology* **2015**, *12* (6), 66015.
19. WItten, T. A.; Sander, L. M., Diffusion-limited aggregates, a kinetic critical phenomenon. *Physical Review Letters* **1981**, *47* (19), 1400.
20. Wang, Q.; Zhang, T., Review of mathematical models for biofilms. *Solid State Communications* **2010**, *150* (21), 1009–1022.
21. Vicsek, T.; Czirok, A.; Ben-Jacob, E.; Cohen, I.; Shochet, O., Novel type of phase transition in a system of self-driven particles. *Physical Review Letters* **1995**, *75* (6), 1226–1229.
22. Nadell, C. D.; Xavier, J. B.; Levin, S. A.; Foster, K. R., The evolution of quorum sensing in bacterial biofilms. *PLOS Biology* **2008**, *6* (1), e14.
23. Xavier, J. B.; Foaster, K. R., Cooperation and conflict in microbial biofilms. *PNAS* **2007**, *104* (3), 876–881.
24. Fozard, J. A.; Lees, M.; King, J. R.; Logan, B. S., Inhibition of quorum sensing in a computational biofilm simulation. *Biosystems* **2012**, *109* (2), 105–114.
25. Head, D. A., Linear surface roughness growth and flow smoothening in a three-dimensional biofilm model. *Physical Review E* **2013**, *88* (2), 032702.
26. Jayathilake, P. G.; et al., A mechanistic individual-based model of microbial communities. *PLOS One* **2017**, *12* (8), e0181965.
27. Farrell, F.; Hallatschek, O.; Marenduzzo, D.; Waclaw, B., Mechanically driven growth of quasi-two-dimensional microbial colonies. *Physical Review Letters* **2013**, *111* (16), 168101.
28. Mabrouk, N.; Deffuant, G.; Tolker-Nielsen, T.; Lobry, C., Bacteria can form interconnected microcolonies when a self-excreted product reduces their surface motility: Evidence from individual-based model simulations. *Theory Biosciences* **2010**, *129* (1), 1–13.

29. Johnson, L. R., Microcolony and biofilm formation as a survival strategy for bacteria. *Journal of Theoretical Biology* **2008**, *251* (1), 24–34.
30. Jang, S. S.; Oishi, K. T.; Egbert, R. G.; Klavins, E., Specification and simulation of synthetic multicelled behaviors. *ACS Synthetic Biology* **2012**, *1* (8), 365–374.
31. Blee, J. A.; Roberts, I. S.; Waigh, T. A., Spatial propagation of electrical signals in circular biofilms: A combined experimental and agent-based fire-diffuse-fire study. *Physical Review E* **2019**, *100* (5-1), 052401.
32. Naylor, J.; Fellermann, H.; Ding, Y.; Mohammed, W. K.; Jakubovics, N. S.; Mukherjee, J.; Biggs, C. A.; Wright, P. C.; Krasnogor, N., Simbiotics: A multiscale integrative platform for 3D modeling of bacterial populations. *ACS Synthetic Biology* **2017**, *6* (7), 1194–1210.
33. Akabuogu, E. U.; Martorelli, V.; Krasovec, R.; Roberts, I. S.; Waigh, T. A., Emergence of ion-channel mediated electrical oscillations in *Escherichia coli* biofilms. *eLife* **2023**, to appear.
34. Zhang, Z.; Igorshin, O. A.; Cotter, C. R.; Shimkets, L. J., Agent-based modeling reveals possible mechanisms for observed aggregation cell behaviors. *Biophysical Journal* **2018**, *115* (12), 2499–2511.
35. Melaugh, G.; Hutchison, J.; Kragh, K. N.; Irie, Y.; Roberts, A.; Bjarnsholt, T.; Diggle, S. P.; Gordon, V. D.; Allen, R. J., Shaping the growth behaviour of biofilms initiated from bacterial aggregates. *PLOS One* **2016**, *91* (3), e0149683.
36. Korabel, N.; Clemente, G. D.; Han, D.; Feldman, F.; Millard, T. H.; Waigh, T. A., Hemocytes in Drosophila melanogaster embryos move via heterogeneous anomalous diffusion. *Communications Physics* **2022**, *5* (1), 269.
37. Lardon, L. A.; Merkey, B. V.; Martins, S.; Dotsch, A.; Picioreanu, C.; Kreft, J. U.; Smets, B. F., iDynoMiCS: Next-generation individual-based modelling of biofilms. *Environmental Microbiology* **2011**, *13* (9), 2416–2434.
38. Bauerle, E.; Zimmermann, J.; Baldini, F.; Thiele, I.; Kaleta, C., BacArena: Individual-based metabolic modeling of heterogeneous microbes in complex communities. *PLOS Computational Biology* **2017**, *13* (5), e1005544.
39. Rana, N.; Ghosh, P.; Perlekar, P., Spreading of nonmotile bacteria on a hard agar plate: Comparison between agent-based and stochastic simulations. *Physical Review E* 2017, *96* (5-1), 52403.

28 Communication between Bacteria

Most bacteria are social organisms and they communicate with one another. A dramatic example is when bacteria form biofilms via *quorum sensing* in which the bacteria vote on the suitability of a new environment to establish new homes by releasing signalling molecules. If the signalling molecule concentration is above a threshold value, a quorum is passed and the bacteria commit themselves to creating a biofilm. This in turn implies new possibilities for the treatment of biofilms are possible, such as disturbing the process of quorum sensing to disrupt the creation of the biofilm i.e. *quorum quenching*.[1]

Quorum sensing is not restricted to the formation of biofilms with bacteria. It also occurs in other forms of gene regulation such as with virulence factors, motility, bioluminescence, nitrogen fixation and sporulation.

Two typical types of small molecules (called autoinducers, since they affect both themselves and their neighbours) used in quorum sensing are *acyl homoserine lactones* for Gram-negative bacteria and *modified oligopeptides* for Gram-positive bacteria.[2] Membrane proteins in a two-component signalling system are used to detect oligopeptide autoinducers in Gram-positive bacteria (Chapter 18), which are subsequently transduced using a phosphorylation cascade.

The *speed of communication* in bacterial systems is normally diffusion limited by the transport of small molecules and ions. Their diffusion coefficients are relatively large, so communication is reasonably quick over small length scales. Bacteria do not appear to directly use luminescence for communication among themselves, although they are involved in luminescent signalling in communities of bobtail squid (a symbiotic effect). Furthermore, the *milky sea* phenomenon due to luminescent bacteria (over areas >100 000 km^2) is not fully understood.[3] Spatial light detection has been demonstrated in single cyanobacteria, but it is used for phototaxis rather than communication i.e. it allows them to respond to spatial gradients in light. Bacteria can harvest electricity using conducting pili and long-range conducting cable bacteria also exist (Chapter 29), but no role in communication has yet been established. Waves of charged potassium ions have been observed with *Bacillus subtilis* and *Escherichia coli* biofilms (Section 29.2). The speed of communication is limited by the diffusion coefficient of the potassium ions, which is only slightly faster than standard quorum molecules (the ions are slightly smaller in size). This discovery of electrochemical signalling using ion channels in bacterial cells is fairly recent (Chapter 18) and big advances are expected in this area, since most bacteria contain many varieties of ions channels, which to date have badly defined physiological roles, including voltage-gated varieties that are implicated in signalling.

Quorum quenching is a strategy in which the quorum sensing signalling processes required to initiate biofilm formation are interfered with.[1,4] Small volatile molecules (e.g. butanol) released by *B. subtilis* were found to inhibit the biofilms of competing *E. coli*.[5] Furthermore, some bacteria eat quorum sensing molecules of rival species. Extracellular quorum sensing is tightly bound to intracellular signalling pathways. For example, cyclic dimeric guanosine monophosphate (c-di-GMP) is thought to play an important role.[6]

In addition to protein (gene) circuits that regulate quorum sensing, small RNA molecules often play an important role e.g. in *Vibrio cholerae* and *Vibrio harveyi*.[7] More developments are expected in this area as the RNA-based circuits are understood in quantitative detail, since it is a large emergent area of systems biology.

Predatory bacteria use chemical signalling systems similar to quorum sensing to coordinate attacks on other organisms and to decide whether to form fruiting bodies.[8]

Optogenetics techniques can allow events in the communication processes of bacteria to be followed at the level of individual cells.[9] The role of communication in the branching patterns formed by lubricating bacteria was highlighted.[10]

Sometimes vesicles are used to transport cargoes between bacteria e.g. with *Pseudomonas aeruginosa*.[11] The cargoes are called exosomes in this case and analogous systems are often observed in multicellular eukaryotic organisms.

Host-bacteria signalling is also important for both symbiosis and parasitism but has not yet received much quantitative research.[12] There is evidence for bacteria-host communication with enterohemorrhagic *E. coli*.[13]

28.1 Quorum Sensing

There is a wide range of *signalling pathways* in bacteria for small molecules.[2] Some examples of molecules specifically used in quorum sensing are shown in Table 23.4.

Historically, quorum sensing was first explored in an exotic marine system i.e. chemiluminescence of bobtail squid due to bacteria (*Vibrio fischeri*).[14] Quorum sensing is now known to occur in both Gram-positive and negative bacteria. Gram-negative bacteria tend to use acetylated homoserine lactones as autoinducers, whereas Gram-positive bacteria use oligopeptides.[15] Communication can occur between different bacterial species that form biofilms, if they use a common signalling molecule. Quorum sensing also occurs in fungal biofilms.[16] Surprisingly, even viruses that infect *B. subtilis* use quorum sensing with small peptides in the bacterial biofilms to coordinate lysogeny and lytic cycles.[17,18] Eukaryotic microorganisms that cause sleeping sickness (parasitic protozoans called trypanosomes) are also found to use quorum sensing at a stage in their life cycle.[19]

Bacteria often use multiple signalling molecules during quorum sensing and have multiple sensing circuits arranged either in parallel or competitively[20] e.g. *B. subtilis*, *P. aeruginosa*, *V. cholerae* and *V. harveyi*. Prokaryote-prokaryote and eukaryote-prokaryote quorum quenching are both well-established mechanisms used to disrupt biofilm formation[1] e.g. biofilm formation is inhibited by the reduction of the concentration of quorum molecules. The opposite process to quorum sensing, *biofilm disassembly*, is triggered by D-amino acids with a range of bacteria.[21] Collective responses of quorum-sensing bacteria have been considered from an evolutionary perspective.[22]

Different genetic pathways in bacteria are switched on depending on the nature of the surface.[23] There is a complex interplay of sensory information with quorum sensing strategies. A mathematical model for quorum sensing is described in Section 18.3.

Voltammetry has detected quorum-sensing molecules that are charged.[24] Other standard molecular biology techniques (nuclear magnetic resonance, mass spectrometry, etc.) are typically used for the analysis of quorum molecules.

Suggested Reading

Forger, D. B., *Biological Clocks, Rhythms and Oscillations: The Theory of Biological Timekeeping*. MIT Press: 2017.

Strogatz, S. H. *Non-linear Dynamics and Chaos with Applications to Physics, Biology, Chemistry and Engineering*, 2nd ed. Westview Press: 2015. Coupling of non-linear oscillators is considered in detail.

References

1. Grandclement, C.; Tannieres, M.; Morera, S.; Dessaux, Y.; Faure, D., Quorum quenching: Role in nature and applied developments. *FEMS Microbiology Reviews* **2016**, *40* (1), 86–116.
2. Camilli, A.; Bassler, B. L., Bacterial small-molecule signaling pathways. *Science* **2006**, *311* (5764), 1113–1116.
3. Miller, S. D.; Haddock, S. H. D.; Straka III, W. C.; Seaman, C. J.; Combs, C. L.; Wang, M.; Shi, W.; Nam, S., Honing in on bioluminescent milky seas from space. *Scientific Reports* **2021**, *11* (1), 15443.
4. Paluch, E.; Rewak-Soroczynska, J.; Jedrusik, I.; Mazurkiewi, E.; Jermakow, K., Prevention of biofilm formation by quorum quenching. *Applied Microbiology Biotechnology* **2020**, *104* (5), 1871–1881.
5. Hou, Q.; Keren-Paz, A.; Korenblum, E.; Oved, R.; Malitsky, S.; Kolodkin-Gal, I., Weaponizing volatiles to inhibit competitor biofilms from a distance. *npj Biofilms and Microbiomes* **2021**, *7* (1), 2.
6. Hengge, R., Principles of c-di-GMP signalling in bacteria. *Nature Reviews Microbiology* **2009**, *7* (4), 263–273.
7. Lenz, D. H.; Mok, K. C.; Lilley, B. N.; Kulkarni, R. V.; Wingreen, N. S.; Bassler, B. L., The small RNA chaperone Hfq and multiple small RNAs control quorum sensing in *Vibrio harveyi* and *Vibrio cholerae*. *Cell* **2004**, *118* (1), 69–82.

8. Munoz-Dorado, J.; Marcos-Torres, F. J.; Garcia-Bravo, E.; Moraleda-Munoz, A.; Perez, J., Myxobacteria: Moving, killing, feeding, and surviving together. *Frontiers in Microbiology* **2016**, *7*, 781.
9. Lindner, F.; Diepold, A., Optogenetics in bacteria – applications and opportunities. *FEMS Microbiology Reviews* **2022**, *46* (2), 1–17.
10. Ben-Jacob, E.; Levine, H., Self-engineering capabilities of bacteria. **2006**, *3* (6), 197–214.
11. Gill, S.; Catchpole, R.; Forterre, P., Extracellular membrane vesicles in the three domains of life and beyond. *FEMS Microbiology Reviews* **2019**, *43* (3), 273–303.
12. Baker, B.; Zambryski, P.; Staskawicz, B.; Dinesh-Kumar, S. P., Signaling in plant-microbe interactions. *Science* **1997**, *276* (5313), 726–733.
13. Sperandio, V.; Torres, A. G.; Jarvis, B.; Nataro, J. P.; Kaper, I. B., Bacteria-host communication: The language of hormones. *PNAS* **2003**, *100* (15), 8951–8956.
14. Fuqua, W. C.; Winons, S. C.; Greenberg, E. P., Quorum-sensing in bacteria the luxr-luxi family of cell density-responsive transcriptional regulations. *Journal of Bacteriology* **1994**, *176* (2), 269–271.
15. Miller, M. B.; Bassler, B. L., Quorum sensing in bacteria. *Annual Review of Microbiology* **2001**, *55*, 165–199.
16. Padder, S. A.; Prasad, R.; Shah, A. H., Quorum sensing: A less known mode of communication among fungi. *Microbiological Research* **2018**, *210*, 51–58.
17. Erez, Z.; et al., Communication between viruses guides lysis-lysogeny decisions. *Nature* **2017**, *541* (7638), 488–493.
18. Whiteley, M.; Diggle, S. P.; Greenberg, E. P., Progress in and promise of bacterial quorum sensing research. *Nature* **2017**, *551* (7680), 313–320.
19. Mony, B. M.; Matthews, K. R., Assembling the components of the quorum sensing pathway in African trypanosomes. *Molecular Microbiology* **2015**, *96* (2), 220–232.
20. Waters, C. M.; Bassler, B. L., Quorum sensing: Cell to cell communication in bacteria. *Annual Review in Cellular Developmental Biology* **2005**, *21*, 319–346.
21. Kolodkin-Gal, I.; Romero, D.; Cao, S.; Clardy, J.; Kolter, R.; Losick, R., D-Amino acids trigger biofilm disassembly. *Science* **2010**, *328* (5978), 627–629.
22. Popat, R.; Cornforth, D. M.; McNally, L.; Brown, S. P., Collective sensing and collective responses in quorum-sensing bacteria. *Journal of the Royal Society – Interface* **2015**, *12* (103), 20140882.
23. Davey, M. E.; O'Toole, G. A., Microbial biofilms: From ecology to molecular genetics. *Microbiology and Molecular Biology Reviews* **2000**, *64* (4), 847–867.
24. Oziat, J.; Cohu, T.; Elsen, S.; Gougis, M.; Malliaras, G. G.; Mailley, P., Electrochemical detection of redox molecules secreted by *Pseudomonas aeruginosa* – Part 1: Electrochemical signatures of different strains. *Bioelectrochemistry* **2021**, *140*, 107747.

29 Bacteria and Electricity

All biological cells (eukaryotic, bacterial and archaeal) have electrical potentials across them due to imbalances between the concentrations of charged ions on either side of their membranes (Chapter 3). A minority of human cell types can actively modulate their potentials and are called *excitable cells* e.g. sensory cells, neurons and cardiac cells. Many bacterial cells also appear to be excitable, and large changes in transmembrane potentials are observed in response to external stimuli. For example, there is good data for excitability in *Escherichia coli*, *Bacillus subtilis* and *Pseudomonas aeruginosa*.[1–4] All of these species have voltage-sensitive potassium (K^+) ion channels, and these provide a molecular mechanism for excitability. A large number of other bacterial species contain similar voltage-gated ion channels (although not all) and a wide range of bacteria are therefore expected to be excitable.

Compared with the electrical phenomena in eukaryotic cells (e.g. sensory cells, neurons and cardiac cells), the study of electrophysiology in bacteria has some advantages and disadvantages. Bacterial genetics are much simpler than that of the eukaryotic organisms (including optogenetics techniques), so many of the original studies with ion channels were made with those from bacterial cells (including McKinnon's Nobel Prize-winning work on the structure of the K^+ channel, Kcs A, from *Streptomyces lividans*[5]). However, bacterial cells are an order of magnitude smaller in size than eukaryotic cells, which means direct measurement of voltages via patch clamping is much harder. Furthermore, a direct connection has not yet been made between a diseased state and an electrical phenomenon with bacteria (in contrast to studies of heart disease, deafness, blindness or epilepsy), so research budgets have been more limited, although recent work on the electrophysiology of the antibiotic tolerance of bacteria may change this situation. In conclusion, bacteria clearly demonstrate a range of electrophysiological phenomena, but detailed research is much less extensive than their eukaryotic counterparts due to the lack of direct applications presently and some experimental challenges.

An ecological guiding principle for the evolution of bacteria is that if there is a source of energy in a biological niche, bacteria will evolve to use it[6] e.g. chemicals, sunlight, thermal energy and electricity are all used to power bacteria. Thus all bacteria appear to be electrochemically powered to use such energy sources, and electrochemical redox reactions have been widely developed in bacteria during their evolution.[7] Energy recovery is thus possible, which has been harvested using bacteria, and bacterial materials show promise for high-performance organic electronics.[8]

Gap junctions are another type of ion channel in eukaryotic cell membranes and they play a crucial role in connecting neighbouring cardiac cells, so they can transmit electrical waves in the heart. Gap junctions are known to occur in filamentous cyanobacteria.[9] Whether they play similar roles to those in cardiac tissue within aggregated bacterial systems still needs more research, and their roles in electrical communication need to be explored.

Although *bacterial ion channels* were the first to have detailed X-ray crystallographic structures available,[5] strong structure-function connections to single cell physiology have been slower to develop than with eukaryotic cells.[10] The textbook attribution is that bacterial ion channels play roles in osmoregulation, pH regulation, mechanosensitivity and motility.[7] The role of ion channels in bacteria for regulating membrane potentials for communication is a relatively new finding.[11] Furthermore, mechanisms of bacterial conductivity and energy harvesting are under intensive research, but the role of ion channels is not well understood.

Assays for determining whether bacteria are *alive* typically use positively charged fluorescent dyes (e.g. propidium iodide or SYTOX Green) that are affected by the membrane potential. Their use in live/dead assays depends on their membrane impermeability in live cells (only cells that have holes due to lysis are expected to be permeable) and thus the lysed cells become fluorescently stained. However, they can thus overestimate the number of dead cells due to the existence of cells with small negative or small/large positive membrane potentials (which do not absorb the dyes).[12] More direct measures of the bacterial membrane potential are expected to provide better assays for cell death e.g. to determine the efficacy of antibiotics.

There are many other reasons that bacteria could have evolved to detect electrical potentials beyond intercellular signalling and one example is found in immunology. Killing of bacteria by neutrophils is ionic (the neutrophils drop the pH) and electrical signalling could provide crucial information for bacteria to evade the neutrophils. Thus, an alternative explanation for the sensitivity of the bacteria to potassium ions, K^+ (beyond that of electrical signalling), may be associated with phagocytosis, since the killing activity of neutrophils is mediated through the activation of proteases by a flux of K^+ ions.[13]

Many metals typically used for electrodes can be toxic for bacteria if they are placed in contact with the culture media e.g. silver and gold. Care must be taken to avoid artefacts due to bacterial stress responses e.g. to coat the electrodes with thin layers of polymers to avoid ionic leaching, such as 10 nm layers of PMMA via spin coating.

Electrochemical potentials in bacterial cells were introduced in Chapter 3. In this chapter, mechanisms that modulate the action potentials are considered together with electrical communication, electrical conductivity and a few miscellaneous applications of electrical fields in bacterial biophysics.

29.1 The Action Potential of Bacteria

It is well understood that two-component signalling and MAPK systems are used for sensory information processing in bacteria (Chapter 18). However, some of the sensory information in bacteria is also transduced by ion channels e.g. the mechanosensitive

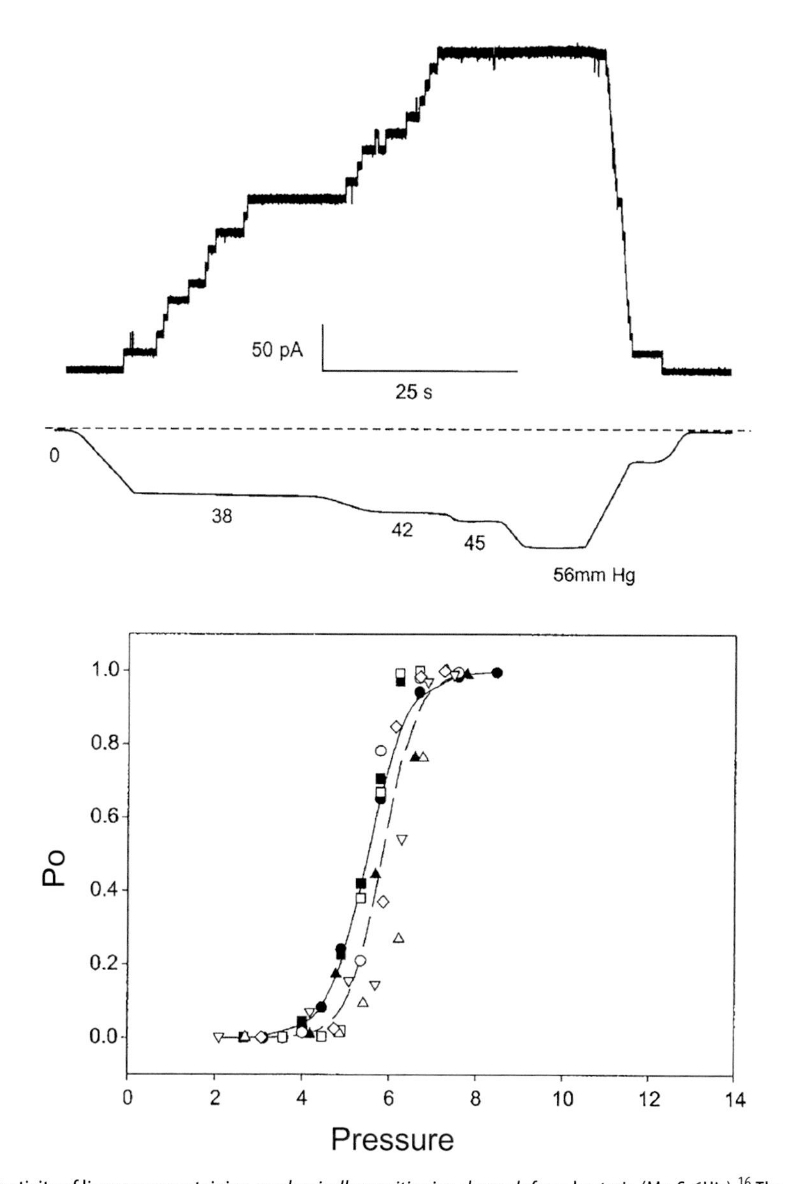

Figure 29.1 (a) Conductivity of liposomes containing *mechanically sensitive ion channels* from bacteria (MscS-6His).[16] The current is shown on the *y*-axis and time on the *x*-axis. Discrete steps in the current correspond to discrete opening/closing events of single ion channels. (b) The opening probability of the ion channels (P_0, Equation (29.1)) as a function of pressure (the *x*-axis). Reprinted from Sukharev, S., Purification of the small mechanosensitive channel of *Escherichia coli* (MscS): the subunit structure, conduction, and gating characteristics in liposomes. *Biophysical Journal* 2002, *83*, 290 with permission from Elsevier.

channels of *E. coli*,[14,15] and stepwise conductivity changes are observed as a function of membrane curvature when cells are mechanically deformed.

Mechanosensitive ion channels show abrupt changes in conduction that is correlated with membrane curvature for mechanically sensitive bacterial channels embedded in liposomes, due to the opening and closing of individual ion channels (Figure 29.1).[16]

The curvature of the liposomes measured with video microscopy could thus be related to the channels' electrical activity.[14] The data from the liposomes could be fit into a simple model based on Boltzmann statistics for thermal equilibrium,

$$\frac{P_o}{P_c} = e^{-(\Delta G - \gamma \Delta A)/kT}, \quad (29.1)$$

where P_o is the open probability of the channels, P_c is the closed probability, ΔG is the free energy change of a channel opening in an unperturbed membrane, γ is the surface free energy of the membrane, ΔA is the change in membrane area due to deformation and kT is the thermal energy.

The measurement of single ion channel activity in intact bacterial cells is challenging, since the small size of the cells makes conventional patch clamping techniques awkward (the pipette is around the size of the bacteria). Some experiments have been made on giant mutant *E. coli* treated with detergents to break up the cells (creating spheroplasts), but question marks remain over how representative the measurements are for wild-type cells. Although the mutant spheroplasts provide a slightly artificial system, they have given valuable information e.g. osmoregulation by MscS ion channels was studied.[15]

Individual electrical spiking events were measured for *E. coli* cells using a genetically expressed voltage indicator protein.[2] Such optogenetics techniques seem very well placed to help revolutionise the field of bacterial electrophysiology.[17] Recent optogenetics work has examined Ca^{2+} signalling (the original work was based on K^{+1}).[18] Furthermore, fluctuations in the conductivity of bacterial ion channels were also seen in response to antibiotic stress using optogenetics.[19]

In a further study with higher-resolution optogenetic fluorescent probes, spiking potentials were seen in *E. coli*, *B. subtilis* and *Salmonella typhimurium* cells.[20] The spikes sensitively depended on the growth media surrounding the cells i.e. frequent hyperpolarisation spikes were triggered by high Na^+ concentration and low K^+ concentration environments in the growth media. Antibiotic tolerance of ampicillin was found to correlate with the membrane potential.

A separate study using *microelectrode arrays* (MEAs) that were originally developed for spatially resolved studies of neuronal tissue found spiking behaviour at short time scales, similar to optogenetic fluorescence measurements,[2] but much faster than the hyperpolarisation phenomena measured with synthetic Nernstian dyes[21] (Figure 29.2). Membrane voltages were directly measured with the MEA. Thus, there may be additional kinetic phenomena broadening the electrochemical responses measured with the Nernstian dyes. How the results of fluorescence measurements integrate over a short time scale behaviour is not clear, although different species of bacteria were used with the MEAs and the action potentials appear to be a generic phenomenon. Specifically, MEAs with 1 824 electrodes were used to map the electrochemical activity across *P. aeruginosa* biofilms.[22] Microelectrode arrays were also used to measure the electrical activity in both biofilms and planktonic *Bacillus licheniformis*, *Pseudomonas. alcaliphila* and *E. coli*[21] (Figure 29.2). Spike amplitudes

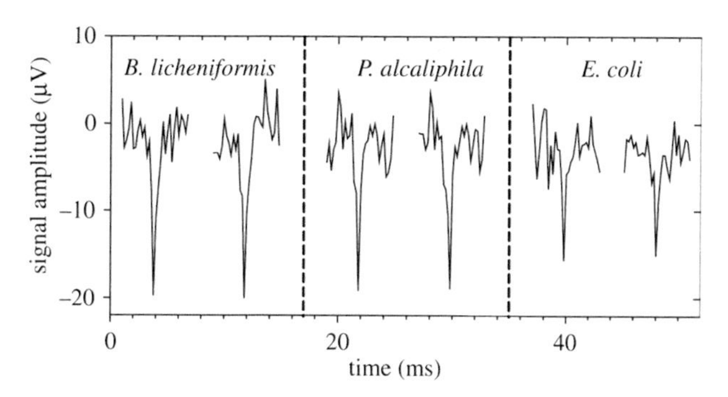

Figure 29.2 *Action potentials* as a function of time recorded with a microelectrode array on three types of bacteria: *Bacillus licheniformis*, *Pseudomonas alcaliphila* and *E. coli*.[21] Two repeats are shown per organism.

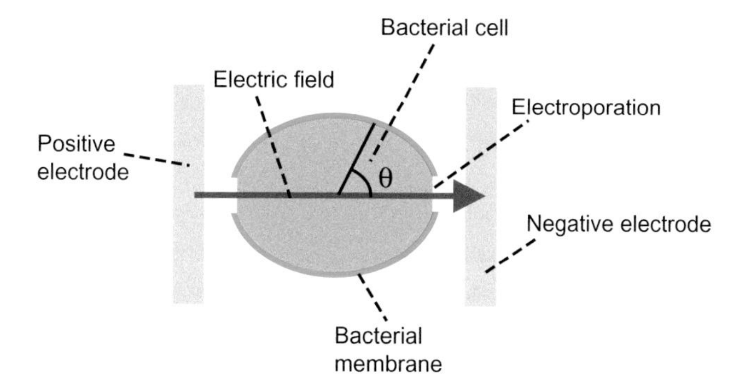

Figure 29.3 Electrode stimulation of a bacterial cell. *Electroporation* will occur on the poles of the bacterium due to the $\cos\theta$ term in the electric field (Equation (29.2)) and small holes are created in the bacterial membrane.

were ~20 μV, the action potential rate was ~1 spike s^{-1} and the spike duration was around 1 s for all three samples.

The *potential distribution* on a cell membrane placed in an external electric field E is

$$V = 1.5aE\cos\theta, \qquad (29.2)$$

where V is the transmembrane potential, a is the cell radius, θ is the angle between the position on the cell membrane and the direction of E.[23] Equation (29.2) is a solution of the Laplace equation with the necessary boundary equations.

Direct *electrode stimulation* of bacterial cells was used in parallel with fluorescence microscopy with Nernstian dyes[24] to measure the viability of cells (Figure 29.3). Work in our own lab indicates that *E. coli* cells may experience reversible electroporation due to such electrode stimulation and AC electro-osmosis of cationic fluorophores could occur (ghost spikes due to motion of the fluorophores), which complicates the interpretation of the phenomena. Furthermore, hyperpolarisation of the cells occurred in response to blue light in our studies as well as due to electrode stimulation, although this lack of agreement may be an issue with sensitivity to the ions in the media, which determines the response of the cells.[20] An interesting further question is to extend these electrical stimulation studies to biofilms e.g. to understand whether biofilms protect bacteria from electric fields during electrode stimulation.

The *Schwann equation* can be used to calculate the membrane potential in a bacterial cell placed directly between two electrodes $(\theta = 0)$ in response to an oscillatory electric field, on the poles of the cell (Figure 29.3),

$$V_{max} = 1.5aE\left(1+\left(2\pi f\tau\right)^2\right)^{-1/2}, \quad (29.3)$$

where V_{max} is the induced membrane potential, a is the cell radius, E is the applied electric field strength, f is the AC field frequency and τ is the relaxation time of the membrane.[25] The threshold of the applied electric field (E) for reversible electroporation is ~300 V cm^{-1}, whereas that for irreversible electroporation is ~2 kV cm^{-1}, although these figures depend on the specific type of membrane considered.[26]

The *Hodgkin–Huxley* model for bacteria was introduced in Chapter 3, and this is the main tool for describing action potentials. Some simplifications of the model to reduce the complexity are *FitzHugh–Nagumo* and, *integrate and fire* models.[27,28] These may be useful intermediate coarse-grained models to speed up simulations e.g. with agent-based models of biofilms.[29]

Another study examined the effects of blue light on the spiking of bacterial potentials[4] using a Nernstian dye (ThT) in *B. subtilis* and *P. aeruginosa* (Figure 3.2). The cells hyperpolarise before they swim away from the surface, but no evidence was found for intercellular communication and the times for departure were independent stochastic events. More recent data with *E. coli* indicate that these bacteria can synchronise their reaction to light stress and a wavefront of cellular hyperpolarisation events (Figure 5.1a) passes through a biofilm (which is not observed with systems of low bacterial concentration).[3]

Wave pulses of cellular hyperpolarisation have been observed to propagate across *N. gonorrhoea* biofilms (Figures 5.1b and 29.4). They are thought to be initiated by oxidative stress.[30]

Fluorescence experiments with *E. coli* show that voltage-gated calcium flux mediates mechanosensation.[18] Calcium-gated voltage fluctuations were also found in response to antibiotics.[19]

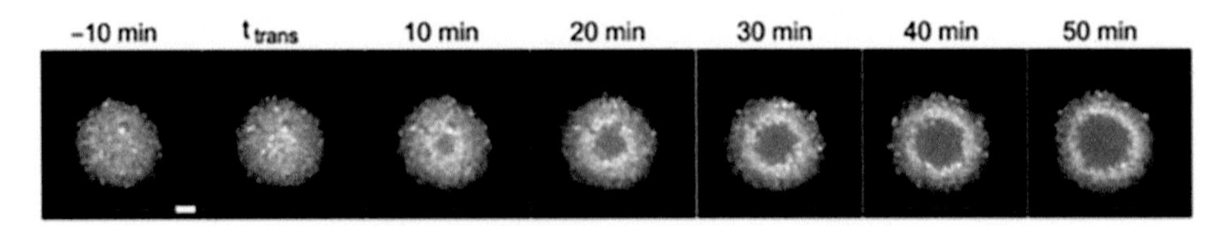

Figure 29.4 *Wave pulses* of depolarisation in *Neisseria gonorrhoeae* biofilms as observed with fluorescence microscopy as a function of time.[30] The process of membrane polarisation correlates with the influx of K^+ ions. A fluorescent ring propagates from the centre of the biofilm.

The study of neuronal electrical oscillations is much better developed than the electrical oscillations of bacteria.[28] It is an open question whether bacteria correspond to type I, II or III neuronal oscillators. Furthermore, are they *integrator oscillators* or *resonance oscillators*[31]? Why has no limit cycle yet been observed with bacteria in which they experience a series of bursting spikes? It may be possible to classify bacteria based on their response to a pulsed electric field via their non-linear phase portraits in a similar manner to neurons.

Action potentials in neuronal cells can be initiated by pulsed magnetic fields.[32] Magnetic field stimulation has not yet been achieved with bacteria or bacterial biofilms. Furthermore, the conductivity of neuronal and cardiac cells is anisotropic (e.g. the axons are long and thin) and it can be explored experimentally using two-dimensional (2D) arrays of electrodes to rotate the direction of the electric field with respect to the cells. Many bacteria are elongated; some form elongated aggregates and aligned liquid crystalline behaviour (Chapter 8) can occur in dense suspensions and biofilms. Thus, it would be interesting to explore the anisotropy of bacterial systems to electric and magnetic stimulation.

29.2 Electrical Communication in Bacteria

The groundbreaking study of Prindle et al. demonstrated electrochemical signalling in 2D bacterial biofilms of *B. subtilis* that was based on wavefronts (Figure 5.1a) of potassium ions (K^+) due to the activity of the voltage-gated ion channel, YugO.[1] Evidence was presented that the signalling was driven by glutamate starvation. However, electrochemical signalling may just be a general mechanism of communicating stress in bacterial biofilms[29] e.g. both centrifugal (outward) and centripetal (inward) wavefronts were also observed in additional experiments on *B. subtilis* (Figure 29.5), which are harder to justify with the glutamate hypothesis and similar waves were observed in three-dimensional (3D) *E. coli* biofilms in response to light stress with no significant nutritional limitations.[3] Wave pulses (Figure 5.1b) of excitation have also been observed in *Gonorrhoeae* biofilms (Figure 29.4), but it was claimed that electrical communication did not occur in this case[30] (just diffusion of ROS). Further evidence has been presented for long-range electrical (K^+) communication between different species of bacteria in biofilms.[33] Specifically, *B. subtilis* biofilms were found to attract *P. aeruginosa* to their surfaces. The electrical signalling is associated with hyperpolarisation of the cells i.e. the cells become more negatively charged as they release potassium. Molecules that block ion channels used in electrical communication could thus be useful for the treatment of bacterial diseases, since they appear to be crucial for the social interactions of bacterial cells.

The *Eikonal approximation* (Equation (5.3)) has been used to qualitatively rationalise the dependence of the wavefront velocity on the curvature of the wavefronts, and it has been used to understand reaction–diffusion processes in a wide range of biological systems e.g. waves across cardiac tissue[27] or CAMP signalling in amoeba.[34] However,

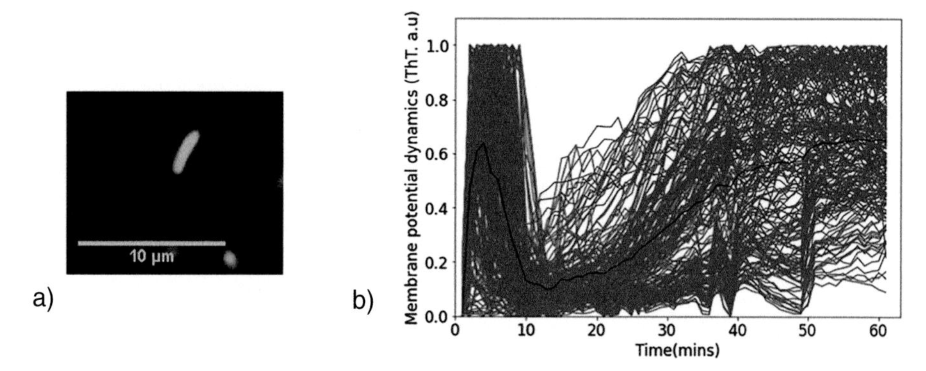

Figure 29.5 Membrane polarisation dynamics of single *E. coli* in response to 440 nm *blue light stress*.[3] (a) A fluorescence microscopy image of a single hyperpolarised *E. coli* cell containing ThT in a microfluidic device showing its hyperpolarisation in response to blue light. (b) Normalised ThT fluorescence intensity from single cells as a function of time.

experiments and agent-based models quantitatively diverge from the predictions of the Eikonal approximation in *E. coli*[3] and *B. subtilis* biofilms, and in general, wavefronts rarely propagate at constant velocities in reaction–diffusion systems, which is an assumption in Equation (5.4). Thus, there exist questions on the effect of biofilm geometry on the propagation of potassium wavefronts in biofilms.

A *fire-diffuse-fire* model was introduced to describe the propagation of 2D and 3D wavefronts of electrical signalling and was solved numerically using agent-based models,[3,29]

$$\frac{\partial Q}{\partial t} = D_Q \nabla^2 Q - \theta Q + \sigma \sum_i \delta(r - r_i)\delta(t - t_i), \tag{29.4}$$

where Q is the potassium concentration, D_Q is the potassium diffusion coefficient, θ is the rate of uptake of potassium, σ is a fixed amount of potassium released per spiking event, r is the space coordinate, t is the time, r_i is the position of the ith bacterium and t_i is the time the ith bacterium experiences a K^+ concentration higher than the threshold value (Figure 5.2). More sophisticated models could replace the delta functions for potassium release[33] e.g. those based on Hodgkin–Huxley models for the membrane channels with spiking potentials. The density of the bacteria sensitively affects the velocity of the wavefront and can determine whether the wavefronts follow sub-diffusive, diffusive, ballistic, super-diffusive or super-ballistic kinetics via Equation (5.3).[3]

Open questions also exist concerning how the electrical wavefronts are affected by microheterogeneities in biofilms e.g. defects in the biofilms (in comparison with the analysis of cardiac infarctions where dead tissue changes patterns of heart contraction,[27] what is the effect of transport channels in bacterial biofilms?) and mixed-species biofilms (is there a percolation transition for the more highly conducting bacteria as a function of volume fraction?) will affect the wavefronts.

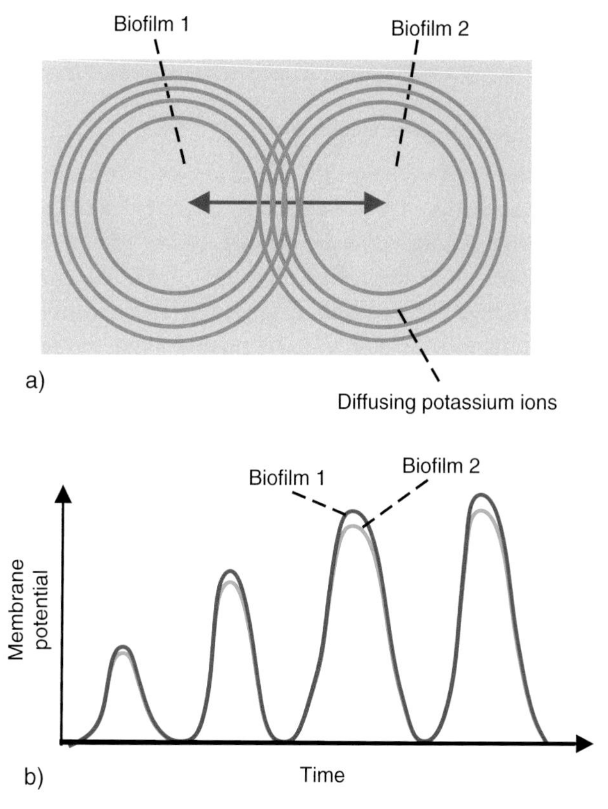

Figure 29.6 Synchronisation of electrical signalling in *B. subtilis* biofilms.[36] (a) Schematic diagram of two isolated biofilms that synchronise their electrochemical oscillations by the transmission of potassium ions. (b) Synchronisation of the membrane potentials as a function of time measured using a fluorescent Nernstian dye (ThT).

A two-channel Hodgkin–Huxley model was used to interpret the double hyperpolarisation phenomena observed when *E. coli* was exposed to blue light (Figure 29.5b). The first hyperpolarisation event was when the bacteria first register the light and the second hyperpolarisation event was due to a habituation phenomenon. The model used was a simple extension of Equation (3.3) using two separate ion channel currents (corresponding to each of the peaks). Mutants with the potassium (K_{ch}) ion channels removed lacked the second hyperpolarisation event.

A non-linear bistable model for the electrical oscillations of neighbouring *B. subtilis* biofilms has been created (Chapter 12). This model predicts the sudden discontinuous emergence of collective electrochemical oscillations in the biofilms and a minimum size of the biofilms for oscillations to occur.[35]

A separate study of *B. subtilis* biofilms also explored the process of communication between two separated communities (Figure 29.6). Electrical oscillations in the biofilms became coupled and thus synchronised (either in antiphase or in phase)[36] due to their non-linear interactions.[37]

29.3 Electrical Conductivity of Bacteria

Conductivity and impedance spectroscopy are standard techniques used in organic electronics to understand the electrical properties of materials.[38] Care must be taken with such electrical measurements, since artefacts are common due to a wide range of competing conduction mechanisms e.g. the contact resistance at the electrodes needs to be controlled. Reproducibility can be challenging with direct electrical measurements on bacteria e.g. making conductivity measurements robust to the choice of culture media and the electrode chemistry. Four, three and two electrodes are typically used (Figure 29.7), which have different advantages and disadvantages e.g. there is an ambiguity in voltage measurement with only two electrodes (a single half-cell in terms of its electrochemistry).[38]

Standard microbiological systems to study extracellular electron transfer (EET) with bacteria are found with *Shewanella oneidensis* and *Geobacter sulfurreducens*.[39] Four-point measurements are the gold standard for conductivity measurements, but they are tricky with living biofilms and two-point measurements are more common. Interdigitated microelectrode arrays (IDAs) are a convenient modern geometry to reduce artefacts. The structure of the conducting *G. sulfurreducens* filaments was measured using cryoTEM.[40] Heme groups from cytochrome OmcS in the filaments were found, disproving the hypothesis that the conductivity was due to pili proteins. Biofilms of *G. sulfurreducens* grown on copper electrodes increased their conductivity as copper was incorporated into their structures.[41]

A reasonably broad range of marine bacteria are thought to have electrically conducting nanowires including *Shewanella oneidensis*, *Synechocystis* and *Pelotomaculum thermopropionicum*.[42] Shewanella also secretes flavins to mediate EET.[43]

The electrical resistivity of bacterial nanowires from *Shewanella oneidensis* MR-1 was found to be 1 Ω cm,[44] which is comparable to synthetic organic materials used in commercial conductivity applications. Atomic force microscopy (AFM) experiments showed the wires had reasonably high Young's moduli at 1 GPa, which are typical of

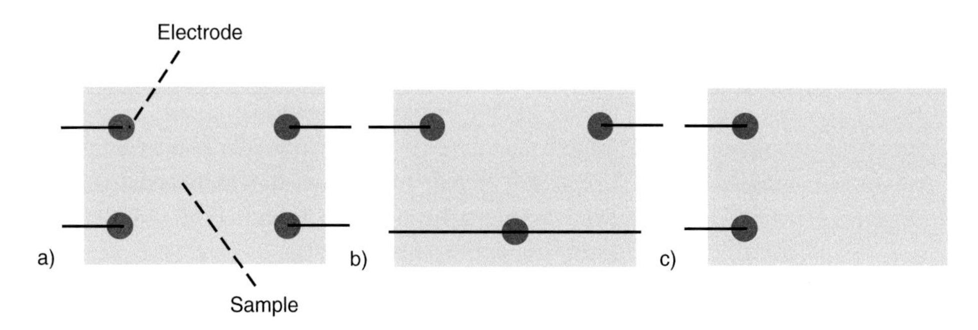

Figure 29.7 (a) Four, (b) three and (c) two electrodes are typically used to measure the electrical properties of biological molecules and cells.[38]

polymeric materials. The conductivity of nanowires produced by *Shewanella oneidensis* was measured using a modified electrically sensitive AFM.[45] The electrical transport behaviour was highly non-linear with a series of local maxima. It is thought to be due to hopping or electron transfer between energy states (either delocalised or localised). *Shewanella oneidensis* interacts with electrodes via flavins, whereas *Geobacter sulfurreducens* makes direct electrical contact via c-type cytochromes.[46] Microelectrode experiments with *G. sulfurreducens* show quantised increases in current as individual cells contribute to long-range charge transport.[47] The currents per cell were 80–200 fA, similar to other experiments with *Shewanella* cells.

Inspiration from bacterial systems may provide methods to create biocompatible electrode contacts and robust self-repairing aqueous electronics.[48] Some evidence exists that the extracellular polymeric substance in the biofilms of *Shewanella oneidensis*, *Bacillus* sp. and the yeast *Scheffersomyces stipitis* (=*Pichia stipitis*) facilitate EET.[49] The biofilm components could thus provide inspiration for organic electronics in addition to molecules in the bacteria themselves.

There is some controversy as to whether *G. sulfurreducens* biofilms do indeed demonstrate metallic conductivity in their pili.[50] Experiments by different groups have demonstrated the existence of long-range charge transport but ascribed it to redox conductivity rather than a metallic mechanism. Evidence of long-distance redox conductivity was also seen in *Marinobacter–Chromatiaceae–Labrenzia* biofilms with a conductivity one order of magnitude greater than *G. sulfurreducens* (which has a similar conductivity to some standard synthetic conducting polymers).[51]

A separate research field was developed to understand charge transport in filamentous species of marine bacteria that form macrocables[52] (Figure 29.8 the macrocables are centimetres in length compared with pili that are ~ 10 μm). Initially, these studies were confined to marine bacteria, but good evidence is now available for freshwater species of cable bacteria from muddy swamps.[53] Bacteria in sediments and mud become anoxic, which makes some metabolic functions more challenging. Conduction of electrons over large distances using macrocables provides a mechanism for them to cope with such anoxic environments.

Evidence was found for the conduction mechanism employed by single cells of cable bacteria using Raman microscopy.[55] Cytochromes were found to play an important role in conduction. Additional work has been performed on the chemotaxis and gliding motility of cable bacteria.[56] The bacteria are thought to glide when there is little oxygen and stop gliding when oxygen is present. The cables in the bacteria have a cartwheel-type structure[57] (Figure 29.8). Cable bacteria were found to interact with marine and freshwater plants e.g. rice. Nickel is found to play an important role in the conductivity of cable bacteria.[58] Both Gram positive and negative cable bacteria have been found.[59]

Microfluidic dielectrophoresis experiments were found to be a sensitive tool to detect EET in bacteria (*G. sulfurreducens*, *S. oneidensis* and *E. coli*).[60] There is evidence that quinones (signalling molecules) allow EET in *Shewanella putrefaciens*.[61]

Another perspective is that electron transfer is a useful strategy that encourages symbiosis in mixed bacterial communities.[62] Electrons can be more easily exchanged between different species than more chemically specialised signalling molecules.

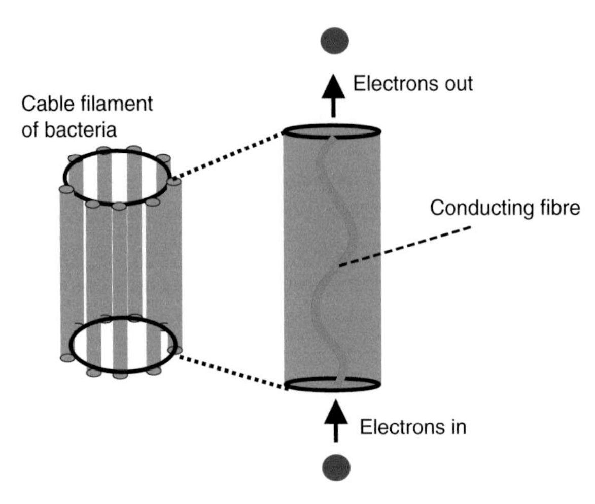

Figure 29.8 Schematic diagram of *cable bacteria* arranged end on end in a conducting filament. Multiple filaments are bundled together in a macrocable (cm in length).[54]

29.4 Other Applications of Electricity

It is possible to harvest electricity from sediments that contain bacteria e.g. 0.016 W can be extracted per square metre of electrode.[63] Graphite electrodes were used.

Application of constant voltages or currents across conducting surfaces can delay the onset of the growth of a biofilm e.g. it delays bacterial cell attachment and reactive oxygen species can both disrupt biofilms and increase the efficacy of biocides.[64] Current carrying electrodes embedded in wound dressings have been investigated over the last 20 years.[65] Currents can encourage wound healing via stimulation of the migration of neutrophils, fibroblast activity and blood flow. In addition, the dressings are found to be antibacterial and antibiofilm.

Application of electrical fields to bacteria increases the effect of antibiotics on *E. coli* biofilms.[66] Results suggest the energy of the signal is the crucial factor determining the improved efficacy, relatively independent of the oscillatory effects (alternating current [AC] versus direct current [DC]). However, electrophysiological measurements of membrane potentials imply a more nuanced understanding is possible e.g. to determine the different activities of charged and non-charged antibiotics. Low-amperage antibiofilm activity has been found in a wide range of bacterial biofilms using 2 – 200 μA currents through conducting surfaces for day-long durations.[67] Whether the disruption of electrical signalling or electroporation to facilitate entry of the antibiotics occurs (Section 29.2) is yet to be determined.

Another perspective on bioelectricity in bacteria is to examine how flagellar motors are powered. External voltage sources have been used to power *E. coli* motors.[68] More recently, the activity of *E. coli* motors has been used to calibrate the measurement of Nernstian potentials using fluorescent dyes.[69] However, the motors can have clutches, which complicates measurements[70] e.g. flagella can be immobile at high potentials if the clutches are engaged. Furthermore, voltage gating of ion channels in the *E. coli* cells was not included in the analysis of the activity of the flagellar motors (a Hodgkin–Huxley analysis is needed).

Application of constant electric fields can modulate the rate of attachment of *Pseudomonas fluorescens* to conducting surfaces.[71] A QCM-D electrochemical module was used to probe the behaviour.

Boron-doped diamond electrodes have provided nanomolar detection of quorum-sensing molecules from *P. aeruginosa* that are electrically active.[72] Three separate quorum-sensing molecules could be simultaneously detected. Electrochemical metabolites can be used to identify bacterial infections,[73] although genetic studies remain the gold standard for species identification.

Alternating electric fields can induce clustering of *E. coli*.[74] Dielectrophoresis effects with a wide range of cell types have shown promise in cell sorting applications.

Electroporation is a standard tool to increase the efficiency of DNA transfection with bacteria.[23,26,75] Separate electric field thresholds occur for *reversible* (smaller) and *irreversible* (larger) *electroporation* for artificial membranes, bacterial and eukaryotic cells.

There are many more outstanding questions that need to be understood with the interactions of electric fields with bacteria. For example, can propagating action potentials be measured in the hyphae of fibrous bacteria (e.g. with *Streptomyces lividans*) in response to electrical stimulation, how to interpret negative capacitances in electrical impedance spectroscopy experiments with bacteria and biofilms[76] (they are expected from the Hodgkin–Huxley model of neurons[77] and were first measured in neurons by K. S. Cole in 1941[78,79], but Joule heating is also implicated) and can anisotropic electrical response be measured for bacteria (e.g. rotating the electric field, so it aligns with the long axis of a rod-like bacterium using a 2D array of electrodes)?

Suggested Reading

Benarroch, J. M.; Asally, M., The microbiologists guide to membrane potential dynamics. *Trends in Microbiology* **2020**, *28* (4), 304–314.

Beyend, H.; Bubaute, J. *Biofilms in Bioelectrochemical Systems: From Lab Practice to Data Interpretation*. Wiley, 2015.

Blee, J. A.; Roberts, I. S.; Waigh, T. A., Spatial propagation of electrical signals in circular biofilms: A combined experimental and agent-based fire-diffuse-fire study. *Physical Review E* **2019**, *100*, 052401.

Blee, J. A.; Roberts, I. S.; Waigh, T. A., Membrane potentials, oxidative stress and the dispersal response of bacterial biofilms to 405 nm light. *Physical Biology* **2020**, *17*, 036001.

Grimner, S.; Martinsen, O. G. *Bioimpedance and Bioelectricity*, 3rd ed. Academic Press: 2015.

Leung, K. M. et al., *Shewanella oneidensis* MR-1 bacterial nanowires exhibit p-type, tunable electronic behaviour. *Nanoletters* **2013**, *13*(6), 2407–2411.

References

1. Prindle, A.; Liu, J.; Asally, M.; Ly, S.; Garcia-Ojalvo, J.; Sudel, G. M., Ion channels enable electrical communication in bacterial communities. *Nature* **2015**, *527* (7576), 59–63.
2. Kralj, J. M.; Hochbaum, D. R.; Douglass, A. D.; Cohen, A. E., Electrical spiking in *Escherichia coli* probed with a fluorescent voltage-indicating protein. *Science* **2011**, *333* (6040), 345–348.
3. Akabuogu, E. U.; Martorelli, V.; Krasovec, R.; Roberts, I. S.; Waigh, T. A., Emergence of ion-channel mediated electrical oscillations in *Escherichia coli* biofilms. *eLife* **2023**, to appear.
4. Blee, J. A.; Roberts, I. S.; Waigh, T. A., Membrane potentials, oxidative stress and the dispersal response of bacterial biofilms to 405 nm light. *Physical Biology* **2020**, *17* (3), 036001.
5. MacKinnon, R., Potassium channels and the atomic basis of selective ion conduction (Nobel Lecture). *Angewandte Chemie International ed. in English* **2004**, *43* (33), 4265–4277.
6. Madigan, M. T.; Bender, K. S.; Buckley, D. H.; Sattley, W. M.; Stahl, D. A., *Brock Biology of Microorganisms*, 15th ed. Pearson: 2018.
7. Kim, B. H.; Gadd, G. M., *Prokaryotic Metabolism and Physiology*, 2nd ed. CUP: 2019.
8. Logan, B. E., Exoelectrogenic bacteria that power microbial fuel cells. *Nature Reviews. Microbiology* **2009**, *7* (5), 375–381.
9. Weiss, G. L.; Kieninger, A.-K.; Maldener, I.; Forchhammer, K.; Pilhofer, M., Structure and function of a bacterial gap junction analog. *Cell* **2019**, *178* (2), 374–384.
10. Kopronski, P.; Kubalski, A., Bacterial ion channels and their eukaryotic homologues. *BioEssays: News and Reviews in Molecular, Cellular and Developmental Biology* **2001**, *23* (12), 1148–1158.
11. Martinac, B.; Saimi, Y.; Kung, C., Ion channels in microbes. *Physiological Reviews* **2008**, *88* (4), 1449–1490.
12. Kirchoff, C.; Cypianka, H., Propidium ion enters viable cells with high membrane potential during live-dead staining. *Journal of Microbiological Methods* **2017**, *142*, 79–82.
13. Reeves, E. P.; et al., Killing activity of neutrophils is mediated through activation of proteases by K+ flux. *Nature* **2002**, *416* (6878), 291–297.
14. Sukharev, S., Purification of the small mechanosensitive channel of *Escherichia coli* (MscS): The subunit structure, conduction, and gating characteristics in liposomes. *Biophysical Journal* **2002**, *83* (1), 290–298.
15. Sotomayor, M.; Vasquez, V.; Perozo, E.; Schulten, K., Ion conduction through MscS as determined by electrophysiology and simulation. *Biophysical Journal* **2007**, *92* (3), 886–902.

16. Markin, V. S.; Martinac, B., Mechanosensitive ion channels as reporters of bilayer expansion: A theoretical model. *Biophysical Journal* **1991**, *60* (5), 1120–1127.
17. Cohen, A. E.; Venkatachalam, V., Bringing bioelectricity to light. *Annual Review of Biophysics* **2014**, *43*, 211–232.
18. Bruni, G. N.; Weekley, R. A.; Dodd, B. J. T.; Kralj, J. M., Voltage-gated calcium flux mediates *Escherichia coli* mechanosensation. *PNAS* **2017**, *114* (35), 9445–9450.
19. Bruni, G. N.; Kralj, J. M., Membrane voltage dysregulation driven by metabolic dysfunction underlies bactericidal activity of aminoglycosides. *eLife* **2020**, *9*, e58706.
20. Jin, X.; et al., Sensitive bacterial V_m sensors revealed the excitability of bacterial V_m and its role in antibiotic tolerance. *Proceedings of the National Academy of Sciences of the United States of America* **2023**, *120* (3), e2208348120.
21. Masi, E.; Ciszak, M.; Santopolo, L.; Frascella, A.; Giovannetti, L.; Marchi, E.; Viti, C.; Mancuso, S., Electrical spiking in bacterial biofilms. *Journal of the Royal Society, Interface* **2015**, *12* (102), 20141036.
22. Bellin, D. L.; Sakhtah, H.; Zhang, Y.; Price-Whelan, A.; Dietrich, L. E. P.; Shepard, K. L., Electrochemical camera chip for simultaneous imaging of multiple metabolites in biofilms. *Nature Communications* **2016**, *7*, 10535.
23. Tseng, T. Y., Electroporation of cell membranes. *Biophysical Journal* **1991**, *60* (2), 297–306.
24. Stratford, J. P.; Edwards, C. L. A.; Ghanshyam, J.; Malyshev, D.; Delise, M. A.; Hayashi, Y.; Asally, M., Electrically induced bacterial membrane-potential dynamics correspond to cellular proliferation capacity. *PNAS* **2019**, *116* (19), 9552–9557.
25. Marszalek, P.; Liu, D. S.; Tsong, T. Y., Schwan equation and transmembrane potential. *Biophysical Journal* **1990**, *58* (4), 1053–1058.
26. Weaver, J. C.; Chizmadzhev, Y. A., Theory of electroporation: A review. *Bioelectrochemistry and Bioenergetics* **1996**, *41* (2), 135–160.
27. Keener, J.; Sneyd, J., *Mathematical Physiology*. Springer: 2009.
28. Gerstner, W., *Neuronal Dynamics: From Single Neurons to Networks and Models of Cognition*. CUP: 2014.
29. Blee, J. A.; Roberts, I. S.; Waigh, T. A., Spatial propagation of electrical signals in circular biofilms. *Physical Review E* **2019**, *100* (5–1), 052401.
30. Hennes, M.; Bender, N.; Cronenberg, T.; Welker, A.; Maier, B., Collective polarization dynamics in bacterial colonies signify the occurrence of distinct subpopulations. *PLOS Biology* **2023**, *21* (1), e3001960.
31. Izhikevich, E. M., *Dynamical Systems in Neuroscience: The Geometry of Excitability and Bursting*. MIT: 2010.
32. Stern, S.; Rotem, A.; Burnishev, Y.; Weinreb, E.; Moses, E., External excitation of neurons using electric and magnetic fields in one- and two-dimensional cells. *Journal of Visualized Experiments* **2017**, *123*, e54357.
33. Humphries, J.; Xiong, L.; Liu, J.; Prindle, A.; Yuan, F.; Arjes, H. A.; Tsimring, L.; Suel, G. M., Species-independent attraction to biofilms through electrical signalling. *Cell* **2017**, *168* (1–2), 200–209.

34. Palsson, E.; Lee, K. J.; Goldstein, R. E.; Franke, J.; Kessin, R. H.; Cox, E. C., Selection for spiral waves in the social amoebae *Dictyostelium*. *PNAS* **1997**, *94* (25), 13719–13723.
35. Martinez, R.; Liu, J.; Suel, G. M.; Garcia-Ojalvo, J., Bistable emergence of oscillations in growing *Bacillus subtilis* biofilms. *PNAS* **2018**, *115* (36), E8333–E8340.
36. Liu, J.; et al., Coupling between distant biofilms and emergence of nutrient time-sharing. *Science* **2017**, *356* (6338), 628–642.
37. Strogatz, S. H., *Nonlinear Dynamics and Chaos: With Applications to Physics, Biology, Chemistry and Engineering*, 2nd ed. Westview Press: 2014.
38. Grimnes, S. J.; Martinsen, O. G., *Bioimpedance and Bioelectricity*. Academic Press: 2014.
39. Beyend, H.; Bubaute, J., *Biofilms in Bioelectrochemical Systems: From Lab Practice to Data Interpretation*. Wiley: 2015.
40. Wang, F.; et al., Structure of microbial nanowires reveals stacked hemes that transport electrons over micrometers. *Cell* **2019**, *177* (2), 361–369.
41. Beuth, L.; Pfeiffer, C. P.; Schroder, U., Copper-bottomed, electrochemically active bacteria exploit conductive sulphide networks for enhanced electrogeneity. *Energy & Environmental Science* **2020**, *13* (9), 3102.
42. Gorby, Y. A.; et al., Electrically conductive bacterial nanowires produced by *Shewanella oneidensis* strain MR-1 and other microorganisms. *PNAS* **2006**, *103* (30), 11358–11363.
43. Marsili, E.; Baron, D. B.; Shikhare, I. D.; Coursolle, D.; Gralnick, J. A.; Bond, D. R., *Shewanella* secretes flavins that mediate extracellular electron transfer. *PNAS* **2008**, *105* (10), 3968–3973.
44. Leung, K. M.; Wanger, G.; Guo, Q.; Gorby, Y.; Southam, G.; Lau, W. M.; Yang, J., Bacterial nanowires: Conductive as silicon, soft as polymer. *Soft Matter* **2011**, *7* (14), 6617–6621.
45. El-Naggar, M. Y.; Gorby, Y. A.; Xia, A.; Nealson, K. H., Molecular density of states in bacterial nanowires. *Biophysical Journal Letters* **2008**, *95* (1), L10–L12.
46. Lovley, D. R., Electromicrobiology. *Annual Review of Microbiology* **2012**, *66*, 391–409.
47. Jiang, X.; et al., Probing single-to multi-cell level charge transport in *Geobacter sulfurreducens DL-1*. *Nature Communications* **2013**, *4*, 2751.
48. Zhang, L.; Lu, J. R.; Waigh, T. A., Electronics of peptide- and protein-based materials. *Advances in Colloid and Interface Science* **2021**, *287*, 102319.
49. Xiao, Y.; Zhang, E.; Zhang, J.; Dai, Y.; Yang, Z.; Christensen, H. E. M.; Ulstrup, J.; Zhao, F., Extracellular polymeric substances are transient media for microbial extracellular electron transfer. *Science Advances* **2017**, *3* (7), e1700623.
50. Yates, M. D.; Strycharz-Glaven, S. M.; Golden, J. P.; Roy, J.; Tsoi, S.; Erickson, J. S.; El-Naggar, M. Y.; Barton, S. C.; Tender, L. M., Measuring conductivity of living *Geobacter sulfurreducens* biofilms. *Nature Nanotechnology* **2016**, *11* (11), 910–913.
51. Yates, M. D.; Eddie, B. J.; Kotloski, N. J.; Lebedev, N.; Malanoski, A. P.; Lin, B.; Strycharz-Glaven, S. M.; Tender, L. M., Toward understanding long-distance

extracellular electron transport in an electroautotrophic microbial community. *Energy & Environmental Science* **2016**, *9* (11), 3544–3558.
52. Pfeffer, C.; et al., Filamentous bacteria transport electrons over centimetre distance. *Nature* **2012**, *491* (7423), 218–221.
53. Risgaard-Petersen, N.; et al., Cable bacteria in freshwater sediments. *Applied and Environmental Microbiology* **2015**, *152*, 122–142.
54. Nielsen, L. P.; Risgaard-Petersen, N., Rethinking sediment biogeochemistry after the discovery of electric currents. *Annual Review of Marine Science* **2015**, *7*, 425–442.
55. Bjerg, J. T.; et al., Long-distance electron transport in individual living cable bacteria. *PNAS* **2018**, *115* (22), 5786–5791.
56. Bjerg, J. T.; Damgaard, L. R.; Holm, S. A.; Schramm, A.; Nielsen, L. P., Motility of electric cable bacteria. *Applied and Environmental Microbiology* **2016**, *82* (13), 3816.
57. Cornelissen, R.; et al., The cell envelope structure of cable bacteria. *Frontiers in Microbiology* **2018**, *9*, 3044.
58. Boschker, H. T. S.; et al., Efficient long-range conduction in cable bacteria through nickel protein wire. *Nature Communications* **2021**, *12* (1), 3996.
59. Powell, L. C.; Abdulkarim, M.; Stokniene, J.; Yang, Q. E.; Walsh, T. R.; Hill, K. E.; Gumbleton, M.; Thomas, D. W., Quantifying the effects of antibiotic treatment on the extracellular polymer network of antimicrobial resistant and sensitive biofilms using multiple particle tracking. *npj Biofilms and Microbiomes* **2021**, *7* (1), 13.
60. Wang, Q.; Jones III, A. A. D.; Gralnick, J. A.; Lin, L.; Buie, C. R., Microfluidic dielectrophoresis illuminates the relationship between microbial cell envelope polarisability and electrochemical activity. *Science Advances* **2019**, *5* (1), eaat5664.
61. Newman, D. K.; Kolter, R., A role of excreted quinones in extracellular electron transfer. *Nature* **2000**, *405* (6782), 94–97.
62. Shrestha, P. M.; Rotaru, A. E., Plugging in or going wireless: Strategies for interspecies electron transfer. *Frontiers in Microbiology* **2014**, *5* (237), 237.
63. Bond, D. R.; Holmes, D. E.; Tender, L. M.; Lovley, D. R., Electrode-reducing microorganisms that harvest energy from marine sediments. *Science* **2002**, *295* (5554), 483–485.
64. Sultana, S. T.; Babauta, J. T.; Beyenal, H., Electrochemical biofilm control: A review. *Biofouling* **2015**, *31* (9–10), 745–758.
65. Roy, S.; et al., Disposable patterned electroceutical dressing is safe for treatment of open clinical chronic wounds. *Advances in Wound Care* **2019**, *8* (4), 149–159.
66. Kim, Y. W.; et al., Effect of electrical energy on the efficacy of biofilm treatment using the bioelectric effect. *npj Biofilms and Microbiomes* **2015**, *1*, 15016.
67. Schmidt-Malan, S. M.; Karau, M. J.; Cede, J.; Greenwood-Quaintance, K. E.; Brinckman, C. L.; Mandrekar, J. N.; Patel, R., Antibiofilm activity of low-amperage continuous and intermittent direct electrical current. *Antimicrobial Agents and Chemotherapy* **2015**, *59* (8), 4610–4615.
68. Fung, D. C.; Berg, H. C., Powering the flagellar motor of *E. coli* with an external voltage source. *Nature* **1995**, *375* (6534), 809–812.

69. Mancini, L.; Tian, T.; Guillaume, T.; Pu, Y.; Li, Y.; Lo, C. J.; Bai, F.; Pilizota, T., A general workflow for characterization of Nernstian dyes and their effects on bacterial physiology. *Biophysical Journal* **2020**, *118* (1), 4–14.
70. Blair, K. M.; Turner, L.; Winkelman, J. T.; Berg, H. C.; Kearns, D. B., A molecular clutch disables flagella in the *Bacillus subtilis* biofilm. *Science* **2008**, *320* (5883), 1636–1638.
71. Gall, I.; Herzberg, M.; Oren, Y., The effect of electric field on bacterial attachment to conductive surfaces. *Soft Matter* **2013**, *9* (8), 2443–2452.
72. Buzid, A.; et al., Molecular signature of *Pseudomonas aeruginosa* with simultaneous nanomolar detection of quorum sensing signally molecules at a boron-doped diamond electrode. *Scientific Reports* **2016**, *6*, 30001.
73. Goluch, E. D., Microbial identification using electrochemical detection of metabolites. *Trends in Biotechnology* **2017**, *35* (12), 1125–1128.
74. Bao, M. M.; Igwe, I. E.; Chen, K.; Zhang, T. H., Modulated collective motions and condensation of bacteria. *Chinese Physics Letters* **2022**, *39* (10), 108702.
75. Chen, C.; Smye, S. W.; Robinson, M. P.; Evans, J. A., Membrane electroporation theories: A review. *Medical and Biological Engineering and Computing* **2006**, *44* (1–2), 5–14.
76. Akabuogu, E. U.; Zhang, L.; Krasovec, R.; Roberts, I. S.; Waigh, T. A., Electrical impedance spectroscopy with bacterial biofilms: Neuronal-like behaviour. *ACS Nanoletters* **2024**, *24* (7), 2234–2241.
77. Bou, A.; Bisquert, J., Impedance spectroscopy dynamics of biological neural elements: From memristors to neurons and synapses. *The Journal of Physical Chemistry B* **2021**, *125*, 9934–9949.
78. Cole, K. S., Rectification and inductance in the squid giant axon. *The Journal of General Physiology* **1941**, *25* (1), 29–51.
79. Cole, K. S., *Membranes, Ions and Impulses*. University of California Press: 1968.

30 Manipulation of Biofilms

A pressing need for *biofilm manipulation* is found in the cleaning business. Specifically, how to remove problematic biofilms from surfaces that cause biofouling during all of their different stages of growth. There are lots of efficient strategies to do so *in vitro* in laboratories (e.g. use an autoclave), but the challenge is to develop minimally invasive procedures that can be performed *in situ* e.g. in a pipe line of an oil rig or in an ice cream maker in a food processing plant, and *in vivo* e.g. on the replacement valve of a heart patient or a replacement polyethylene hip joint for someone with osteoarthritis.

Anti-adhesion therapies to coat surfaces have had mixed results clinically. One crucial problem is due to the multiple adhesion strategies employed by single strains of bacteria[1] (Chapter 4). For example, maltose has shown some promise to block *Escherichia coli* adhesion in urinary tract infections (UTIs), but it only blocks one of the bacteria's adhesion strategies and they have multiple alternatives.

Numerous methods have been developed to modify surfaces so that they kill bacteria. These tend to be successful in deterring colonisation at very low concentrations of bacteria over short time scales, but at slightly higher concentrations, the dead bacteria tend to shield the surface effects for the next wave of colonisation and the surface modifications quickly become ineffective.

Many biological surfaces in contact with bacteria are constantly shed e.g. epithelial cells on human skin or mucin in the lungs. This is a useful protection mechanism but also an issue for the dissemination of bacteria e.g. people's homes are often covered in a layer of *Staphylococcus aureus* attached to dead skin. Pealing of thin surfaces thus offers a useful approach to the removal of biofilms.

Active release strategies from surfaces promise an improved longevity over passive surface chemistries.[2] A statistical study found 4.3% of 2.6 million orthopaedic implants annually become infected. Cardiovascular implants are even higher at 7.4%. A critical 6-hour post-implantation period is a crucial time window to reduce infections and allow tissue integration. Active release of antibiotics during this critical time window can play a crucial role. Examples include embedding antibiotics, silver and nitrogen oxide in a range of implant coatings e.g. in polyurethane.

Nanostructured surfaces can lead to bactericidal activity[4] and it sensitively depends on their topography. Textured substrates, found on the surfaces of the wings of cicadas surface structures (Figure 30.1) and their biomimetic analogues, can rupture Gram-negative cells, but Gram-positive cells often need additional irradiation to kill them.[3] In general, surfaces with micron-sized roughness are thought to reduce bacterial attachment.[5] However, very low <30 nm and very high >1 – 2 μm roughness encourages adsorption on metallic surfaces, so some contradictory evidence exists

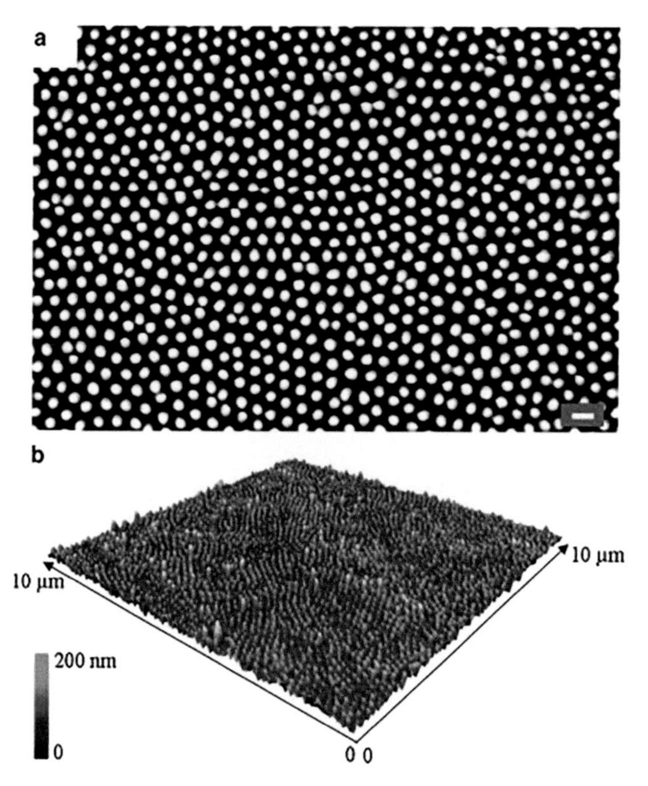

Figure 30.1 The topography of the cicada (*Psaltoda claripennis*) wing[3] measured with (a) scanning electron microscopy (scale bar is 200 nm) and (b) atomic force microscopy. Reprinted from Pogodin, S.; Hasan, J.; Baulin, V. A.; Webb, H. K.; Truong, V. K.; Nguyen, T. H. P.; Boshkovikj, V.; Fluke, C. J.; Watson, G. S.; Watson, J. A.; Crawford, R. J.; Ivanova, E. P., Biophysical model of bacterial cell interactions with nanopatterned cicada wing surfaces. *Biophysical Journal* 2013, *104*, 835–840, with permission from Elsevier.

in the literature (additional parameters beyond the roughness are expected to be important). Unidirectional textures on surfaces also led to increased adsorption of bacteria. The surface interactions were found to be specific to the bacterial species e.g. nanopillars kill Gram-positive bacteria due to cell membrane stretching but are less effective with Gram-negative bacteria. Shark scale topographies have also been investigated for their antibiofilm properties.

Coating concrete with a layer of slime produced by bacteria is found to protect it from sulphate attack from other bacteria.[6] Perhaps it encourages the proliferation of non-sulphate-attacking bacteria or presents a physical barrier for the bacteria. Artificial slime coatings could have many further applications e.g. mimicking mucus on aquatic animals, such as dolphins, and the intestines of mammals. The slime would need to be continually secreted for it to provide a long-term solution to bacterial adhesion.

Small peptides have been used to block the adsorption of bacterial adhesins onto teeth (the peptides could be used as additives in tooth paste or mouth wash).[7] Components in human milk inhibit the adhesion of bacteria and thus help with fighting infections.[8] Cranberry juice (which contains maltose) has been extensively studied as an anti-adhesive to help treat UTIs.

There is evidence that blue light can inhibit biofilm formation *in vivo* i.e. *Pseudomonas aeruginosa* biofilms in skin wounds of mice.[9] It is also found to have significant effects on the membrane potentials of bacteria *in vitro*[10] (Chapter 3), presumably since the light stresses the cells via the production of reactive oxygen species (ROS).

Enzymes for the degradation of *extracellular polymeric substance* (EPS) have been discussed in previous chapters (Chapters 23 and 24) and represent an effective mechanism for the treatment of biofouling, if the chemistry of the EPS is well characterised (the enzymes tend to be specific to the type of polymer treated). More generally, plasma and radiation treatments are excellent methods to sterilise the majority of bacterial samples (e.g. in foods or on surgical devices) and were discussed in Chapter 24, but are impractical *in vivo*, since they can also damage human cells. Bacteria can be killed via photothermal effects using a combination of lasers with gold nanoparticles.[11] Practically, such a mechanism could only be envisaged in specialised applications e.g. to treat challenging *in vivo* infections, due to the cost.

Magnetic nanoparticles have shown some success to disrupt biofilms. Magnetic methods tend to be less invasive than lasers, electric fields or radiation-based methods, so they have reasonable prospects for *in vivo* use. Curie heating in the magnets or oxidation by nanoparticle chemistry can be used to kill the bacteria.[13] Nanorobots have been made from magnetic particles and can be conveniently produced using three-dimensional (3D) printing with iron oxide nanoparticles[12] (Figure 30.2). Different manipulators can be used to translate the nanobots around biofilms to destroy them, such as permanent magnets attached to a robotic arm, high-frequency oscillated electromagnets and 3D-controlled electromagnets.[12]

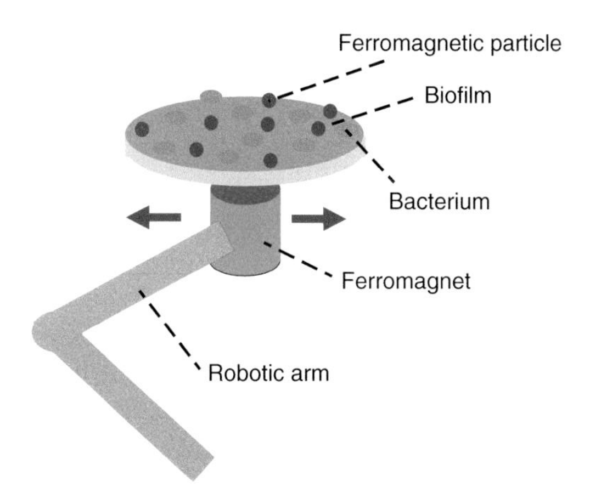

Figure 30.2 *Magnetic nanobots* have been used for the cleaning of biofilms from surfaces.[12] A robotic arm is used to move a ferromagnet around a biofilm treated with ferromagnetic particles. Curie heating of the ferromagnetic particles is also possible using oscillatory magnetic fields.[13]

Another novel approach was to create magnetic micron-sized pillars that beat in response to an external electromagnetic field.[14] These allow bacteria to be shed from surfaces, similar to cilial transport in the lungs. The pillars were used to coat catheters, since their surfaces are prone to biofilm fouling.

Modification of the surface chemistry of biomaterials has been studied in detail to combat biofilms. Unfortunately, it has received relatively limited success in long-term infections, although it can provide crucial additional inhibition times to help the immune system to combat infections. Bacteria find a way to stick to hydrophobic/hydrophilic surfaces and neutral/anionic/cationic surfaces, so modulation of the physical forces experienced by these surfaces has not been overly successful for persistent infections. Steric stabilisation with polymeric brushes has shown a little more promise, since it tends to provide a reasonably robust repulsive potential to a range of nanoparticle chemistries and over a range of nanoparticle sizes.[15] Antifouling coatings have a wide variety of applications to inhibit the adsorption of both biomolecules and microorganisms e.g. in marine biofilms.[16] Polyethylene glycol (PEG) coatings are commonly used to reduce protein biofouling and bacterial adsorption (the higher-molecular-weight polymer brushes tend to perform better). Polyelectrolyte multilayers have been created on surfaces that include antibiotics in their structures. The active prevention of biofilm formation via such multilayers has been studied in a micro chemostat.[17]

Suggested Reading

Tripathy, A., et al., Natural and bioinspired nanostructural bactericidal surfaces. *Advances in Colloid and Interface Science* **2017**, *248*, 85–104.

References

1. Ofek, I.; Bayer, E. A.; Abraham, S. N., Bacterial adhesion. In *The Prokaryotes*, Rosenberg, E., DeLong, E. F., Lory, S., Stackebrandt, E., Thompson, F., Eds., Springer: **2013**; pp. 107–123.
2. Hetrick, E. M.; Schoenfisch, M. H., Reducing implant-related infections: Active release strategies. *Chemical Society Reviews* **2006**, *35* (9), 780–789.
3. Pogodin, S.; et al., Biophysical model of bacterial cell interactions with nanopatterned cicada wing surfaces. *Biophysical Journal* **2013**, *104* (4), 835–840.
4. Tripathy, A.; Sen, P.; Su, B.; Briscoe, W. H., Natural and bioinspired nanostructured bactericidal surfaces. *Advances in Colloid and Interface Science* **2017**, *248*, 85–104.
5. Linklater, D. P.; Baulin, V. A.; Juodkazis, S.; Crawford, R. J.; Stoodley, P.; Ivanova, E. P., Mechano-bactericidal actions of nanostructured surfaces. *Nature Reviews Microbiology* **2021**, *19* (1), 8–22.

6. Yang, K. H.; Lim, H. S.; Kwon, S. J., Effective bio-slime coating technique for concrete surfaces under sulfate attach. *Material* **2020**, *13* (7), 1512.
7. Sharma, S.; Lavender, S.; Woo, J.; Guo, L. H.; Shi, W. Y.; Kilpatrick-Liverman, L.; Gimzewski, J. K., Nanoscale characterization of effect of L-arginine on *S. mutans* biofilm adhesion by atomic force microscopy. *Microbiology-SGM* **2014**, *160* (7), 1466–1473.
8. Sharon, N., Carbohydrates as future anti-adhesion drugs for infectious diseases. *Biochimica et Biophysica Acta* **2006**, *1760* (4), 527–537.
9. Rupel, K.; et al., Blue laser light inhibits biofilm formation *in vitro* and *in vivo* by inducing oxidative stress. *npj Biofilms and Microbiomes* **2019**, *5* (1), 29.
10. Blee, J. A.; Roberts, I. S.; Waigh, T. A., Membrane potentials, oxidative stress and the dispersal response of bacterial biofilms to 405 nm light. *Physical Biology* **2020**, *17* (3), 036001.
11. Zharov, V. P.; Mercer, K. E.; Galitovskaya, E. N.; Smeltzer, M. S., Photothermal nanotherapeutics and nanodiagnostics for selective killing of bacteria targeted with gold nanoparticles. *Biophysical Journal* **2006**, *90* (2), 619–627.
12. Hwang, G.; et al., Catalytic antimicrobial robots for biofilm eradication. *Science Robotics* **2019**, *4* (29), eaaw2388.
13. Li, J.; Nickel, R.; Wu, J.; Lin, F.; Lierop, J. V.; Liu, S., A new tool to attack biofilms: Driving magnetic iron-oxide nanoparticles to disrupt the matrix. *Nanoscale* **2019**, *11* (14), 6905–6915.
14. Gu, H.; Lee, S. W.; Carnicelli, J.; Zhang, T.; Ren, D., Magnetically driven active topography for long-term biofilm control. *Nature Communications* **2020**, *11* (1), 2211.
15. Murata, H.; Koepsel, R. R.; Matyjaszewski, K.; Russell, A. J., Permanent, non-leaching antibacterial surfaces-2: How high density cationic surfaces kill bacterial cells. *Biomaterials* **2007**, *28* (32), 4870–4879.
16. Banerjee, I.; Pangule, R. C.; Kane, R. S., Antifouling coatings: Recent developments in the design of surfaces that prevent fouling by proteins, bacteria, and marine organisms. *Advanced Materials* **2011**, *23* (6), 690–718.
17. Balagadde, F. K.; You, L.; Hansen, C. L.; Arnold, F. H.; Quake, S. R., Long-term monitoring of bacteria undergoing programmed population control in a microchemostat. *Science* **2005**, *309* (5731), 137–140.

31 Other Applications

The economic significance of bacteria and biofilms is huge. Here some applications in the food industry, petrochemical industry, optics, water purification, ecology and wound healing are considered.[1]

31.1 Food Industry

Listeria monocytogenes is a bacterium that accounts for 28% of foodborne illnesses.[2] Other common examples of foodborne illnesses with a microbial origin are: *Campylobacter* (bacteria), *Cryptosporidiosis* (eukaryote), *Cyclosporiasis* (eukaryote), *O157:K7 Escherichia coli* (bacteria), scombroid fish poisoning (bacteria), *Giardiasis* (eukaryote), *Salmonella* (bacteria), *Toxoplasma gondii* (eukaryote), *Vibrio parahaemolyticus* (bacteria) and *Yersinia* (bacteria).

A surprisingly wide range of food products require bacteria for their creation.[3] Bacteria are needed to make coffee, chocolate, Weiss beer, kimchi, sour kraut and yoghurt. Sometimes, bacterial cultures are used in tandem with fungi.

Extracellular polymeric substances (EPS) in slimes and biofilms are important in industrial settings as a source of food e.g. those produced from lactic acid bacteria.[4] Bacterial biofilms also play an important role in food spoilage e.g. *Lactobacillus brevis* is the majority cause of beer spoilage.[5]

Cleaning is a big issue in a huge range of chemical engineering processes (Chapter 30) and the disruption of biofilms tends to be one of the most challenging areas. It is particularly important in bioprocessing applications, such as the food industry and pharmaceuticals. Lipid A from rogue bacterial membranes is a common problem that can cause unpleasant side effects to pharmaceuticals, such as sepsis in extreme cases.

Biofilms commonly occur in the human gut, so their interaction with food components is important. The appendix is postulated to be a place to house biofilms of commensal bacteria for reinoculation of the colon to maintain well-balanced communities in this organ.[6,7]

31.2 Petrochemical Applications

The souring of oil wells is an important issue for the trillion-pound petrochemical industry, which is often driven by marine bacterial biofilms. For example, poisonous hydrogen sulphide (H_2S) gas is produced by *Desulfobacteraceae* that can corrode steel pipes.[8] Fluid transport in porous media is required to extract oil and it is affected by biofilm growth.[9] Furthermore, biofilm growth needs to be encouraged to clean up oil slicks. The breakup of oil slicks can be accelerated by spraying with microorganisms (in addition to surfactants).[10,11]

31.3 Optical Properties

The *optical properties* of biofilms control their light absorption, which is important for photosynthesis and the protection of bacteria from ultraviolet (UV) light (parasols for bacteria). Biofilm formation occurs with a wide variety of bacteria, including cyanobacteria (purple bacteria), i.e. bacteria that photosynthesize. The biofilms of photosynthetic bacteria have interesting optical properties as a result.[12]

Excitons are quasi-particles (bound electron/hole pairs) and are involved in photosynthesis. There is evidence from depolarised resonant light scattering for excitons in propine and chlorophyll, which are both molecules known to be involved in photosynthesis.[13] The adsorption of a photon by bacteriochlorophyll is found to be completely delocalized over the ring structure in low-temperature single-molecule fluorescence spectroscopy experiments.[14]

Quantum coherence of bacterial chlorophyll is still a hotly debated subject. There is ultrafast spectroscopy evidence for coherent excitation effects in the antenna complex of *Rhodopseudomonas acidophila* (a type of purple bacterium).[15] It is perhaps surprising that the complex molecular nature of the antenna at room temperature does not disrupt this process of coherent control. Evidence for coherent effects in excitons is also found in 2D Fourier transform electronic spectroscopy[16] and the coherence is remarkably long-lived. However, recent work has questioned the experimental evidence for long-lived coherence and suggests that dissipative phenomena occur instead, which have been optimised to create efficient energy flow.[17] Coherence due to entanglement has also been claimed in the photosynthetic apparatus of green sulphur bacteria.[18]

31.4 Water Purification

The issue of biofilm contamination in water purification has many factors in common with generic biofouling problems.[19] In general, substantial issues with biofouling are observed with ion exchangers, paper production, heat exchangers, the treatment of waste water, membrane separation technology, ship hulls, ship fuel systems, piping and sea chests, marine sensors, drinking water systems, marine aquaculture, washing machines, dish washers, cultural heritage and air conditioning systems.

A key step in water purification is the use of multivalent (e.g. cationic polymers) flocculants to improve water quality. These can modify the interparticle potentials of bacteria (Section 4.1.3) and the resultant aggregated bacteria can be conveniently removed with filtration.

More specialist concerns for water purification are found in environmental microbiology. Bacteria can clean poisonous heavy metals from polluted water e.g. via uranium sequestration.[20] Similar gene circuits that promote heavy metal tolerance can lead to the resistance of bacteria to silver ions when treated with silver-impregnated wound dressings, so metal sequestration is of general interest.

31.5 Ecology

Much of the *environmental conditions* on planet Earth have been perturbed by the work of bacteria over the 4.5 billion years of its existence. Bacteria were around for most of Earth's history, starting ~3.7 billion years ago. Of the six major elements H, C, N, O, S and P that constitute the major building blocks for all biomolecules, the first five are driven by microbially catalysed redox reactions[21] (Figure 31.1).

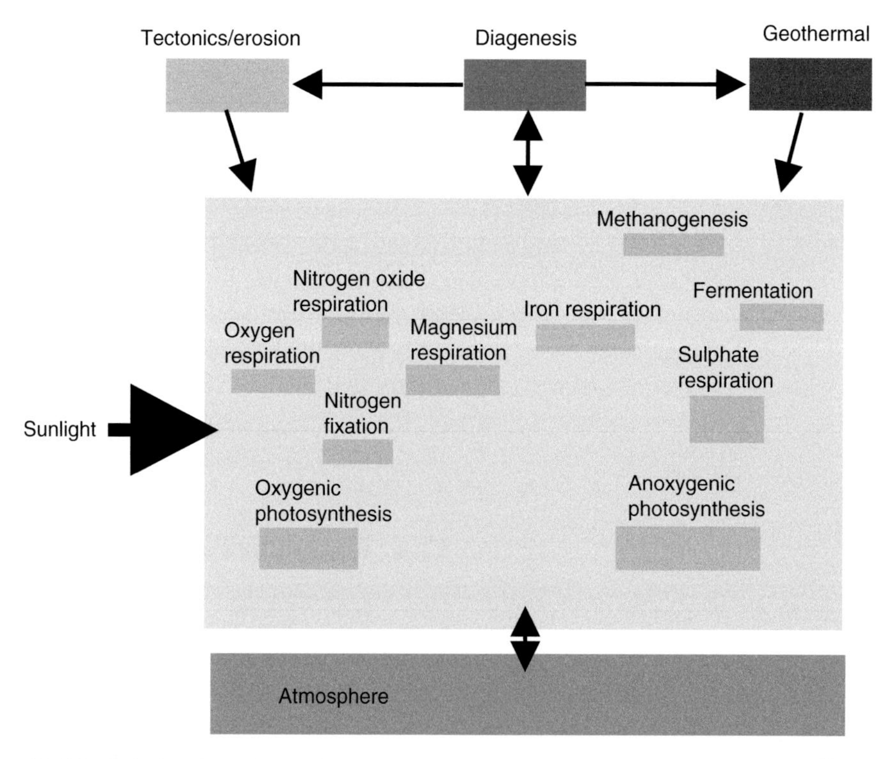

Figure 31.1 A model of the Earth's biosphere showing the inputs and outputs of redox processes controlled by bacteria.[21]

Microbial life has been extremely robust in its performance between the ice ages, during volcanic activity and during asteroid impacts.

31.6 Wound Healing

The development of *infections in wounds* delays the healing process and increases the chances of death. Serious wounds are classified as *acute* (e.g. surgical, traumatic or burns) or *chronic* (e.g. diabetic, foot ulcers, venous leg ulcers, arterial ulcers or pressure ulcers). Both types of infected wounds can lead to systemic infection in which sepsis leads to death. Common wound infections are caused by *Staphylococcus aureus*, *Pseudomonas aeruginosa* and *Streptococcus pyogenes*. In addition to the virulence components released by the bacteria that actively attach to the host (e.g. those that cause degradation of fibrinogen and destruction of all types of cell), the bacteria can interfere with wound healing in a targeted manner e.g. *S. aureus* secretes anti-angiogenic and anti-inflammatory factors and can stop leukocytes from functioning properly.

Chronic wounds tend to be populated by multiple species of microorganisms and have an increased microbial mass compared with acute wounds that can be due to single bacterial species. Standard treatments for critically infected chronic wounds include silver, iodine, honey and polyhexamethylene biguanide that are impregnated into wound dressings. Such wounds also need to be cleaned and debrided to remove foreign organic matter, such as biofilms. Swabs are taken of wounds to culture microorganisms for diagnosis. Acute wounds can be successfully treated with antibiotics, when they have been correctly diagnosed, provided they have not developed biofilms. A very simple continuum reaction–diffusion model (Chapter 5) has been shown to be reasonably successful in describing hypoxic and heterogeneous growth in microbial biofilms from chronic infections.[22]

In general, *wound healing* is an exceedingly complex process and robust quantitative models are scarce. Wounds are extremely variable (e.g. they can be very moist, exuding puss or very dry) and robust data to allow an accurate quantitative comparison of different commercial products (e.g. a honey dressing versus a silver dressing) rarely exists. This is compounded by commercial issues, e.g. intellectual property, complex multi-component formulations and opaque governmental regulatory processes.

Suggested Reading

Flanagan, M., ed. *Wound Healing and Skin Integrity: Principles and Practice*. Wiley: 2013.

Pepper, I. L.; Gerba, C. P.; Gentry, T. J. *Environmental Microbiology*. Academic Press: 2014.

Ray, B.; Bhunia, A. *Fundamental Food Microbiology*. CRC Press: 2014.

References

1. Camara, M.; et al., Economic significance of biofilms: A multidisciplinary and cross-sectoral challenge. *npj Biofilms and Microbiomes* **2022**, *8* (1), 42.
2. McLandsborough, L.; Rodriguez, A.; Perez-Conesa, D.; Weiss, J., Biofilms: At the interface between biophysics and microbiology. *FOBI* **2006**, *1* (2), 94–114.
3. McGee, H., *McGee on Food and Cooking*. Hodder and Stoughton: 2004.
4. de Vuyst, L.; de Vin, F.; Vaningelgem, F.; Degeest, B., Recent developments in the biosynthesis and applications of heteropolysaccharides from lactic acid bacteria. *International Dairy Journal* **2001**, *11* (9), 687–707.
5. Riedl, R.; Dunzer, N.; Michel, M.; Jacob, F.; Hutzler, M., Beer enemy number one: Genetic diversity, physiology and biofilm formation of *Lactobacillus brevis*. *Journal of the Institute of Brewing* **2019**, *125* (2), 250–260.
6. Bollinger, R. R.; Barbas, A. S.; Bush, E. L.; Lin, S. S.; Parker, W., Biofilm formation in the gut. *Journal of Theoretical Biology* **2007**, *249* (4), 826–831.
7. Donaldson, G. P.; Lee, S. M.; Mazmanian, S. K., Gut biogeography of the bacterial microbiota. *Nature Reviews Microbiology* **2016**, *14* (1), 20.
8. Cheng, Y.; et al., Microbial sulfate reduction and perchlorate inhibition in a novel mesoscale tank experiment. *Energy Fuels* **2018**, *32* (12), 12049–12065.
9. Seymour, J. D.; Gage, J. P.; Codd, S. L.; Gerlach, R., Anomalous fluid transport in porous media induced by biofilm growth. *Physical Review Letters* **2004**, *93* (19), 198103.
10. Prince, R. C., Petroleum spill bioremediation in marine environments. *Critical Reviews in Microbiology* **1993**, *19* (4), 217–242.
11. Prasad, M.; et al., *Alcanivorax borkumensis* biofilms enhance oil degradation by interfacial tubulation. *Science* **2023**, *381* (6659), 748–753.
12. Roeselers, G.; Van Loosdrecht, M. C. M.; Muyzer, G., Phototrophic biofilms and their potential applications. *Journal of Applied Phycology* **2008**, *20* (3), 227–235.
13. Parkash, J.; Robblee, J. H.; Agnew, J.; Gibbs, E.; Collings, P.; Pasternack, R. F.; de Paula, J. C., Depolarized resonance light scattering by porphyrin and chlorophyll a aggregates. *Biophysical Journal* **1998**, *74* (4), 2089–2099.
14. van Oijen, A. M.; Ketelaars, M.; Kohler, J.; Aartsma, T. J.; Schmidt, J., Unraveling the electronic structure of individual photosynthetic pigment-protein complexes. *Science* **1999**, *285* (5426), 400–402.
15. Herek, J. L.; Wohlleben, W.; Cogdell, R. J.; Zeidler, D.; Motzkus, M., Quantum control of energy flow in light harvesting. *Nature* **2002**, *417* (6888), 533–535.
16. Engel, G. S.; Calhoun, T. R.; Read, E. L.; Ahn, T. K.; Mancal, T.; Cheng, Y. C.; Blankenship, R. E.; Fleming, G. R., Evidence for wavelike energy transfer through quantum coherence in photosynthetic systems. *Nature* **2007**, *446* (7137), 782–786.
17. Cao, J.; et al., Quantum biology revisited. *Science Advances* **2020**, *6* (14).

18. Sarovar, M.; Ishizaki, A.; Fleming, G. R.; Whaley, K. B., Quantum entanglement in photosynthetic light-harvesting complexes. *Nature Physics* **2010**, *6*, 462–467.
19. Fleming, H. C., Biofouling and me: My Stockholm syndrome with biofilms. *Water Research* **2020**, *173* (1), 5576.
20. Wall, J. D.; Krumholz, L. R., Uranium sequestration. *Annual Review of Microbiology* **2006**, *60*, 149–166.
21. Falkowski, P. G.; Fenchel, T.; Delong, E. F., The microbial engines that drive Earth's biogeochemical cycles. *Science* **2008**, *320* (5879), 1034–1039.
22. Stewart, P. S.; Zhang, T.; Xu, R.; Pitts, B.; Walters, M. C.; Roe, F.; Kikhney, J.; Moter, A., Reaction-diffusion theory explains hypoxia and heterogeneous growth within microbial biofilms associated with chronic infections. *npj Biofilms and Microbiomes* **2016**, *2*, 16012.

Index